161 388215 3

D1585393

# MICROBIAL ECOLOGY

# Microbial Ecology

## An Evolutionary Approach

**Dr. J Vaun McArthur**

University of Georgia
Savannah River Ecology Laboratory
Aiken, South Carolina

ELSEVIER

AMSTERDAM • BOSTON • HEIDELBERG • LONDON
NEW YORK • OXFORD • PARIS • SAN DIEGO
SAN FRANCISCO • SINGAPORE • SYDNEY • TOKYO
Academic Press is an imprint of Elsevier

| | |
|---|---|
| Senior Acquisitions Editor: | Nancy Maragioglio |
| Project Manager: | Heather Furrow |
| Associate Editor: | Kelly Sonnack |
| Marketing Manager: | Linda Beattie |
| Cover Design: | Eric DeCicco |
| Composition: | SNP Best-set Typesetter Ltd. |
| Cover Printer: | Phoenix Color |
| Interior Printer: | Courier Westford |

Academic Press is an imprint of Elsevier
30 Corporate Drive, Suite 400, Burlington, MA 01803, USA
525 B Street, Suite 1900, San Diego, CA 92101-4495, USA
84 Theobald's Road, London WC1X 8RR, UK

This book is printed on acid-free paper. ∞
Copyright © 2006, Elsevier Inc. All rights reserved.

Cover image: "Three common types of bacterial morphology," copyright Dennis Kunkel, Microscopy, Inc.

No part of this publication may be reproduced or transmitted in any form or by any means, electronic or mechanical, including photocopy, recording, or any information storage and retrieval system, without permission in writing from the publisher.

Permissions may be sought directly from Elsevier's Science & Technology Rights Department in Oxford, UK: telephone: (+44) 1865 843830, fax: (+44) 1865 853333, e-mail: permissions@elsevier.com. You may also complete your request on-line via the Elsevier homepage (http://elsevier.com), by selecting "Support & Contact" then "Copyright and Permission" and then "Obtaining Permissions."

**Library of Congress Cataloging-in-Publication Data**
McArthur, J Vaun.
  Microbial ecology: an evolutionary approach/J Vaun McArthur.
     p.   cm.
  Includes bibliographical references and index.
  ISBN-13:  978-0-12-369491-1
  ISBN-10:  0-12-369491-4
  1. Microbial ecology.  2. Surface chemistry.  I. Title.
  [DNLM:  1. Microbiology.  2. Ecology.  3. Evolution.  QW  4
M4778m  2006]
QR100.M387  2006
579'.17—dc22

                                                        2005030991

**British Library Cataloguing in Publication Data**
A catalogue record for this book is available from the British Library

ISBN 13: 978-0-12-369491-1
ISBN 10: 0-12-369491-4

For all information on all Elsevier Academic Press Publications
visit our Web site at www.books.elsevier.com

Printed in the United States of America
06  07  08  09  10    9  8  7  6  5  4  3  2  1

Working together to grow
libraries in developing countries

www.elsevier.com | www.bookaid.org | www.sabre.org

ELSEVIER   BOOK AID International   Sabre Foundation

# Contents

# Preface

Microbial ecology is an exciting and growing discipline with broad interest and potential impacts. Our understanding of the many beautiful and detailed processes that are controlled all, or in part, by microorganisms is expanding daily. With that understanding has come an appreciation for things microbial. For many years, microbes got, at best, second billing to higher organisms when they were even mentioned and they had incredibly bad press. To most people all microbes, especially viruses and bacteria, were germs and as such something to eradicate. Through the dedicated efforts of numerous scientists a microbially dominated world has come into view. This new view shows a world where life as we know it could not exist without the processes and functions of these tiny organisms.

Over the years there have been a few microbial ecology textbooks that have been produced and used to greater or lesser extents. Writing a textbook on a subject that is growing rapidly is dangerous business. New material that is important will probably be published while the textbook is being readied for printing. Therefore, to be of value to the student, a textbook in microbial ecology must capture general concepts that can be used to understand what we have already discovered and all or most future microbial discoveries. To this end, in this textbook I have sought to present microbial ecology through the lens of evolutionary ecology. The principles of evolution and ecology have been, with varied success, applied to higher organisms but few if any attempts have been made to apply these concepts to microorganisms. However, these basic biological principles should help us understand various microbial strategies and predict others that may not be fully appreciated yet. In addition, microbes provide a unique tool for understanding basic evolutionary and ecological principles because of their relatively short generation times that allow thousands of generations to be reared and observed—a luxury not available to scientists who study higher organisms. The power of evolutionary and ecological theories also can be tested using the organisms that began all ecology and evolution. There is much to learn from these oldest of all creatures.

Microbial ecology from a taxonomic perspective is an incredibly broad topic. It includes most of the living diversity of the planet. While the textbook title suggests a very broad sweep through all of the microbes, including eukaryotes and prokaryotes, the actual presentation spends a greater proportion of time discussing bacterial evolutionary ecology. This bias is recognized and greater emphasis to other microbial groups could be added in future additions if professors choose to use this approach (textbook) in their teaching microbial ecology.

I am especially grateful for the administrative assistance of Juanita Blocker who worked tirelessly on helping with the graphics for this book. Dr. Paul Bertsch, Director of the Savannah River Ecology Laboratory, and Dr. Carl Strojan, Associate Director, provided encouragement and the opportunity to write. Dr. Ramunas Stepanauskas provided comments on sections of the manuscript and many worthwhile discussions on microbial ecology. Angela Lindell and Catherine King kept the lab going while I was engaged in writing. Dr. R. Cary Tuckfield was a sounding board for ideas, concepts, and approaches.

Many thanks to my family, especially Jackie Lynne, for her support and belief in this project and more importantly in me.

J Vaun McArthur

# Ecology and Evolution

1

# Core Concepts in Studying Ecology and Evolution

The study of microbiology, ecology, and evolution is extremely broad. Entire textbooks and sets of textbooks have been written on each subject. There are numerous scientific journals that continue to expand our knowledge of each of these subjects. Numerous annual, semi-annual, and special scientific meetings are held in which cutting-edge science is presented and discussed. Specialists spend their careers examining and describing subsets of only one of these disciplines. However, an understanding of basic underlying principles of biology requires that efforts be made to see linkages among disciplines and to meld the theories and insights of one discipline with those of another. It is beyond the scope of this text to present in detail many of the aspects of each of these disciplines. In this introductory chapter we will briefly cover some of the major topics belonging to each discipline as an introduction to the material that will follow.

## The Beginnings of Microbiology

Although van Leeuwenhoek described microbes during the 16th century with his ingenious microscopes, little was done to understand these very small organisms for many, many years. Van Leeuwenhoek's observations using his microscopes surpassed all other observations for over 400 years. Although not formally trained as a scientist, van Leeuwenhoek had a natural propensity to ask questions about various substances and samples and then examine many different types of samples. He studied river, pond, well, and seawater. He described the microbiota of spittle and teeth tartar, and he was the first to see sperm cells.

Unfortunately he left no descriptions of his apparatus with which he was able to observe both protozoa and bacteria. His descriptions of various organisms were so exact that we can almost assign taxonomic affiliations based just on the descriptions. His observations are so good that he must have had novel means of illuminating his subjects and he probably discovered dark field microscopy, a practice still used today. Although van Leeuwenhoek made observations of many different types of samples for over fifty years and wrote over 150 letters to the Royal Society of London little attention was paid to these marvelous little creatures for a considerable length of time.

It was not until several centuries later that scientists began to ask questions about these organisms. Among the first questions was whether these microorganisms could arise through spontaneous generation. Although Redi had proved that maggots in meat arose from flies, there was no definitive proof that microbes could not come into existence spontaneously. Redi had excluded the adult flies using gauze with a weave that prevented the adult flies from *ovipositing* their eggs on the meat. However, although they did not know how microorganisms came into being, no gauze would have been fine enough to prevent microbes from colonizing the meat. It was not until Pasteur that the source of many microbes in contaminated substances was understood. Pasteur's experiments were a masterpiece of logic and experimentation. However, his results did not completely destroy the idea of spontaneous generation. It was not until Tyndall demonstrated that sterilization of various materials required different means that spontaneous generation was laid to rest. Tyndall was able to completely sterilize the starting broths or infusions from a variety of substances and by so doing eliminated all subsequent growth.

Formal microbiology has its origin in the discoveries of the 19th century that began to reveal the existence and importance of many organisms too small to see. Today through the creative insights and persistence of numerous scientists the role these tiny organisms play in disease, various nutrient cycles, the food industry, and their nearly global distributions have been worked out. Microorganisms were found to be everywhere the scientists were creative enough to sample. However, outside of medicine, few other studies were done to understand what role these creatures played and more exactly who the organisms were that were involved. Microorganisms were obviously too small to be of much importance and they were extremely difficult to study.

Fortunately, people do not like to be sick. It is this disposition that gave impetus to the study of disease causing organisms. In the process, many capable scientists have contributed to our understanding of microorganisms. They have formulated and developed methods and resources for the systematic study of microorganisms and, in so doing, have provided a wealth of information on the genetics, physiology, and to some degree the behavior of disease causing organisms. They have discovered ways to isolate, characterize, and recognize various microbes. The isolation of disease causing organisms has been the mainstay of microbial taxonomy. Although much effort has been made in understanding disease-causing organisms in a host, very little effort has been made to study the ecology of disease causing organisms outside of the hosts. This is an area waiting for clever scientists to study.

Understanding both ecological and evolutionary concepts is critical in attempting the study of organisms in nature. Unfortunately, microorganisms have not been the subjects of ecological or evolutionary studies to any great degree. Perhaps our inability to really cure microbially induced diseases is a function of our lack of understanding of the basic *natural history* of these organisms. In one sense the realization that the strains of viruses that cause influenza differ among themselves and may differ even within a strain over time is a first cut at considering the evolution and ecology of these organisms.

This chapter presents a brief natural history of two major classes of microbes: viruses and bacteria. These groups are too broad and diverse to present much information in detail but this information will form a basis for much of what follows. Although there is a rich literature on microbial eukaryotes, little attention will be paid to these organisms except in their interactions with viruses and bacteria. Exceptions will include some discussion of some fungi.

# Viruses

Viruses are curious things. They have been the subjects of much debate as to whether or not they are even alive. The reader is asked to review any beginning biology textbook to get the basic arguments for or against viruses being alive. Regardless of what the textbooks say, viruses have interesting life histories, and some of the main properties of living things (i.e., genetic material and genetic change) are integral to the biology of viruses. Interactions between the host and a virus are essential in understanding the ecology and evolution of viruses. This is because viruses are obligate intracellular parasites. Outside of a host, viruses do not fit the classic definition of an organism because there is no metabolism, they do not reproduce themselves, they do not take energy from the environment, and they do not respond to stimuli. However, within a host, viruses demonstrate many of these properties.

Viruses are small. They range in size from 0.02 to 0.2 μm and are composed of a protective protein covering that surrounds a very small amount of nucleic acid (DNA or RNA) in quantities ranging from a few to several tens of kilobases. This genetic information is enough to code for only a handful of proteins. Even though they have such little genetic information, they are highly adaptable and maintain high levels of genetic diversity. Despite their extremely small size, there is a diversity of shapes found among the viruses (Figure 1.1).

The basic steps in viral replication are similar for all viruses whether they infect plant, animal, or bacterial cells. Viruses can only infect cells in which there are chemical structures on the surface of the virus and the host cell, which permit attachment and penetration. How many types of cells (or the host range) a particular virus can infect is determined by these receptor-dependent interactions.

The first step in a viral infection is for the viral nucleic acid to get into a cell through penetration of the host cell surface. For some viruses penetration is facilitated by release of an enzyme located in the phage tail that degrades a small part of the cell wall. The nucleic acid is then literally forcibly injected into the cell. Then replication of viral nucleic acids and the synthesis of viral constituents take place using the cell's anabolic machinery.

There are three different outcomes that can come from an infection: the infecting viruses multiply into many progeny which kill and lyse the host cell releasing the newly made viruses into the environment; the viruses multiply into many progeny within the host but the host cell does not lyse and the cell survives; and a stable condition is established with little or no viral multiplication and where the virus is integrated into the cell's genome or it remains as a separate entity within the cytoplasm of the cell.

Virus or virus-like particles can reach very high densities in the environment. These densities are probable sufficient to expose many bacteria and eukaryotes to viral infection. As will be seen in later discussion viruses may actually regulate microbial populations in nature to some extent.

(a) Polydedral, naked

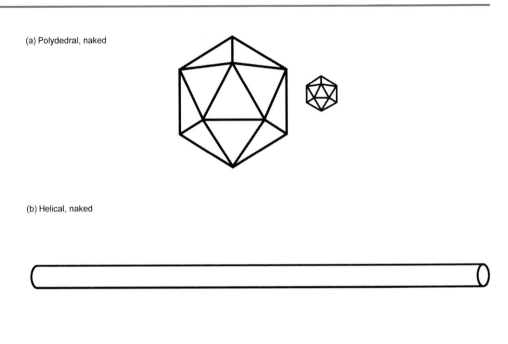

(b) Helical, naked

(c) Enveloped

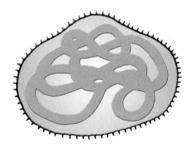

(d) Combination of polydedral and helical, naked

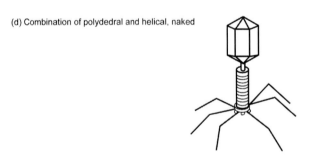

**Figure 1.1**  Different shapes of viruses or virus-like particles: polyhedral (**a**); helical (**b**); enveloped (**c**); and combination (**d**). (From Nester EW, Roberts CE, McCarthy BJ, Pearsall NN. *Microbiology: Molecules, Microbes, and Man.* New York, Holt, Rinehart & Winston, 1973.)

## Bacteria

Bacteria are among the most diverse if not the most diverse groups of organisms found on the planet. Attempts to classify these organisms using culturing techniques have resulted in more than 3,000 named bacteria. Because there are more than 600,000 species of beetles, this statement of being the most diverse group seems a little rash. This number of species is woefully small, and research using culture-independent techniques have demonstrated considerable diversity (discussed later). The purpose of this text is not to describe the various groups of microorganisms in any definitive manner. Because the main text does not discuss specific groups of bacteria except as they relate to specific topics, the major groups of bacteria are briefly mentioned here. Table 1.1 summarizes some of the differences found among the major groups of bacteria. Much more complete descriptions are given in many microbiology textbooks and these should be referred to for additional information. Our purpose here is to review these groups as an indicator of the wide diversity of bacteria found on the earth. Although the source for these comparisons and description comes from a work written in the 1970s, most of these groups were observed in the 1600s, and little new morphological information can be found. Rather considerable differences have been found in subcellular constituents in recent years.

## Photosynthetic Bacteria

All major morphological types of bacteria are represented among the photosynthetic bacteria: rods, cocci, and helical forms suggesting that photosynthesis among the bacteria is derived from diverse evolutionary origins. Some of the oldest fossils resemble blue-green or cyanobacteria. The details of bacterial photosynthesis are presented in the following section. Photosynthetic bacteria are found in wet soil, and both freshwater and marine habitats that lack oxygen but where light can penetrate.

## Gliding Bacteria

These bacteria, as their grouping suggests, have the ability to glide or slide over solid surfaces. They lack flagella. There are several interesting taxa associated with this group that are found in a wide variety of habitats. Some of these habitats include the inside of the mouth, polluted rivers, sulfur springs, black mud, manure, decaying organic matter, and soils. As one might expect when a grouping is done on a phenotypic trait like gliding many different and distantly related organisms will be grouped together. This is true of the gliding bacteria.

Some of the species can form long filaments composed of chains of cells that are encased in a common wall. These filaments are not mobile; however, short segments can be. Other multicellular forms can be found including species found in the mouth and on decaying organic matter.

## Sheathed Bacteria

The sheathed bacteria are another group of multicellular organisms that live in filaments. These filaments differ from those discussed in the previous section in that the filaments are enclosed in a sheath of lipoprotein-polysaccharide that is chemically different from bacterial cell walls. These organisms can increase to very high densities, especially below sewage outfalls.

**Table 1.1** The Major Groupings of Bacteria and Their General Characteristics

| Bacterial Group | Mode of Motility | Morphology | Nutrition | Staining | Other Features |
|---|---|---|---|---|---|
| Photosynthetic | | | | | |
|   Purple, sulfur | Flagella, if motile | Rods, cocci, and helices | Phototrophic; autotrophic | Gram-negative | Sulfur granules deposited intracellularly |
|   Green, sulfur | Flagella, if motile | Rods | Photosynthetic, autotrophic | Gram-negative | Sulfur granules deposited extracellularly |
|   Purple, nonsulfur | Flagella, if motile | Rods, helices | Phototrophic, heterotrophic | Gram-negative | No sulfur granules |
| Gliders | | | | | |
|   Filamentous sulfur | Glide on solid substratum | Multicelled filaments | Oxidize reduced sulfur compounds | Gram-negative | Sulfur granules deposited intracellularly |
|   Nonfruiting mycobacteria | Glide on solid substratum | Long rods | Heterotrophic | Gram-negative | |
|   Fruiting myxobacteria | Glide on solid substratum | Short rods that form microcysts | Heterotrophic | Gram-negative | Forms elaborate fruiting structures |
| Sheathed | Flagella | Multicelled filament enclosed in sheath | Heterotrophic | Gram-negative | Sheath |
| Prosthecate and budding | Flagella, if motile | Unicellular rods, vibrios, cocci, some have appendages and some divide by budding | Heterotrophic | Gram-negative | |
| Spirochetes | Axial filaments | Helical; flexible wall | Heterotrophic | Too thin to stain well | Cell is flexible and bends easily |
| Spiral or curved | Flagella, if motile | Bent rods or helical; rigid wall rods | Heterotrophic | Gram-negative | Cell is rigid and does not bend |
| Strictly aerobic, Gram-negative rods | Polar flagella if motile | | Heterotrophic, nonfermentative | Gram-negative | |
| Facultatively anaerobic, Gram-negative rods | Polar or peritrichous flagella if motile | Rods, some very short | Heterotrophic | Gram-negative | Many pathogens, enterobacteria |
| Strictly anaerobic, Gram-negative rods | Polar flagella if motile | Curved and straight rods, some spindle shaped | Heterotrophic | Gram-negative | |
| Nonphotosynthetic autotrophs | Flagella | Rods, spheres, helices | Use $CO_2$ as major carbon source | Gram-negative | Reduced inorganic compounds supply energy |
| Gram-negative cocci | Nonmotile | Spherical found in pairs | Heterotrophic. Parasitic and pathogens | Gram-negative | Includes causative agents of gonorrhea and bacterial meningitis |
| Gram-positive cocci | Nonmotile | Spherical, chains, packets or clusters common | Heterotrophic | Gram-positive | Includes important pathogens |
| Endospore formers | Peritrichous flagella if motile | Rods | Aerobic, facultative or anaerobic heterotrophs | Gram-positive | Spore is very resistant |
| Non–spore-forming, Gram-positive rods | Peritrichous flagella if motile | Rods | Heterotrophic | Gram-positive | Diverse group |
| Branching | Flagella, if motile | Branching rods and nonseptated filaments; some produce spores on aerial hyphae | Heterotrophic | Some acid fast | Includes important pathogens |
| Mycoplasmas | Nonmotile | Irregular shape due to absence of cell wall | Heterotrophic; most require sterols | Gram-negative | Only prokaryotes with sterol-containing cytoplasmic membranes |
| Obligate intracellular | Nonmotile | Small, short rods | Heterotrophic. Must grow intracellularly | Gram-negative | Some have leaky cytoplasmic membrane |

Adapted from Nester EW, Roberts CE, McCarthy BJ, Pearsall NN. *Microbiology: Molecules, Microbes, and Man.* New York, Holt, Rinehart & Winston, 1973.

## Budding and Prosthecate Bacteria

These organisms have appendages that project out from the bacterial cells. The pros-thecae are the stalks by which some of these bacteria attach to surfaces by adhesion by a holdfast located in the tip of the appendage. The prosthecae are actual exten-sions of the bacterial cell and contain cytoplasm. Budding bacteria may or may not attach to surfaces. Their unique characteristic is the ability to reproduce by budding.

## Spirochetes

The spirochetes are also grouped based on morphology. The specific morphology is helical or wave shaped. These organisms have flexible cell walls and are capable of movement using axial filaments. These organisms are found in aquatic habitats and as parasites of warm-blooded animals. Van Leeuwenhoek probably observed some of these organisms as indicated by his drawings.

## Spiral and Curved Bacteria

We might expect that a spiral shaped bacterium be grouped with the spirochetes but such is not the case. These organisms are much larger than the spirochetes and they have a rigid cell wall. Some of these organisms are predators of other bacteria and will be discussed in more detail in later chapters.

## Strictly Aerobic Gram-Negative Rods

This is one of the large "cover-everything" type groups. These organisms are Gram-negative rods. In fact it contains one of the best-known groups: the pseudomonads. This group by itself is interesting because the name literally means false unit (Latin *pseudo* = false, *monad* = unit). These organisms are characterized as being motile by polar flagella. They live in just about every habitat that has ever been sampled. Some members of these strictly aerobic organisms can fix nitrogen while others are known parasites of humans and animals.

## Facultative Anaerobic Gram-Negative Rods

This group is characterized by the ability to ferment carbohydrates. Many bacteria found in the guts of other organisms are within this group (e.g., *Escherichia coli*, many plant and animal pathogens). Some of these pathogens are responsible for plague.

## Strictly Anaerobic Gram-Negative Rods

These organisms are important both as free-living forms and as pathogens and par-asites of man. The upper respiratory, gastrointestinal and lower urogenital tracts usually have high numbers of these organisms. In nature members of this group perform part of the sulfur cycle.

## Nonphotosynthetic Autotrophic Bacteria

Nonphotosynthetic autotrophic bacteria generate energy by oxidizing inorganic compounds rather than by using organic matter. As with all man-made groupings, some of these organisms can be classified with other groups. This includes members of this group that are also gliding bacteria. Some of these bacteria are involved in the sulfur cycle and actually produce sulfuric acid. Others can oxidize reduced iron and fix carbon dioxide, and still others of this group can reduce carbon dioxide to methane and are found in the digestive tracts of cattle or other animals or in the anaerobic mud of marshes and wetlands.

## Gram-Negative Cocci

Gram-negative cocci can be commonly found associated with animals but rarely in the external environment. Members of this group cause gonorrhea and meningitis. They can be either aerobic or anaerobic.

## Gram-Positive Cocci

The Gram-positive cocci are important in industry and food processing, as well as extremely difficult to control disease-causing bacteria. Some members of this group have developed multiple resistances to a wide variety of antibiotics.

## Endospore-Forming Bacteria

Endospores are mostly formed by Gram-positive rod-shaped bacteria that are commonly found in soils. Some require anaerobic conditions while others can grow in air. The endospore is a structure that allows survival during adverse or harsh conditions (discussed later).

## Non–Spore-Forming, Gram-Positive Rods

The non–spore-forming, gram-positive rods are very widespread and have representatives from a number of bacterial taxonomic groups. Some are found in the mucous membranes of humans. Others inhabit soils where they are capable of breaking down complex man-made compounds such as insecticides and herbicides.

## Branching Bacteria

Some bacteria grow by extending mycelia into their growth medium. They produce many types of antibiotics and are important in the degradation of lipids and waxes in plants and animal tissues. Others are widespread in soils and give soil its distinctive odor. This group includes both Gram-negative and Gram-positive organisms and several novel and interesting modes of reproduction.

## Obligate Intracellular Bacteria

This is a group that is receiving considerable attention recently. Some of these bacteria are capable of altering the behavior or the gender of the organisms they inhabit (discussed later). These organisms are transmitted during reproduction of the host and do not have a free-living form.

We have just barely touched on some of the wonderfully unique aspects of these amazing microorganisms. In so doing, it is important that we consider why and how this diversity in form and function arose and is maintained. Organisms live somewhere. In the process of living, they are affected by their environment, and in most cases, they are capable of modifying their environment to a greater or lesser degree. Evolutionary science is the study of how organisms change over time. Ecology is the study of organisms in their environments. We briefly review the sciences of ecology and evolution as preface to our in-depth study of microbial evolutionary ecology.

## Ecology Becomes a Science

Ecology or the study of the factors that control the distribution and abundance of organisms began in the early 20th century. Ecology has its roots in *natural history*. Natural history is the knowledge of organisms in their environment. Where, when, and how to find various creatures were the questions that formed the basis of natural history. Ancient human tribes were well acquainted with the movements of various creatures, and where and how to find important plants. Plagues of insects and other organisms were frequently encountered and required explanation. However, the possible explanations were beyond the scope of experimentation and often included supernatural explanations.

As man became more interested in why certain plants or animals were found in specified locations and not elsewhere *biogeography* began to be studied. Biogeography was initially concerned with distributions of organisms over broad geographical scales—like continents or islands. Both botanists and zoologists were concerned about such patterns but they differed in their approach and interests. Students would be advised to study the history of ecological thought as it developed among botanists and zoologists. However, an exhaustive history of the science of biogeography is beyond the scope of this introductory material. Suffice it to say that there was major controversy between the two disciplines.

Early in the past century and accelerating during the 1950s and later, various theoretical concepts began to be developed by botanists and zoologists. Unfortunately, most of this thought was geared toward trying to understand the distribution and abundance of organisms that could be easily, or without too much trouble (e.g., insects), be observed without aid. Microbes, if they were considered at all were thought of as decomposers and lumped together as a single taxonomic and functional unit. Even this limited acknowledgment was flawed as scientists failed to grasp the magnitude of microbial processes or the level of microbial diversity. Exceptions were found in medicine in which scientists and health professionals sought to eliminate worldwide epidemics such as polio and smallpox. The recognition of modes of dispersal of pathogens and the observation that these same organisms were being encountered on a worldwide basis required significant improvement in isolation and identification techniques.

Ecologists have made significant strides in describing factors and processes that determine the distribution and abundance of the earth's biota, especially the macrobiota—plants and animals. In contrast to other scientists, who can perform most of their experiments in the laboratory under tightly controlled conditions, ecologists study perhaps the most variable of all subjects—nature. Nature is inherently messy, and variance is the rule. No two ecosystems are exactly alike. No two samples, even if taken very close to each other either in space or time, are exactly alike. The beauty of extremely controlled and replicated experiments is difficult if not impossible to obtain in ecological studies. Exceptions include greenhouse studies and *microcosm* or *mesocosm* studies in which the variability of nature is controlled through simplification. In these types of experiments, researchers are able to control some of the variability by reducing the complexity of the experiment. Single-species responses to various treatments are found in many greenhouse studies on plants. Mesocosm and microcosm studies often involve a small subset of species that may interact and under a much narrower set of environmental conditions than those found in nature.

Any field study has various levels of uncertainty associated with every variable measured. Most of these variables change both spatially and temporally. Nevertheless, ecologists have through various methods and approaches successfully completed numerous studies that detail processes and events that occur in nature. The beauty and intricacies of these processes and events is exceptional. The cleverness of the scientists is profound at times. These scientists have given incredible insight into the workings of nature for many organisms in many parts of the world.

It has been said that ecology is the painstaking description of the obvious. In other words, almost anyone can see patterns in nature but most are unable to explain why these patterns should exist. For example, the grasslands of the United States have few trees—why? The observation is open for everyone to see who passes through grassland. Ecologists attempt to answer such questions.

Ecology as a discipline spans scales from molecules to the biosphere and everything in between. Ecology may be concerned with chemical and organismal interactions or organism-to-organism interactions. Ecologists may seek to understand a single species or try to explain why certain groups of organisms are found together. Some ecologists measure the dynamics of important plant nutrients as a means of understanding the roles plants, animals, and the abiotic environment have on these substances. In reality, only ecologists who work with higher, easily seen organisms are explaining the obvious. Many ecologists study things that most people have never imagined or observed.

Over the past century, it has become clear that much of what determines distributions and abundances of even higher organisms is not readily visible. Such things as nutrient availability, temperature, pH, redox potential, moisture, and many other variables also affect where organisms are found. Variability within each of these abiotic variables has resulted in incredible levels of biological diversity both within taxa and among taxa. Biological variability includes such things as behaviors, mating systems, competitive ability, predation, population dynamics, community interactions, and many others.

Ecology is not just a descriptive science but involves carefully planned and executed experiments that have helped tease apart some of the relationships mentioned in the previous paragraphs. Ecology has a rich body of theory and an ever-expanding corps of researchers that have interests in all types of organisms and habitats in pristine, extreme, and heavily perturbed environments.

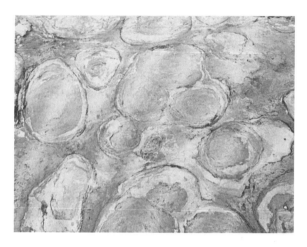

**Figure 1.2** Fossilized cyanobacterial-like microorganisms. (From the National Audubon Society Collection/Photo Resesearchers and K. and B. Collins, Visuals Unlimited.)

# Evolution

The oldest fossils discovered resemble bacteria that are found today. These fossils look remarkably like certain cyanobacteria (Figure 1.2). Other fossil microbes bear a strong resemblance, at least in outward morphology, to other bacteria that are present among us. These earliest known organisms are over 3 billion years old. Microbes have not changed in their physical outward appearance for a very long time.

Evolution in its simplest definition is change. However, that definition can be applied to many different systems, including living and nonliving systems. In biology, evolution is change that is transmitted through generations or, as stated by Futuyma, "descent with modification and often with diversification." Evolution is driven by a process called *natural selection* that is discussed later. Natural selection acts on the differences found among individuals that affect the rate of survival and reproduction.

Evolution by natural selection is based on four points that were made by different people but eventually summarized by Charles Darwin. These four points are

1. The offspring of an organism is more like the parents than any other organism. In other words, *like begets like*.
2. In every population of organisms, there are variations that have occurred by chance, and these variations are heritable or can be passed on through reproduction. These chance differences are the products of mutations that occur in the genes. However, Darwin did not know about DNA, RNA, and protein synthesis.
3. Most species produce more offspring than can actually survive and reproduce. Compared with the number produced, the number of organisms that actually reproduce may be very small.
4. Which organisms reproduce and produce viable offspring is in part determined by the action of their environment on chance variations produced by mutations. Over many generations, these favorable mutations will accumulate, and the population will be different from the original population and, on average, made up of more individuals with the favorable mutation.

Anyone who has been to a natural history museum can readily see that for some organisms, there is an incredible and often dramatic change in form and often function over the millions of years of the fossil record. Most species that have existed on this planet are now extinct and the extant species are a paltry subset of the diversity that has at times been found. However, where lineages are known or suspected, the visual representation of evolution is impressive and awe inspiring.

We began this section by observing that the microbial fossil record appears to show that for these organisms form has changed little over billions of years. Now it must be pointed out that fossilization of organisms is a chancy business at best. Even for large, hard-bodied creatures, the chance of being fossilized requires the organism to die in the right place; a place where geological processes can capture an image of the creature. For soft-bodied organisms that decompose rapidly, fossilization is a very chancy thing. For microbes, scientists must be able to recognize a microscopic fossilized body from the other components of the rocks. This is no small task (no pun intended). However, the fact that outwardly the known microbial fossils resemble the living organisms is interesting. For some microbial life, outward morphology has not changed for billions of years.

If we cannot observe physical differences among most fossilized microbes, why should we study the evolution of microbes? That is the rest of the story! Microbial evolution in the past and microbial evolution today have resulted in an incredible array of organisms, functions, and abilities to survive. Knowing something about how that diversity in species, functions, and abilities arose is important for understanding microbial life. Such knowledge can aid researchers in predicting various interactions that may aid in the clean-up of toxic wastes, in helping increase soil fertility, in industrial applications. Because microbes have been evolving longer than all other organisms put together perhaps we can learn more of the process of evolution by studying microorganisms.

## Natural Selection

Not all individuals are able to reproduce at the same rates or levels in every generation and leave copies of their genes in the next generation. Nonrandomness in reproduction is natural selection. If there is a random effect, each individual, regardless of whether it was better adapted than another individual, may not reproduce simply by chance. Random effects could eliminate the best-adapted organisms simply by chance. However, if some organisms can leave more copies of themselves in the next generation than another organism then they will have somewhat of an advantage in the next generation.

Natural selection can act on all phases of an organism's life, including fertilization, development, growth, and sexual maturity. At every stage of life, various selective agents can act on the individual through disease, predation, and parasitism, as well as developmental or physiological problems. Differences in the ability to withstand or avoid these selective agents are a function of the genetic make-up of the organism. Mutations cause changes in alleles, and if favorable, they are retained; if not, the mutation and the individual are eventually eliminated.

Any individual or line of individuals that increase in number relative to others in the population will have more copies of their genes and will affect natural selection.

*Fecundity* is the number of offspring surviving into the next generation. Sometimes it is advantageous not to produce the most offspring. If producing more offspring means less maternal investment, fewer offspring may result in increased survival. For example, many species of birds spend considerable amounts of time finding food and then feeding their offspring. If these birds had too many chicks, they would not have the time or the energy to find the required food items, and all the chicks could perish. Having fewer offspring in this example means a higher likelihood that some of the offspring survive.

There seems to be a negative relationship between the numbers of offspring and the overall chances of survival. In general this relationship can be stated as high fecundity = low survival and low fecundity = high survival rates of offspring. For example, humans produce few offspring with fairly high levels of survival, but fruit flies produce many offspring, each with a low probability of survival.

In sexually reproducing organisms, not all of an organism's genes are passed on to the next generation. All of the genes are found somewhere in the gametes but not together. Sexually reproducing organisms leave only a haploid version of their genome in their offspring. Consider a heterozygote that has two different alleles, *T* and *t*. If every gamete has a 50:50 chance of getting one of the alleles, there is the same 50:50 chance that one of the alleles will not be passed on to the next generation. If a population is large, these losses by one individual are compensated by another such that the overall frequency of alleles remains fairly constant, though not an exact replica, between generations. In small populations there is an increased chance of what is called a sampling error or *genetic drift*. When the populations are small one of the alleles may be completely lost from the population by chance alone. In other words, natural selection does not bring about the change in gene frequency but rather random events do. The effects of genetic drift may be considerable especially if we consider that many populations of higher organisms are small and isolated from each other (isolation is discussed later).

## Patterns of Selection

Because an organism can survive in a particular location, it is not an indication that natural selection has favored that particular genotype. The organism must be able to reproduce in the habitat. If the organism survives only as an individual and never reproduces, it is in reality poorly adapted. Natural selection does not measure the survival of individuals, but rather the ability of various organisms to leave the most offspring that can continue to reproduce. *Fitness* is a measure of the number of offspring produced by one individual genotype relative to that produced by another genotype.

There are basically three ways that selection can act within a population that result in a change in allele or gene frequencies as observed as changes in phenotypes. These three ways are *directional selection, stabilizing selection,* and *disruptive selection.* Each of these patterns is shown graphically in Figure 1.3. In general, directional selection favors one extreme phenotype; stabilizing selection occurs if an intermediate phenotype is favored, and disruptive selection occurs when two or more phenotypes are fitter than the intermediate phenotypes. The relationship between fitness and the phenotype is called the *selection regime.*

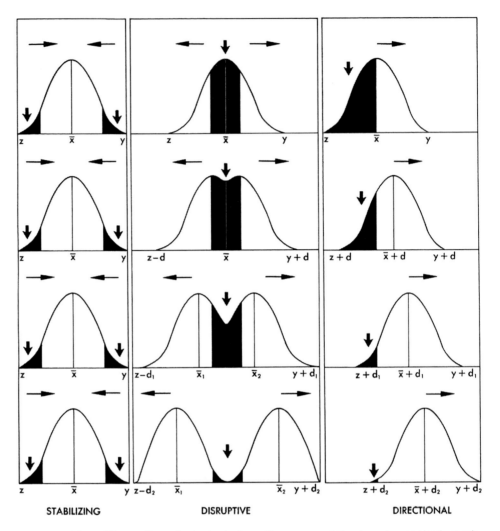

**STABILIZING**          **DISRUPTIVE**          **DIRECTIONAL**

**Figure 1.3**  Three different effects of natural selection acting on a population. In each example, the x-axis is the ordering from low to high of some phenotypic trait, and the y-axis is the frequency of the trait. Stabilizing selection is selection that acts on both tails of the distribution and maintains the mean characteristic. In disruptive selection, the mean phenotype is selected against that resulting over time in two separate populations with different mean characteristics. In directional selection, the effect is on one of the tails of the distribution, which results in a shift in the mean characteristic. (From Solbrig OT. *Principles and Methods of Plant Biosystematics*, 1st edition, ©1970. Reprinted by permission of Pearson Education, Inc., Upper Saddle River, NJ.)

Futuyma (1998) points out that fitness is most easily conceptualized for an asexual organism in which all individuals reproduce at the same time and then die. There are no overlapping generations. A more extensive discussion of fitness is provided by Futuyma (1998).

Evolutionary biology mostly has been restricted to academia. Application of evolutionary thinking has been applied to agriculture (both crops and animals) and to some extent in disease control. However, applied evolution (Bull and Wichman, 2001) is being used in a variety of contexts including design of biotechnology protocols that result in new drugs and industrially important enzymes, development of computer technologies, and the avoidance of resistant microbes and pests.

# Evolutionary Ecology

The study of organismal variability is the purview of evolution. Evolution is change. However, this definition is much too broad. Biological evolution involves the modification and diversification of organisms over generations. Evolutionary thought and studies have prevailed since Darwin opened the window and exposed a process that seems to explain much of biology. Over the past century, enormous strides have been made in our understanding of the theoretical and empirical processes that drive evolution. The discovery of DNA and genes and the development of molecular biology have opened up previously unknown processes, expanding our understanding of how evolution operates. However, it has become clear that evolution alone is not sufficient to understanding much of the pattern seen in nature. Ecology and evolution became united in the subdiscipline of evolutionary ecology, which sought to understand how evolution affects the ecology and how ecology affects the evolution of organisms. Much insight into the biology of many organisms has been obtained because of the union of these two disciplines.

Microbiology has been, in general, a science outside of both evolution and ecology even though many important observations that have advanced evolution have come from microbiology (e.g., *transformation*) and lateral transfer of genetic information. In the 1960s, a few microbiologists began to leave the laboratory and sought to understand some of the incredible processes that were occurring in nature that seemed to be controlled by microbes. Microbial ecology became a discipline.

Many of the studies performed by microbial ecologists have been describing the distribution or prevalence of various types or strains of microorganisms in a variety of habitats. Some of these habitats included the mouth, digestive system, various plants, sewage, oceans, rivers, and extreme environments such as hot pools, deep ocean vents, and Antarctic ice. The ability to sample and then describe the microorganisms found requires careful planning and an understanding of the environmental conditions that maintain the health and survival of the organism.

Microbial ecology was strongly influenced by clinical and medical microbiology, which required that microbes be cultured to pure culture so that a name could be attached. Over and over, it has been shown that there are millions on millions of microbes in environmental samples, but only a few of these organisms can be made to grow on laboratory media. Clinical and medical microbiologists have been successful in identifying the causes of many microbiologically induced diseases. Treatments of these diseases required knowledge of the basic biology of these organisms and the discovery of various antimicrobials that were sometimes species specific. Unfortunately, most environmental microbes defied isolation.

Were the microbes that grew on laboratory media representative of all the other unculturable microbes or an exclusive subset? To answer questions like these, researchers had to know something about relatedness. Comparisons required some level of knowledge of differences or similarities. In medical microbiology, disease-causing organisms were grouped into divisions based on various phenotypic properties. Classification of microbes in medicine is essential for effective treatment, but what about environmental studies? Do we really need to know the phylogenies of the microbes taken from nature? Species designations have clearly defined meanings for higher organisms (discussed later). Microbes live life differently from higher organisms and so we ask do such designations have any meaning for microbes?

In 1953, Watson and Crick successfully described the double helix structure of one of the most important molecules—DNA. DNA and its related molecule RNA have been found in every living organism. There are no exceptions. Many wonderful experiments have shown that DNA is the molecule of inheritance. The early evolutionists were hampered in their quest for understanding by not knowing the fundamental units of inheritance (i.e., the gene). These scientists knew there had to be something physical that contained and transferred genetic information, but they did not know what that material was. Discovery that DNA was the information molecule has propelled biology through the last half of the past century and into this new century. Why did life select this molecule over other molecules to carry the information of life?

The molecules of life are important in any discussion of evolution and especially evolutionary ecology. The history of those molecules forms the basis of evolution. A basic understanding of those molecules is necessary if we desire to understand evolution and, more importantly, if we seek to understand evolutionary ecology. In many ways, microbial ecology is essentially gene ecology because of the high level of environmental responsiveness expressed by microbes and the fact that there is not much more to a microbe than a bag of genes and a few gene products. Life has surrounded microbes with a cell membrane and, in some cases, a cell wall to protect and facilitate the life of the microbe. However, the incredibly small size of most microbes puts them in constant interaction with their environment. Unlike many higher organisms that are able to moderate their internal and external environments, microbes are a product of their environment. Microbes can in some ways alter their environment and make it more suitable for growth; however, they are very responsive to changes in the physical conditions to which they are exposed.

A basic understanding of two concepts (i.e., the molecules of life and the species concept) is essential before we begin any discussion of microbial evolutionary ecology. A thorough discussion of both of these concepts would take volumes. However, we must cover the topics sufficiently to understand the theoretical and empirical data that are presented in much more detail later. Much of ecology and evolution is based on the species concept, and most of evolution is based on changes and modifications of the molecules of life. We begin with a discussion of the molecules of life and present some models on which of these molecules were most important to earth's earliest inhabitants. To understand natural selection and evolutionary change, a basic knowledge of possible and plausible scenarios concerning the origin of life is needed. However, these models of the origin of life are likely scenarios and are not definitive.

Because all meaningful discussions of evolutionary ecology are based on the concept of a species, we will spend some time describing the various species concept models and their misapplication to microbes. As with the origin of life models, the concept of a species is difficult to pin down, especially for microbes.

Our presentation of microbial ecology sometimes follows paths known to most microbiologists, but at other times, we will discuss theoretical and empirical evidence developed for higher organisms and seek to find whether the concepts and predictions are applicable to microorganisms. Sometimes, we will be successful in application of evolutionary ecology principles to microbes; at other times, it may seem that we are trying to force concepts on an unyielding subject. If the approach we are taking is successful in helping microbial ecology students to see the world differently and to ask questions differently results in greater insight into the workings of nature, our task has been worthwhile.

# Molecules and Origins of Life

2

Ecology seeks to explain patterns in species abundance and distributions. Evolutionary ecology considers how the ecology of an organism affects its evolution and more specifically the transfer of genes from one generation to the next. The genealogy of an organism is the complete uninterrupted linking of beginnings to the present. Life on this planet had some beginning. All organisms that are alive today can trace their genealogy to some time when there was no life. This is an extremely disconcerting concept. It conjures up a scenario in which, at one moment, the chemicals of life are floating around and, at the next moment, these same chemicals are working together inside some primordial life form that can reproduce and make copies of itself. That concept takes a leap of faith, but life did originate. It did begin somehow and somewhere. Although the pathways which led to life may or may not still be among us, it seems appropriate that we discuss some of the numerous models of the origin of life. These discussions are essential in helping us understand the basis of life and to understand what natural selection can act on that has resulted in the diversity of life we have before us today.

All early life was microbial but not necessarily prokaryotic. By this I mean that the complex prokaryotic organisms that microbiologists study today were probably not the first living things. Prokaryotes are much simpler than many higher organisms but nevertheless the diversity of metabolism and function within the prokaryotes is immense and very complicated and complex. The first organisms had to be much simpler and easier to construct than even the simplest microbes of today.

All models on the origin of life are based on there being a distribution of various molecules that either are required for life or promote the existence of life. It is very important that microbial ecologists understand the chemicals of life as a precursor to understanding life. The absence of some of these molecules would have made life either nonexistent or perhaps very different from the life forms we see today. After our discussion of the molecules of life, we will discuss various models that seek to describe the origin of life.

These models and concepts are at best guesses about how life originated, because no one knows the exact conditions that prevailed at the site where life first began. However, based on our knowledge of the chemicals of life, these models give us insight into primordial conditions and show us some of the interactions that had to occur before a replicating life form could arise.

# Chemistry of Life

## Water

The first molecule that is important in the evolution of life is water. This is truly a remarkable molecule that has properties that allow life to form and survive. All life, regardless of where it is found, requires liquid water. Fortunately, water is the most common liquid found on the earth, and it has several properties that have allowed life to evolve and to perpetuate.

Water is made up of only two elements: one atom of oxygen and two atoms of hydrogen. Each atom of hydrogen is linked to the oxygen atom by a *covalent bond*. Because the molecule has equal numbers of electrons and protons, it is on paper neutral. However, oxygen, because of its mass, has a greater attraction for electrons than the hydrogen that makes the region around the hydrogen atoms slightly positive and the region around the oxygen slightly negative. The water molecule is *polar*. When two water molecules come close to each other, the oppositely charged regions form a weak bond known as a *hydrogen bond*. Each water molecule can form four hydrogen bonds. It is the hydrogen bond (hydrogen bonds are not found exclusively in water) that gives the water molecule properties that are important to life. In the following paragraphs, we discuss some of the properties of water affected by hydrogen bonding.

*Surface tension* is brought about by the *cohesion* of water molecules. Water holds together at its surface because of its attraction to itself. Water striders and other insects can walk across the surface of water. Many people have placed a needle on the surface of water and observed the float. Surface tension or the hydrogen bonds between the water molecules prevents the needle from sinking. Water forms spherical drops because of surface tension. As soon as a drop of water breaks free from a stream of water it immediately forms a spherical shape. Because water is a polar molecule it will be attracted to any charged surface. The ability of water to wet a surface is due to *adhesion* or the attraction of two different substances. In this case it is the attraction of water molecules to either positive or negatively charged surfaces.

Water can move up tiny tubes or spaces because of cohesion and adhesion. This movement is termed capillary action and it is important in the movement of water in plants but also in the movement of water through various inorganic and organic matrices like soils. The water is attracted to the surface of the tube by adhesion and pulls other water molecules up through cohesion.

Hydrogen bonds in water are also responsible for the high *specific heat* of water. The specific heat is the amount of heat required to raise the temperature of a substance a given amount. It requires one calorie of heat to raise the temperature of 1 cubic centimeter (=1 gram) of water by 1°C. There are not many other substances with as high a specific heat as water. This is important biologically because it means it takes a lot of heat to raise the temperature of water. This means that organisms living in water will experience fairly constant temperatures especially in large bodies of water like oceans and lakes. Also large organisms are composed of significant amounts of water, changes in the temperature of those organisms will be a function of the water content and surface to volume ratios of the organism.

Water also has interesting properties as it freezes. As water gets colder the density increases to around 4°C. At that temperature water molecules are moving so slowly that they can form the maximum number of four hydrogen bonds but to do so the molecules have to move apart. This moving apart continues as the water freezes making frozen water less dense than the water at 4°C. This is an incredibly important property because it means that ice floats. If ice were denser than liquid water the ice would sink to the bottom of a body of water, accumulate, and eventually fill the body of water with solid ice. This is a condition that would make life difficult.

Water is also a good solvent because of its polar nature. Many important substances in living organisms are in solutions. These substances include gases, nutrients, and food. The polarity of water facilitates the separation of ionic molecules. Many molecules that are important to life, such as sugars, are also polar, and they attract water molecules and dissolve in it, making their distribution possible.

## Biological Elements

Six elements make up nearly 99% of all living tissue. These six elements are carbon, nitrogen, phosphorous, hydrogen, oxygen, and sulfur. Considering there are 92 naturally occurring elements, this number seems quite small. These six elements are not among the most abundant elements at the earth's surface. Life did not evolve to take advantage of other extremely abundant elements like silica. Why is life made up of such a few elements and why these six? Part of the answer is that each of these elements requires an addition of electrons to complete the outer energy levels, and they all are able to form covalent bonds. These elements are also relatively small, and that means that the bonds they form result in tight stable molecules. Each of these elements (except hydrogen) is also able to form bonds with more than one atom. The possible number of combinations among these six elements is immense and diverse. Nature has produced thousands of compounds based on these six elements and man has synthesized many more.

Carbon in particular is able to bond with other carbon atoms in a variety of configurations and sizes. This diversity in form results in diversity in function. Millions of organic compounds (i.e., contain carbon) have been identified. These organic compounds can also include hydrogen, oxygen, nitrogen, sulfur, phosphorus, and many other elements and salts. All of these complex compounds made by living organisms are the products of specific genes and/or other gene products such as enzymes.

Every living cell contains a variety of molecules including both organic and inorganic. Many of these molecules are present as charged ions. Considering that 99% of all living tissue is made up of six elements the remaining 1% of biological mass is principally composed of inorganic ions. The positively charged ions are mostly $Na^+$, $Ca^{2+}$, $Mg^{2+}$, $K^+$, and $Fe^{2+}$, and the negatively charged ions are $SO_4^{2-}$, $PO_4^{3-}$, and $Cl^-$. The positively charged ions are important in many enzymatic reactions and functions.

Microorganisms have evolved ways to capture and sequester these important molecules. Living things have many other types of molecules that are essential for life including lipids, fatty acids, proteins, numerous enzymes, and vitamins to name a few. Each of these molecules of life, while fundamental to the survival of living organisms, is the product of cell metabolism. Any model that seeks to describe the origin of the simplest life form must be based on chemical interactions before the metabolism of these molecules (i.e., the constituent molecules need to exist prior to life originating).

# Early Atmosphere and the Beginnings of Life

Geological evidence suggests that the atmospheric chemistry of today is very different from that found 4 billion years ago. Most of the building block chemicals of life including oxygen, nitrogen, hydrogen, sulfur, and carbon were present in this earliest of atmospheres but not in the forms that much of life today uses. The actual forms of each of these elements is still being debated but it is generally accepted that free oxygen ($O_2$) was basically unavailable and that the other elements that make up life were present in the atmosphere or in the waters that covered the earth in simple molecules. Much of what follows is modified from Casti (1989).

In 1922, the Russian biochemist A. I. Oparin came up with the first testable hypothesis about the conditions and events that preceded the first living things. Many people had ideas of how life came to be found on this earth, but few of these ideas were testable. Scientific hypotheses must be testable and generate questions that can be answered through experiments. Oparin reasoned that the primordial atmosphere was reducing rather than oxidizing and as such was filled with methane, ammonia, hydrogen and water vapor. If energy in the form of lightening, volcanic heat, ultraviolet light and other sources of radiation were introduced into mixtures of these gases, he hypothesized that organic molecules would form. Not just any organic molecules but amino and nucleic acids, the basic building blocks of living organisms. Given time and the absence of oxygen these organic compounds could accumulate in the oceans until sufficiently concentrated that the first living organisms could form. In England, J. B. S. Haldane a few years later formulated a similar hypothesis and called the resulting mixture a "hot dilute soup" which has been modernized into the Primordial Soup Theory.

The Oparin-Haldane hypothesis was presented in the early to mid 1920s but was not seriously tested until the 1950s. Why? Remember that spontaneous generation had been laid to rest by Pasteur in 1864. Embodied in this hypothesis was the essence of spontaneous generation. Life could arise from inorganic materials that had been converted into organic compounds. Who wanted to go down that path again and face a scientific audience that had finally accepted that life came from life? In the early 1950s, a graduate student at the University of Chicago was willing to test the hypothesis. This young student was Stanley Miller. Miller was a student of Harold Urey, who had argued in a much more convincing and thorough manner than Oparin that the earth's early atmosphere was reducing and a good place to synthesize the molecules of life.

## Miller Flask Experiment

The basic experimental design Miller used to test the hypothesis is illustrated in Figure 2.1. The primordial atmosphere was simulated using ammonia ($NH_4$), methane ($CH_4$), hydrogen ($H_2$), and water vapor ($H_2O$). An electrode attached to a power supply supplied the energy required. The sparks created by the electrode were meant to mimic lightning. The whole mixture was cycled through a cooling tube that condensed the gases and resulted in a simulated rainfall. The water was slightly heated to promote evaporation. After one week, Miller analyzed the water and found significant amounts

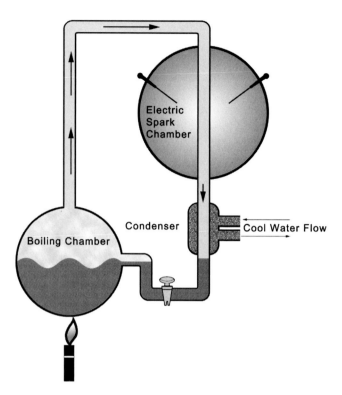

**Figure 2.1** Diagrammatic representation of the experimental apparatus used in the Miller-Urey experiments.

of various amino acids—specifically glycine and alanine, two of the basic building blocks of proteins. Since this experiment was initially conducted, numerous other studies have been performed altering the sources of energy, the temperatures of the water, the starting mixture of gases and each experiment has produced slightly different organic molecules. These studies are extremely important because they demonstrate that the basic molecules of life could originate through totally abiotic means. The studies remain theoretical because we are unable to show that this is the way these molecules came into existence. There is a tremendous amount of evidence in support of the process and most biochemists agree that something akin to these reactions took place in the early earth's history.

Once the chemicals of life were formed we still did not have life. Life is more than the sum of the chemicals that make it up. Every organism that dies still contains, at death, the molecules of life in pretty much the same proportions and concentrations and yet there is not life. Therefore just because certain molecules can be formed through these amazing processes does not mean that we understand how life came into being.

## Which Molecule Came First?

The origin of life is a chicken-or-egg type of problem. Certain molecules are needed to catalyze or code for the formation of other molecules, which are catalyzed or coded for by the other molecules. Let us examine this problem in more detail, because evolutionary ecology is based on these molecules, their synthesis, and regulation.

DNA is divided into short sections that code for certain specific proteins or code for the activation or inhibition of various chemical activities within the cell. These sections are often referred to as *structural* or *regulatory genes* and together they contain the information necessary to construct the organism. This information can be passed on to offspring. Protein synthesis occurs at specialized combinations of RNA and proteins called *ribosomes*. Each group of three base pairs that has been *transcribed* is called a *codon*. Each codon is associated with one of the twenty amino acids that make up the proteins of life or they code for a stop signal that ends translation. Because there are four different bases that make up DNA and there are three bases in each codon, there are a possible 64 different codes. However, there are only 20 amino acids used by all living things. Why are there not more amino acids? The genetic code contains some redundancy, a fact readily observed in Figure 2.2. Some amino acids are coded for by several codons, whereas others have only a single codon. This redundancy prevents serious conformational problems because the simple base substitutions often result in the same amino acids being coded.

We can simplify an extremely beautiful and complex process down to the following schematic designated the *Central Dogma of Molecular Biology* by Francis Crick, the co-discoverer of DNA.

$$\text{DNA} \xrightarrow{\text{transcription}} \text{RNA} \xrightarrow{\text{translation}} \text{Protein}$$

As seen from the diagram protein synthesis appears to be one directional (i.e., from the genetic information of the DNA to the formation of proteins). Genes code only for proteins, a fact that many students fail to comprehend. Genes code only for proteins!

Do proteins ever code for DNA? There are examples of RNA being back coded (reverse transcription) but there are no examples of transfer of information from proteins back to either genetic molecule (i.e., RNA or DNA).

| | A | G | C | U | |
|---|---|---|---|---|---|
| A | lys | glu | gln | stop | A |
| | lys | glu | gln | stop | G |
| | asn | asp | his | tyr | C |
| | asn | asp | his | tyr | U |
| G | arg | gly | arg | stop | A |
| | arg | gly | arg | trp | G |
| | ser | gly | arg | cys | C |
| | ser | gly | arg | cys | U |
| C | thr | ala | pro | ser | A |
| | thr | ala | pro | ser | G |
| | thr | ala | pro | ser | C |
| | thr | ala | pro | phe | U |
| U | ile | val | leu | leu | A |
| | met | val | leu | leu | G |
| | ile | val | leu | phe | C |
| | ile | val | leu | phe | U |

**Figure 2.2** The genetic code consists of four nucleotides that code for 21 amino acids and stop frames.

A one-time competing concept to evolution by natural selection was that championed by Lamarck and known as the *inheritance of acquired characteristics*. In this concept any alteration or mutation in any cell of an organism could be passed on to subsequent generations. The classic example is of a giraffe constantly reaching its neck to feed on higher leaves, which results in a longer neck, and the passing on of the long neck genes to offspring. The proteins that build longer necks can alter the genes that produce them and the reproductive cells that pass the trait on to offspring.

We bring up this concept at this time because there are two competing models of how life evolved. One of these models suggests that genes or DNA came first, followed by proteins. The other model proposes just the opposite (i.e., that proteins evolved first and then genes). The first model is the basis of natural selection as understood today. The second model is very reminiscent of lamarckian evolution. We examine each model as a means of understanding evolutionary ecology.

Casti (1989) identifies three evident facts about life that a theory of its origin must meet:

1. There is life on earth.
2. All life operates according to the same basic mechanisms.
3. Life is very complicated.

Any theory about the origin of life must show how the early conditions of the earth allowed or promoted the rise of life forms. After life had arisen, explanations of facts 2 and 3 would require logical and plausible paths for how primitive organisms were able to evolve the complicated gene-protein linkage seen in all living organisms. All living organisms use the same genetic code and they all use the same small set of basic constituent molecules. All means all. Bacteria and sequoia trees, cheetahs and snails, college students and amoeba all use the same information molecules of life.

The gene-protein linkup is the fundamental problem in the origin of life. Before a living organism can synthesize proteins the genetic material must be read and translated into the appropriate amino acids. On the other hand the genetic code cannot be translated without special proteins (replicases) that facilitate the copying process. In addition to the gene-protein linkup problem there are a few other problems that must be solved before a model on the origin of life is sufficient. Casti identifies three additional hurdles:

1. *Genetic code/protein structure.* There are thousands of possible amino acids and nucleic acids. So why are there only five nucleic acids and twenty amino acids used by all living things?
2. *Chirality.* All molecules in Nature have a mirror image. That means that all amino acids and nucleic acids have both left- and right-handed molecules which are identical in composition but which twist in opposite directions. The direction of twist determines function or chemical action of the molecule. The molecules produced in Miller-type experiments have equal amounts of right- and left-handed molecules. Interestingly all life forms on earth use only right-handed nucleic acids and only left-handed amino acids. Any origin model should explain why these forms of the respective molecules are used exclusively.

3. *Junk DNA.* With the exception of bacteria and viruses, all DNA contains long sections that do not code for any proteins. These junk segments have to be edited out before a protein can be made. Why is this material even here? If bacterial-like creatures were the first living things to come into existence then we need to understand how this extra DNA was incorporated into higher organism's genome.

## Genes-First Models

Let us consider the genes-first model. The essence of the gene-first model is that the first living things were not real organisms but rather replicators of random origin formed from the chemical constituents found in the primitive oceans. There were no proteins and therefore no early replicase enzymes. The major problem of this model is how the replicators form and how they replicate.

Some researchers have performed experiments that demonstrate that RNA can act as an autocatalyst by cutting out a central portion of itself and then resealing the cut ends. It has been shown that some RNA acts as an enzyme by cutting up RNA molecules that are different from it. This self-catalytic RNA can join several short strands of RNA together into chains under conditions that mimic the early earth.

Six major steps are involved in the gene-first model:

1. Start with a primordial soup that contains randomly constructed proteins and lipids (fatty acids) to be able to construct fragments of cell membranes. Nucleotide units must also be available for the construction of nucleic acids.
2. At least one self-catalytic replicating RNA molecule forms by chance. This molecule is not a gene because no proteins are formed. There is no unique nucleotide sequence. The RNA develops a range of enzymatic activities.
3. The RNA molecule evolves in self-replicating patterns and learns to exert control over proteins. The new proteins are better "enzymes" than the RNA was.
4. A series of interactions that are both complex and cooperative occur between nucleic acids and proteins.
5. DNA eventually appears which gives a stable, error-correcting information molecule.
6. RNA is no longer the premier molecule having been replaced by DNA as the information molecule and by proteins which perform the earlier enzymatic functions more effectively.

The biggest question in this scenario is the emergence of the first replicator. This is a random event and presupposes that a subset of right-handed nucleic acids happened to come together and exert control over the other molecules. How difficult is it to randomly assemble even a small strand of RNA? For this molecule to provide continuity between generations the molecule must replicate with a fairly high level of exactness. Unfortunately if the error rate associated with replication is greater than $1/N$ where $N$ is the number of nucleotide bases in the chain the population will probably go extinct. The "proof-reading" step in replication is performed by enzymes but enzymes would not have been made yet because they require longer—much longer strands of nucleic acids. Small RNA strands would not be long enough to code for enzymes and specifically replicases. If you cannot code for the enzyme, it would be impossible to keep replication exact, and long strands of nucleic acids would all be different after just a few generations.

Siefert et al. (1997), using the complete genome sequences of five bacteria, suggest that features shared by these genomes must have arisen early in the evolutionary history of bacteria. While gene order is generally not preserved among these bacteria, there are at least 16 gene clusters of two or more genes whose order remains the same among the eubacteria. Many of these clusters are known to be regulated by RNA-level mechanisms in *E. coli*. This suggests that this type of regulation (i.e., RNA) might have arisen very early during evolution, and although the last common ancestor of these specific bacteria might have had a DNA genome, was likely preceded by progenotes with RNA-based genomes.

## Proteins-First Models

Let us now consider the models that presuppose that proteins came first and were subsequently followed by nucleic acids. There are two main models: that of Oparin and that of Fox. Both of these models have received considerable attention and critique. The first model is based on a series of observations and experiments by A. I. Oparin, the author of the original primordial soup recipe. Oparin has shown that when certain oily liquids are mixed with water that they form small droplets that are called *coacervates*. He added an enzyme that converted sugars into starch and found that the enzyme accumulated in the coacervates. When glucose was added the sugar diffused into the droplets where the enzyme proceeded to convert the sugar into starch and the droplet began to grow. At a certain size the droplet split apart and these "daughter" droplets would also grow and split as long as there was enzyme present. The critical link in this whole scenario is that the enzyme has to be present. Oparin felt that as more diverse molecules accumulated inside the droplet "metabolism" would become more diverse and life would begin. The formation of the initial enzyme from random processes is the weak link in this scenario. Furthermore, coacervates do not have any hereditary mechanism so natural selection could not act on them. The overall summation of Oparin's model is as follows:

Primitive cells (coacervates) → Enzymes or proteins → Genes

Oparin's observations and experiments took place before the discovery of genes and DNA so his model says little about inheritance.

In the 1960s, Sidney Fox came up with another model based on proteins first. This model was based on an observation that when amino acids in certain mixtures that included lysine, aspartic acid, or glutamic acid were heated under dry heat, they formed polymers. These polymers were different from anything found in biology, and Fox labeled them *proteinoids*. When the proteinoids were dissolved in water they formed millions of small spheres that had some nonspecific enzymatic capabilities. In the Oparin studies, the enzyme was very specific but added to the mixture by the researcher. In the Fox scenario, nothing was added, but some enzymatic activity was found. In summary, the model is as follows:

Amino acids → Protenoids → Cells → Genes

**29**

This model has been attacked by many scientists because the dry heat conditions seem difficult to come by and because, like the Oparin model, there is nothing for natural selection to act on.

## Dual-Origin Models

We have spent some time developing theories that from all appearances seem to be deficient in one or more points. These theories cannot be proved or disproved, but they can be supported with evidence from carefully designed experiments. All the theories presented are based on a "chicken or the egg" scenario or that one molecule came first. An alternative approach might be to consider a dual origin. Both proteins and nucleic acids are needed for life to propagate along the lines that all living things seem to follow today.

It is particularly difficult to see how nucleic acids with enough base pairs could have been made in the earth's early conditions. Remember that DNA and RNA have three major components: bases, phosphate, and a sugar. Using Miller flask-type experiments, researchers have been able to synthesize nucleotide bases in the laboratory but only under much colder environmental conditions. Sugars have been synthesized using formaldehyde under very restricted conditions. Phosphate is a natural component of oceans and rocks. Even if the scenarios of how the components can be made are correct or nearly so, there is the problem of how to get the component parts together. Not only do they have to be put together, they have to be put together in the right sequence every time. Nucleotides tend to dissolve in water, a condition that does not promote life.

Proteins carry a sort of genetic code. The order of the amino acids is directly related to the genes that code for them. The order of the amino acids contains the information found in the genes. This information in the absence of nucleic acid genes may have served as a template for making similar protein molecules. This template would require some structure that would support the protein in a fashion that allows the information (i.e., amino acids sequences) to be read, something akin to a ribosome. The basic gist of these models is as follows:

$$Cells \rightarrow Proteins \rightarrow \rightarrow RNA \rightarrow DNA$$
Much later

There are origin of life models that include clays and silica as the support structures that allow translation of proteins to occur. In these models, the crystalline substance would grow through natural abiotic means and any molecule attached through surface charges or otherwise would or could grow as well. We will not spend any more time on this subject, other than to point out that there is evidence or support for these notions from a wide variety of sciences, as summarized by Casti. These include the following ideas:

1. *Biology.* Genes are pure form and not substance. Evolution can act only on this type of replicable form.
2. *Biochemistry.* Nucleic acids, including RNA and DNA, are complex molecules that are fairly difficult to make. They were probably late arrivals geologically and evolutionarily speaking.

3. *Construction industry.* Materials can be added or subtracted during evolution that can lead to mutual dependencies, such as the components of the major biochemical pathways.
4. *Structure of ropes.* Gene fibers can be added or subtracted without adversely affecting the continuity of the gene line. Casti suggests that this may be one way organisms based on a single genetic material could evolve into organisms based on an entirely different genetic material (i.e., proteins to nucleic acids).
5. *History of technology.* Primitive machines have to be made from available resources in the immediate region, and they have to work with little effort being put into them. As such, these primitive machines have a different design and construction compared with more advanced machines, which do not have to be easy to assemble or made from simple parts. In other words, the first organisms were probably very "low tech" compared with the organisms of today.
6. *Chemistry.* The formation of crystals is a low-tech mechanism that may have acted as a primitive genetic code.
7. *Geology.* Inorganic clay crystals are everywhere and continue to form through naturally weathering. Because of their net charge, they can attract and keep various molecules that have the opposite charge.

Let us consider a model proposed by the physicist Freeman Dyson that provides a quantitative prediction about the nature of primitive cellular metabolism that would favor a jump from disorder to order or life. The model has three main parameters: $a$, which represents the number of distinct amino acid or nucleic acid building blocks that were found in the original organism; $b$, the number of distinct chemical reactions that a primitive organism could catalyze; and $N$, the size of the molecular population in a chain of amino or nucleic acids that makes up such a life form. Dyson discovered that there were certain ranges of these parameters that produced interesting behaviors in his model. The ranges of interest were

$a$: 8 to 10
$b$: 60 to 100
$N$: 2,000 to 20,000

What do these values mean in the real world or, more importantly, in the primitive world where life was forming? All life on the earth today uses the same 20 amino acids, but the value of $a$ suggests that life could evolve with as few as eight amino acids. The Miller flask type of experiments produce most of the simple amino acids but not the more complex ones. In other words, 10 or so amino acids could form plenty of diverse proteins before the other amino acids came into being. Alternatively, the model fails if $a$ is less than 4, which implies that there is not enough chemical diversity in four nucleic acids to go from disorder to order. With $b$ in the range of 60 to 100, the predicted chain size of primitive proteins, the model can support a fairly high replication error rate.

Figure 2.3 summarizes aspects of the Dyson model. The transition from disorder to order is predicted to occur in the transitional zone that is highlighted. However, the most meaningful conditions are those near the cusp of the transition zone. The region labeled Dead Zone are states of the model where only disorder is found. This

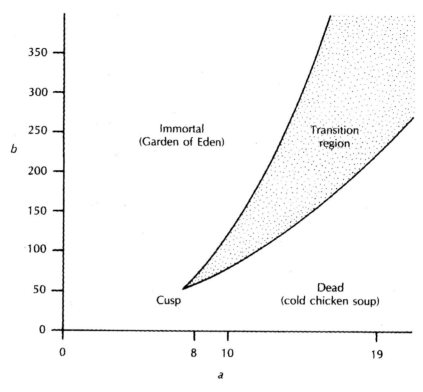

**Figure 2.3** Freeman Dyson model of the origin of life, in which *a* is the number of different amino acids or nucleic acids and *b* is the number of unique catalytic chemical reactions. Dyson predicts that life can originate at the cusp, where the number of amino acids is small and the number of reactions is around 50. (From Figure 4 in Dyson F. *Origins of Life.* Cambridge, UK, Cambridge University Press, 1999; reprinted with the permission of Cambridge University Press.)

occurs because there is too much chemical diversity and too little catalytic capability. The region labeled "immortal" has too little chemical diversity and too much catalytic activity to allow a disordered state to exist (i.e., no death).

Life is found on the earth and the chemistry and biochemistry of that life is very similar and based on the same basic molecules. It is important to consider how these molecules came into existence because evolution is based on changes to these same molecules. Evolutionary ecology then becomes the study of how these molecules of life are modified by their environment and in turn affect the organism. The most primitive fossils have the appearance of bacteria and it seems likely that they were the first living things or closely related to them. Although the fossils suggest that the form of these early organisms is similar to that of some of today's bacteria, we cannot determine whether their cellular biology was similar, even though it seems likely based on the theoretical considerations given previously.

We have not provided an answer to the origin of life. The exact conditions of the ancient atmosphere, the salinity of the seas, the availability of important molecules are almost impossible to determine. However, because all of life uses the same basic blueprint, it seems logical to make some of the assumptions previously described. Life is more than the chemicals that can be analyzed. Immediately on death, all the chemicals of life are present but the organism is dead. Understanding what makes those chemicals interact and function together in the orchestrated way of living things is still a major question. For microbes, the question has been around longer than for any other organism.

# Species Concepts and Speciation

## 3

Much thought has been given to the concept of a species. Darwin entitled his book *The Origin of Species*. By so naming his book, Darwin acknowledged that an entity called the species does exist. However, agreeing on exactly what a species is has been the subject of numerous and continued debate. Living organisms, including bacteria, show definable differences that can be observed and measured. Naturalists have used these differences to aid in the classification of life. The notion of classification is based on the assumption that observable differences are discreet in space and time. If we consider that evolution has been a continuous process since life originated, should we expect a discontinuous product such as a species to be formed, or should we expect continuous variation? Are the apparent discontinuities among living organisms due to random extinctions, or would we be able to identify unique groups even if there had been no extinctions and all life was before us? Variation both within and between species is recognized and either dealt with directly or tacitly acknowledged in discussion of species.

Early taxonomists placed new specimens into discreet hierarchical categories. The lowest and most restrictive category is the species, which is composed of a genus and species designation, the *binomial system of nomenclature* (Box 3.1). In the cataloging of species, a monumental undertaking that involved thousands of scientists, additional relationships among creatures were seen. Some organisms differing in hardly perceptible ways were classified as the same species but were never observed to mate and produce viable, fertile offspring. Other organisms differed in size, coloration, and shape and were mistakenly classified as separate species but later found to be the same species.

The species concept is ancient and originally linked to Providence, at least in Western philosophy. Groups or kinds of organisms were considered divinely created and immutable. A set of essential characteristics defined a particular organism and other observers could, based on those essential characters, identify the organism. Every living thing could, in theory, be placed within a specific group based on sets of observations. With the advent of evolutionary thought, the species concept became central to biology for a different set of reasons than the early naturalists were working under. Instead of Providence, natural selection and other evolutionary processes produced unique and wonderful products, which at the lowest discernible level were called *species*. However, the question remains: Are species real?

> **Box 3.1  The Binomial System of Nomenclature**
>
> Carolus Linnaeus developed a scheme to classify all known living organisms. His scheme was basically based on the form of the organism. Before the development of his system, the names of organisms were Latin descriptions that often were more than one word long. The simplicity and clarity of the linnaen system have been its strongest attributes. Each organism is given two names that correspond to a genus and a specific epithet. A specific organism is always called by both names because many organisms may have the specific epithet but are in entirely different genera (e.g., *Escherichia coli* and *Campylobacter coli*). Both of these bacteria have the species name of *coli*, but they are in different genera. The beauty of the system is that no matter the native language of the scientist, the organism being studied has the same name, decreasing confusion. In writing a binomial name, the genus is always written first and capitalized. A genus is a group of closely related species. The genus name can be used without the species modifier when speaking of this closely related group, such as all *Pseudomonas* bacteria. Taxonomy recognizes categories or levels of biological relatedness above that of the genus. Genera are grouped into families, families into orders, orders into classes, and classes into phyla or divisions. However, all classifications above the species level have no real biological meaning.

Is there a grouping of organisms such that no lower grouping exists? If so, what are the criteria on which the grouping is based? A species, by definition, is the fundamental unit of nature. The word *nature* comes from the Indo-European word *gene*, which is the same root word for genealogy. Identifying the fundamental unit of nature is literally finding the products of descent. Evolutionary descent is particularly difficult to determine. Based on numerous techniques, descent and origin of species can be inferred, but total reconstruction of lineage and relationships is not possible.

Within the scientific community and among nonscientists, species have a practical application and are the basic units of conservation and biodiversity. All attempts to preserve diversity or promote conservation are directed at species. For example, in the United States, legislation assumed that species were real and easily identified with the passing and enforcement of the Endangered Species Act. Much attention has been focused on the levels of biodiversity regionally, nationally, and globally and whether this diversity is decreasing. The concept of biodiversity has meaning only at the species level and presupposes that species exist. Higher-level diversity is confusing and misleading. Some higher-level taxonomic designations such as genus may exist as single groups within the next higher taxon but have numerous species within. Measurements of diversity based on the genus would be much lower than that at the species level.

The criteria used to designate a species, even in the earliest attempts, were threefold:

1. The organism had to have been described from nature by a taxonomist.
2. The organism must be recognizable to others (i.e., characters are constant over generations).
3. There must be fertility when crossed with like organisms.

However, many specimens were dead and had been removed (collected) from their habitats with little ancillary information on the biology of the organism available. Most species designations were based on observed morphological differences, and the third criterion was never observed. Naturalists continued to use the earlier criteria, whereas taxonomists working in museums relied more and more on morphology. It was not until the great synthesis of genetics, systematics, and evolutionary biology in the 1930s and 1940s that a broadly defined concept of a species began to be developed.

# Universal Species Concept

Let us consider the problems associated with defining a universal species concept. In science, there are three common criteria a concept must meet in order to stand (Hull, 1997). These are universality, applicability, and theoretical significance. Most of the species concepts developed today have trouble with being general or universal. Asexual organisms present an intractable problem for these concepts and greatly reduce their generality. If we consider that most of the species diversity on this planet is probably microbial or parasitic, failure to include these groups makes the concept less than universal.

As observed by Hull (1997), the more theoretically significant a concept is, the more difficult it is to apply the concept. Ease of application is important to taxonomists seeking to describe species and species relationships. Operational guidelines, such as morphology, therefore can be formalized and used to describe species. When morphology is not sufficient to discriminate closely related organisms, other operational guidelines, such as genetic relatedness, can be used to define species boundaries. However, operational designations of species define the products of evolution, not the process.

The concept of the species has an evolutionary and a taxonomic meaning, which can be independent of each other. How is membership in a species to be determined? Should it be based on morphology, physiology, reproduction, or some other criterion? Is a species a natural kind or a set of organisms or individuals? If the term *species* is or can be based on morphology, physiology, or reproduction, is there meaning in the term? What is the importance of having a species concept? If we consider that evolution has been acting continuously for eons, why do we expect discontinuous products of evolution? Do species really exist in nature? The concept of the species is fundamental to all aspects of biology and especially to ecology and evolution. Variability among and between organisms is the starting material for evolutionary change. However, much of this variability is plastic– under different conditions organisms have different responses.

Many species concepts have been developed and championed over the years. Although there is overlap and after careful examination *synonymy* among some of the concepts, each has its own set of assumptions, theory (at times) and applicability. Among the many concepts currently being discussed we will choose four that are representative: the biological species concept, the phenetic species concept, the evolutionary species concept, and the phylogenetic species concept. We briefly describe each of these concepts, but the referenced articles provide for a more complete comparison and description of each concept and the many others not discussed in this chapter.

# Biological Species Concept

One of the oldest and most developed concepts is the *biological species concept* (BSC). This concept was championed by Earnst Mayr in 1942 and subsequently developed, discussed, and applied by many of the early, prominent evolutionary biologists. It is still being refined and applied today, and it is the concept of species most widely used in biology by botanists, zoologists, politicians, resource managers, and others concerned about biodiversity.

In the BSC, species exist as part of a reproductive community. Individuals within a reproductive community must be ecologically accessible to others within the same environment, and gene flow must be actually or potentially occurring. Mating maintains the gene pool that "regardless of the individuals that constitute it, interacts as a unit with other species with which it shares its environment." Because the definition is based on reproduction, selection must favor the acquisition of mechanisms that promote breeding with *conspecifics*. *Reproductive isolation* becomes a mechanism for the protection of genotypes. *Speciation* under this concept is the process of achieving reproductive isolation.

The BSC excludes much of the life on this planet, including all *uniparental* species, as well as parthenogenic and self- or sib-mating species. Proponents of the concept sometimes call these other organisms *pseudospecies*, or as Gheslin (1987) posited, the organisms are not species. Hull (1997) said, "It should be kept in mind that very little in the way of gene exchange occurred during the first half of life on earth, and meiosis evolved even later. According to the biological species concept, no species existed for at least the first half of life on earth. Evolution occurred but in the absence of species." This is an important observation, and it relates directly to the application of the BSC concept to bacteria and other asexual organisms. Is it true that most of the world's biota are not species? This seems rather restrictive. It seems likely that if the products of evolution are species, the organisms that were living out their lives under the pressure of natural selection during the first half of the period of life on the planet were probably species.

The BSC has difficulty explaining hybrids because the concept is based on reproductive isolation. Hybrids, especially fertile hybrids, indicate that species designations for the parents are not restrictive. However, many species can hybridize and continue to maintain temporally ecological and genetic identities. Under the BSC, individual parasites and bacteria could be considered a species and each egg or fission a speciation event.

Although the BSC is widely used practically and theoretically, there are several problems associated with the concept. Mayden and Wood (1995) identified 10 elements of this concept that they considered counterproductive for understanding biological diversity:

1. Absence of a lineage perspective
2. Nondimensionality (With this concept, species exist in a brief segment of time with no linkage to the past. Spatially, the concept applies to other organisms that come in contact. Although the BSC has been extended to potentially interbreeding populations, the application is restricted to specific locations and times.)
3. Erroneous operational qualities used as definition

4. Exclusion of non–sexually reproducing organisms
5. Indiscriminate use of a reproductive isolation criterion
6. Confusion of isolating mechanisms with isolating effects
7. Implicit reliance on group selection
8. Its relational nature (i.e., A is a species relative to B and C because it is reproductively isolated from them)
9. Its teleological overtones
10. Its employment as a typological concept, no different from morphological species concepts

# Phenetic and Related Species Concepts

The phenetic species concept is based on *numerical taxonomy*. Species are defined on the basis of overall similarity or divergence in characters that can be given a numerical score, which is usually the presence or absence of the trait. The lowest taxonomic unit in phenetics is the *operational taxonomic unit* (OTU). The OTU is defined in terms of covarying characteristics using various statistical methods. This concept is nondimensional. Individuals for whom variance in characteristics is lower within a group than between groups are considered a distinct taxon. Species do not exist as lineages under this concept. Many molecular and morphological species concepts are basically phenetic. Multiple characters are measured or observed and the responses used to describe relationships. This concept is primarily operational.

# Evolutionary Species Concept

Being dissatisfied with the nondimensionality of the BSC, Simpson (1961) developed the evolutionary species concept (ESC), which extends the BSC through time. Simpson defined evolutionary species as groups that evolve separately from other such lineages and possess their own unitary roles and tendencies. Wiley (1981) made a few modifications to the concept and reworked it so that unitary roles and tendencies were replaced with evolutionary tendencies and historical fates. Wiley and Mayden (1997) argue that the ESC is the only available concept with the capacity to accommodate all known types of biologically equivalent diversity. The ESC is not an operational concept. As described by Mayden (1997), the concept "accommodates uniparentals, species formed through hybridization, and ancestral species." Reproductive isolation is considered a derived characteristic.

# Phylogenetic Species Concept

The phylogenetic species concept is based on the idea that a species is "the smallest diagnosable cluster of individual organisms within which there is parental pattern of ancestry and descendent" (Cracraft, 1983). A species is easily observed as the terminal organism in a lineage (Moreno, 1996) that has a common ancestor to other terminal organisms. The phylogenetic species concept does not rely on higher taxonomic and probably meaningless ranks above the genus level. Avise (1994) pointed out that the problem with this approach is that of determining the difference between gene phylogenies and pedigrees and how to recognize monophyly.

The approach is based on two very different types of data. The first uses comparisons of characters observed between fossil and living or extant species. The second is based on comparisons of molecular sequences obtained from different organisms.

## Bacterial Taxonomy

Many scientists hold to a five-kingdom system that includes plants, animals, fungi, protists, and monerans (i.e., bacteria). With the rise of modern molecular techniques, techniques that can determine differences in the building blocks of genes, an interesting system has been put forward. This system (Figure 3.1), based on differences in the patterns of individual building blocks of a certain type of DNA, suggests that there are only three major groups of living things. Two of the three major groups are microbial. All other living things fall into a single category. In other words, differences among puny, nondescript bacteria are greater than differences between cypress trees and humans.

A major problem with comprehending this immense genetic diversity in microbes is the inability to remember that they have been around for a very long time. Processes that select for certain traits have had plenty of time to fine tune and modify bacterial genes—far longer than all other organisms added together!

We have been considering the genetic diversity that exists among bacteria; let us return to the species problem. All taxonomic categories above that of species exist only in the imagination of man. In theory, the species designation should be the most "real" category. Microbes have been assigned to various species groups. In the past, this was done based on the ability of the microbes to perform various enzymatic and

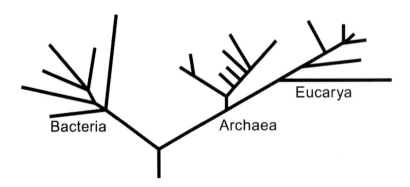

**Figure 3.1** The tree of life as determined by 16S rRNA sequences. This is a simplified version of the tree. Considerable detail has been added to each of the branches. Notice that most of the diversity of life forms is microbial. (Adapted from Woese CR, Kandler O, Wheelis ML. 1990. Toward a natural system of organisms: proposal for the domains Archaea, Bacteria, and Eucarya. *Proc Natl Acad Sci USA* 87:4576–4579, 1990.)

metabolic activities. Using modern molecular biological techniques, scientists can show the degree of relatedness by examining the similarity between DNA from one organism and another. This method of species designation has its own set of ambiguities. For example, at what level of similarity are two organisms the same species? There are no clear boundaries. Let us muddy the waters a little more.

Some bacteria are *promiscuous*, not in the moral sense of the word, but in the sense of the strict scientific meaning. Promiscuity is not restricted to one set or class. The term is most frequently used to describe a person who has multiple sexual relationships. In this sense, a promiscuous person is not restricted to a single individual. In a larger sense, promiscuous could be used to describe businesses that broaden their base and are not restricted to one customer. However, the term does fit bacteria. They are promiscuous. Bacteria can share their genes with other bacteria through several unique evolutionary mechanisms. Bacteria can even share their genes with totally unrelated "species" of bacteria, and there is evidence that they can share genes with eukaryotes.

What is a bacterial species? Many bacterial genes are mobile and can be passed between unrelated organisms. The oldest organisms on the planet are the most difficult to assign to a group. Over 3.8 billion years of evolutionary experimentation, they have broken down barriers that seem to exist among other organisms. They have developed a more cosmopolitan approach to life.

# Bacterial Species Concepts

We know remarkably little about the taxonomy of bacterial species. Even today, the characterization of bacterial species is swayed by the perceived need. Bacteria that are industrially important are better characterized than are medically important species and many times more so than ecologically important species. More than 3,000 species of *Streptomyces* have been identified and patented by the pharmaceutical industry.

Based on various newly devised molecular biological techniques, most known bacterial diversity falls within distinct *phenetic* clusters. These technique-driven methods provide some basis to classify bacteria, but they, as Goodfellow et al. (1997) state, "overlook the fact that species are the product of biological processes." The cookbook nature of these techniques makes them readily accessible to many, and the procedures have been used to describe bacterial species across many habitats. Bacterial species designations are usually assigned to groups of strains that show high levels of similarity in biochemistry, genetics, morphology, nutrition, and structure (Goodfellow et al., 1997). This approach to describing bacterial species is operational but not evolutionary sound, and it relies on the product and not the process to define a species. Even studies that seek to describe evolutionary relationships among bacteria and other taxa (e.g., Woese) use an operational method. Comparisons of 16S rRNA sequences can and do show relationships among groups of organisms, but they do not provide a clear definition of what a species is or how it came to be (discussed later). Numerous studies describe the microbial diversity found in soil, marine, estuarine, freshwater, acid mine drainage, thermal springs, and many other habitats using an operational definition of a species. These studies are important and underscore our need to develop an evolutionary definition of bacterial species.

Most bacterial species names in use today are *taxospecies*. The species designation comes from some application of numerical taxonomy (Sneath and Sokal, 1973). This procedure assigns a numerical code to a series of phenotypic observations of a particular isolate. Numerical taxonomy is the foundation of diagnostic test strips or plates that contain various organic compounds or that indicate the presence of particular degradative or metabolic capabilities. These methods are quick. However, classification to a specific species requires that a complete numerical description of the species be in a database. Because relatively few bacteria can be cultivated and most environmental isolates have not been carefully and thoroughly examined, the databases are restrictive. A number can be generated for an isolate, but classification of that isolate to a species may or may not be valid. Numerical scores are often unique and can be used to designate an unknown species. Levels of microbial diversity can be estimated using the frequency of the numerical scores. Numerical scores that are identical are considered to identify the same organism. Subtle differences in scores can be resolved using multivariate statistical analyses.

A second bacterial species concept is that of the *genomic species*. Genomic species are strains that show DNA:DNA relatedness values greater than some specified value and thermal denaturation values less than some specific rating (Box 3.2). From previous work with enteric bacteria, genomic species were recognized when individuals had 70% or more DNA:DNA relatedness with a difference of 5°C or less in thermal stability.

Neilsen et al. (1995) showed that there was good agreement between DNA:DNA hybridizations and data obtained through numerical taxonomy or chemotaxonomy. This correspondence gives some support to the taxospecies designations and does support the idea that metabolic and functional differences are maintained at the genetic level. The results from numerical taxonomy seemed to suggest that microbial responses are continuous. However, what appeared to be a continuum of responses under numerical taxonomy resolved into defined groups of bacteria when DNA:DNA pairings were made. Exceptions to these clearly defined groups have been found and draw into question the species designations or the universal application of this technique to answer bacterial species questions.

For example, within the genus *Xanthomonas*, DNA:DNA pairings range from 0% to 100% between various pathovars (Hildebrandt et al., 1990), although these pathovars appear indistinguishable when compared biochemically. Such incongruities need further investigation but suggest that the genus designation may be too inclusive. Bac-

**Box 3.2   DNA:DNA Hybridization and Bacterial Species**

Ward et al. (2002) pointed out that the concept of a bacterial species based on DNA:DNA hybridizations is fraught with difficulties. Among these difficulties is the arbitrary nature of the selection criterion: more than 70% homology. The danger of setting a number for species designations is that researchers may feel these numbers are truly thresholds that, once crossed, identify a species. However, true bacterial species may have hybridization percentages that are higher than 70% or that show greater than 97.5% sequence similarity. On the other hand, these types of data describe a continuum of possible relatedness among bacteria and suggest that species designations may be meaningless for some or all groups of bacteria. For many higher organisms, homologies and similarities at these levels would make many recognized species disappear. For example, most of the angiosperms would end up as a single group, and most primates would be designated as a single species. Given enough data, what appears now as a continuum of similarities may resolve into disjunctive unique groupings.

teria that have the same functional capacity as evidenced by the biochemical test but that differ genetically is suggestive of *ecological equivalents*. Ecological equivalents are species found in similar but isolated habitats that look and behave the same although they are unrelated phylogenetically. Examples from higher organisms include the mix of mammals found in North America and those found in Australia. Although no placental mammals were originally found in Australia, equivalents of dogs, mice, and deer can be recognized. The complete overlap in biochemical capacity of bacteria that differ genetically may echo the similarity of habitats colonized by these bacteria.

DNA:DNA pairings do not reflect the actual degree of sequence homology among groups of bacteria. Stackebrandt and Goebel (1994) estimated that bacteria with 70% DNA similarity have 96% to 98% DNA sequence identity. Embley and Stackebrandt (1997) observed that strains that show less than 98% sequence similarity rarely have DNA:DNA homology above 60%. From this observation, they suggest that any environmental sequence that shares 98% 16S rRNA sequence or less similarity to known sequences potentially represent new species. These levels of divergence are not trivial and probably reflect deep evolutionary differences at DNA:DNA pairing values less than 70%.

For bacteria that cannot be isolated or cultured from environmental samples, the amplification by PCR of 16S rRNA has been effective in describing some of the hidden diversity of microbial communities. The use of this technique is based on at least two major assumptions (Goodfellow et al., 1997): that there has been no lateral transfer of 16S rRNA genes and that the amount of evolution or dissimilarity between sequences is representative of the entire genomes of bacteria. If bacteria share rRNA genes between taxa, the use of the method to characterize groups would be compromised. Differences would not be due to evolution acting on these genes but to random mixing of the genes through lateral transfer and recombination. Given that these genes code for the maintenance and tertiary structure of ribosomes, it seems unlikely that they are not highly conserved. Fox et al. (1992) showed that these genes may be too conserved to serve as a means of resolving species differences. Bacteria that take up and incorporate these genes would run the risk of having nonfunctional or functionally reduced ribosomes. This would place these bacteria at a distinct evolutionary disadvantage. Although the possibility exists for transfer of these genes, it has low probability of occurring.

If divergence of bacterial rRNA genes is not tightly correlated with divergence within the genome, the technique cannot be used to differentiate species. Although the sequence of the ribosomal genes is highly conserved, differences do exist among species. These differences are presumably from point mutations and the degree of divergence related to the time since divergence of the species. If other molecules found in organisms do not show similar patterns of divergence, this marker would be considered suspect. Various other molecular markers such as 23S rRNA, ATPase subunits, elongation factors, and RNA polymerase genes, when used to discriminate among bacteria, show a high degree of correspondence with the 16S rRNA results.

Although this and other similar techniques can differentiate bacteria, the resulting designations are operational. We cannot directly infer evolutionary process in the divergence.

Species-specific differences in 16S rRNA genes are primarily found in certain hypervariable regions of the gene. If only part of the 16S rRNA gene is sequenced similarity cannot be determined. Only by sequencing the entire gene can comparisons between studies be effectively made. Relationships constructed from partial sequences are not stable and change with additional sequence data. To compare environmental samples from various sites and times, it is necessary to sequences the entire gene.

## Application of the Phenetic Species Concept to Bacteria

Most, if not all, bacterial species designations are based on phenetics. The compendium of bacterial nomenclature, Bergey's *Manual*, is based almost entirely on the notion of a phenetic species. Morphological distinctions separate bacteria into basic units such as rods, cocci, and spirochetes, but these classifications are subject to observer error and may change with the physiologic state of the bacteria being observed. Other coarse phenotypic divisions being used include the Gram stain, terminal electron acceptor, aerobe/anaerobe tests, pathogenicity, and various chemical and metabolic properties. To be recognized as a unique species, an isolate must be cultured and the following morphological and biochemical descriptions made: general morphology as observed by light and electron microscopy; various physiological and biochemical tests, including growth on various organic compounds; nitrogen source use; ability to fix nitrogen; pH range; various enzymatic tests; analysis of fatty acids; determination of G + C content; DNA:DNA hybridizations with closely related organisms; and phylogenetic analysis. This list is not comprehensive, but it does give an indication of the work necessary to describe a new species of bacteria.

Metabolic diversity can be estimated and scored using diagnostic strips or multiplates on which the response of an isolate to various organic compounds can be visualized. The metabolic phenotype of an isolate is used to characterize the organism. Theoretically, organisms with increasing similarity in metabolic response are considered the same species. Divergence is an indicator of species separation. Unfortunately, there really are no effective ways to determine degree of similarity or dissimilarity. Potential responses end up along a continuum of responses and no clear divisions are present. Researchers have performed these tests on many medical and industrial bacteria. Levels of discrimination can be quite high in the cases where numerous bacteria have been screened and some estimate of within "species" variation is known. However, environmental isolates present considerable difficulty to these methods because little if any information is available on the metabolic variance of free-living bacteria.

Falkwell and others have shown that bacteria isolated from the deep subsurface are metabolically indistinguishable with various aerobic pseudomonads. However, based on DNA:DNA pairings and 16S rRNA analysis of these isolates, they were found to be very different organisms from the pseudomonads. Another example involves the infamous "Jack-in-the-box" *Escherichia coli* strain. This bacterium has been implicated in the deaths of several people. Based on metabolic tests the organism is very similar to *E. coli* K12. Genetic comparisons between the two strains show little overlap. Is the Jack-in-the-box strain the same or a different species from the well-known *E. coli* strains?

On the other extreme, some bacteria have been shown to have incredible variance in their metabolic potential with low genetic differences. Without knowledge of this variance, specific isolates collected from the extremes of the metabolic distribution would be considered different species.

Phenetic species are functional in that distinct classifications can be made, but they fail to show evolutionary relationships and give little information on modes of speciation and change.

## Application of the Phylogenetic Species Concept

Moreno (1996) made several observations about the application of the phylogenetic species concept to bacteria. Remember that the phylogenetic species concept is based on two types of data: fossil and molecular sequences. Microbial fossils are rare and provide little discriminatory power. The evolutionary history of microbes has been inferred from molecular data that are expressed as trees or often bushes. Do molecular trees or gene trees represent species trees? This topic is still being debated, but there appear to be some genes that give support to the notion of a nearly one-to-one correspondence between genes and species. These genes often code for essential cell functions, and they are highly conserved sequences. Genes for ribosomal RNA, cytochrome $c$, elongation factor, chaperone proteins, ATPsynthase, and others are often used. The properties of these genes make them good for phylogenetic analysis at levels above the species, but they are usually not very good at discriminating between closely related sister species. Sometimes, the genes of interest may be duplicated in the same chromosome or be found on different chromosomes in the same organism and potentially increase the diversity within the same group. Sequence differences can be used to discern between closely related bacteria and determine if they are species but there are no clear guidelines on what the limits should be. At some levels of resolution, every individual can be classified as a species, and the concept becomes devoid of any meaning and usefulness in the study of ecology or evolution.

Figure 3.2 is an example of a tree constructed based on the 16S rRNA gene sequences. These trees are ubiquitous in the literature. Unknown sequences are compared with and against known sequences and degree of relatedness determined. In many such examples from the literature, major groups of bacteria are identified (e.g., α-Proteobacteria, β-Proteobacteria, γ-Proteobacteria). From these studies, it is fairly clear that most bacteria fall into the known major groups of bacteria but that specific designations are not as easily determined. Because all bacterial DNA in a sample is seldom sequenced, the true taxonomic diversity is unknown.

# Speciation

We have been considering the products of evolution and trying to determine which species concepts are most applicable to bacteria. We have seen that the evolutionary species concept is the most applicable and theoretically sound concept whereas the phenetic species concept is the most frequently applied to bacteria. The process of forming species (i.e., speciation) is an exciting aspect of bacterial evolutionary biology. Bacteria do things differently from other organisms. To understand how the unique evolutionary mechanisms of bacteria affect their speciation we must first discuss speciation as applied to higher organisms.

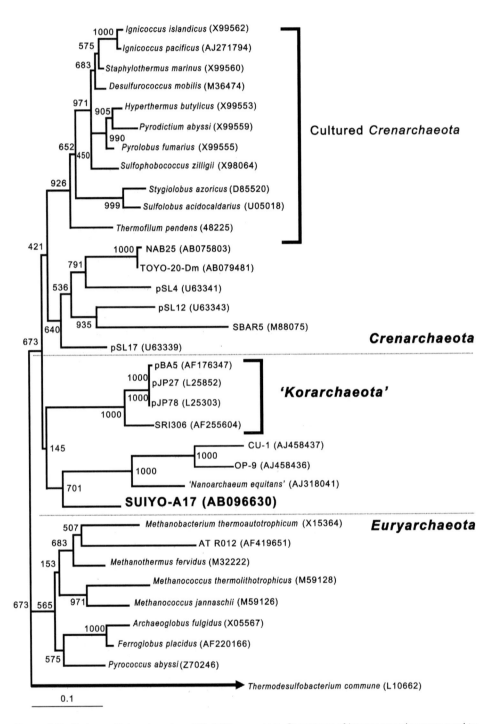

**Figure 3.2** Phylogenetic tree based on 16S rRNA sequences. Sequences of known organisms are used to determine putative relationships among unknown and usually uncultured sequences derived from environmental DNA samples. (Adapted from Figure 6 in Nakagawa T, Ishibashi J, Maruyama A, Yamanaka T, Morimoto Y, Kimura H, Urabe T, Fukui M. Analysis of dissimilatory sulfite reductase and 16S rRNA gene fragments from deep-sea hydrothermal sites of the Suiyo Seamount, Izu-Bonin Arc, Western Pacific. *Appl Environ Microbiol* 70:393–403, 2004.)

Speciation is part of the theoretical basis of the BSC. Most of the aspects of speciation have their genesis in this concept and its development. For most organisms, speciation occurs because there is heritable variation on which natural selection can act and there is some sort of isolation of populations. There can be considerable variation in morphology, physiology, and behavior among individuals found in widely dispersed populations. Variation that is observed due to geography is an indicator of differences in environmental selection acting on the local genotypes. Two patterns of geographic variation are the *cline* and *geographic isolate*.

The cline is a continuous pattern of change in traits and genotypes that results from the mixing of populations along some continuum or along some environmental gradient. This can occur across continental scales for widely dispersed species or over longitudinal gradients with much reduced geographic scales (e.g., rivers). Geographic isolates are populations that have been isolated by some barrier and are not able to share genes with other populations of the same species. Usually, the restriction in gene flow is not complete.

When isolation occurs spatially it is termed *allopatric* or *geographic speciation*. Allopatric situations can arise through *geographic barriers* or *founder effects*.

Allopatric speciation occurs when a population is split apart by the establishing of some barrier that prevents or interrupts gene flow between the two isolated populations. Because each population is under different selective pressures they will accrue unique genetic differences. Given enough time, these accrued differences may be large enough that if the barrier is removed, reproduction between the two populations fails because reproduction cannot occur or the hybrids formed have much lower fitness than either population. The differences accumulated will result in various *isolating mechanisms*, which are characteristics that prevent gene exchange from occurring. If diversification continues these mechanisms can become completely exclusive and the two populations are effectively new species. This definition of a species is based on the BSC. Smith (1980) lists the characteristics of species that are susceptible to allopatric speciation. These characteristics include

1. Species that have low reproductive rates
2. Species that produce few offspring
3. Species that have a long life span
4. Species that have late sexual maturity
5. Species that have high competitive ability
6. Species that have high vagility

The second form of geographic speciation is that brought about by *founder effects*. The name gives some indication of how this process occurs. A founder effect is observed when a single gravid female or a relatively few number of individuals or founders, colonize a new area. In contrast to allopatric speciation, organisms that are susceptible to founder effects demonstrate

1. High reproductive rates
2. Early sexual maturity
3. Large numbers of offspring
4. Short life spans
5. Low competitive ability

Founders are often found on the edge of a species range and experience little gene flow from the center of the population. Founder populations find new suitable habitat that is removed from the main population. These populations generally have much lower genetic diversity and in diploid organisms are more homozygous. As with the geographic barriers once a founder population is established in a new location the population can accumulate different adaptive changes and develop isolating mechanisms.

There are other nongeographic speciation scenarios. Principal among these is *sympatric speciation* or speciation without geographic isolation. Sympatric speciation takes place not on the periphery but in the center of a population living in a patchy environment. Sympatric speciation requires a *stable polymorphism* and *assortative mating*. Assortative mating means that organisms that are adapted to a particular patch or niche tend to mate with one another. Sympatric speciation is thought to occur in plants and in insect parasites of both plants and animals.

## Bacterial Speciation

Sympatric speciation may occur in parasite population without the need to invoke reproductive isolation through adaptive polymorphism and habitat preference (Meeûs et al., 1998). Sympatric speciation may be much more prevalent in parasitic species because hosts provide ample opportunities for niche diversification. This is another argument about the effect of scale in evolutionary and ecological processes. The potential habitats within a single host are immense. Parasites living in one part of a host may in reality be "geographically" isolated from other parasites in the same host. A similar argument could be made for both pathogenic and free-living bacteria. Although bacteria may be living in the same environment as measured by our available methods, they may be isolated from other individuals in both space and time.

Through three unique evolutionary mechanisms, bacteria can and do take up novel DNA and incorporate some or all of this material into their own genomes. These mechanisms are *conjugation*, *transformation*, and *transduction*. Each of these mechanisms is discussed briefly, but extensive reviews of the mechanisms are available. Genetic recombination of this exogenous DNA provides a significant source of genetic variation within bacteria. Not all bacteria are capable of recombination. The potential to recombine novel DNA may be a useful tool to separate bacteria into species.

## Mismatch Repair as a Speciation Mechanism

Vulic et al. (1999) described a mechanism for delimiting bacteria into species based on a specific mutation that affects the ability of bacteria to repair mismatched DNA. Many question whether bacterial species concepts are valid because they do not engage in sexual reproduction like other organisms. However, bacteria do engage in genetic exchange between individuals as mediated by plasmids, viruses, and the uptake of naked DNA from the environment. Based on these mechanisms, a potential exists for recombination in bacteria to be one useful metric for placing bacteria into species groups, although this approach may not be sufficient in all cases.

Vulic et al. (1999) selected one pathway, the methyl-directed mismatch repair (MMR), as a mode of speciation in bacteria. In the MMR pathway, genetic defects increase mutations and recombination rates, and this pathway may be important in speciation of bacteria. Even if speciation does not occur, this and similar pathways with defective repair genes may promote rapid adaptive evolution.

To test whether defects in MMR altered recombination rates, these researchers used a strain of *E. coli* as the founding strain and only source of genetic variation so the only source of variability would be from random mutations. They then allowed the strain to reproduce for nearly 20,000 generations. Some of the lines retained the MMR gene function, but others became defective. From these functional and nonfunctional gene lines, they constructed both donor and recipient genotypes so they could observe recombination. Based on pairwise matings between these independent lines, they found that the effect of mismatch repair systems on recombination rates was greatest in those lines that had evolved nonfunctional repair. This was probably the result of the lines being more sensitive to the recombination-inhibiting effect of a functioning repair system. Most importantly for our discussion, they demonstrated that an incipient barrier (i.e., reproductive barrier) can evolve rapidly during only 20,000 generations (<10 years under their experimental conditions) and influence speciation. The greater the inferred (based on time since evolving and mutation rate) DNA sequence divergence, the higher the rate of recombination (Figure 3.3).

The importance of reproductive isolation in the formation of species is strengthened when we consider that gene flow of any magnitude can swamp genetic

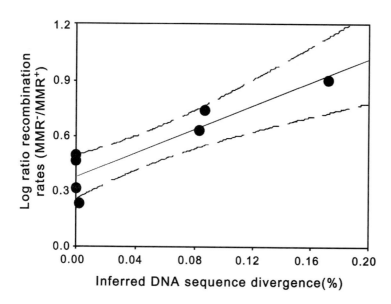

**Figure 3.3** Relationship between the inferred DNA sequence divergence and the rate of recombination. (Adapted from Figure 4 in Vulic M, Lenski RE, Radman M. Mutation, recombination, and incipient speciation of bacteria in the laboratory. *Proc Natl Acad Sci USA* 96:7348–7351, copyright 1999 National Academy of Sciences, USA.)

divergence between sexual populations. Understanding speciation in higher organisms requires that we understand the factors that prevent gene flow and allow genetic divergence. An understanding of speciation in bacteria would similarly involve knowing what drives genetic divergence between bacterial populations and what prevents recombination between divergent genomes.

Sniegowski (1998) considered the MMR hypothesis and found what he identified as two serious problems. First, the effects of genetic divergence in laboratory crosses cannot be directly equated to genetic (sexual) isolation between evolving populations in nature. Sniegowski discussed five steps needed for a successful recombination event to occur in a natural bacterial population:

1. Donor DNA must be taken up by the recipient.
2. The DNA must escape the recipient's restriction enzyme system that cleaves foreign DNA.
3. The donor and recipient DNA must form a *heteroduplex*.
4. The heteroduplex must escape the mismatch repair system, which will abort recombination between divergent sequences.
5. The donor gene product must function in the recipient genetic background. Vulic's hypothesis is concerned with the fourth step only.

Second, rates of recombination measured in prokaryotes cannot be directly related to rates of recombination in eukaryotes. Recombination rates in prokaryotes are controlled by the opportunities for recombination to arise. This is affected by microhabitat distributions of potential recombination genomes. Recombination rates in nature are fairly low, $1 \times 10^{-9}$ for *E. coli* and $1 \times 10^{-8}$ for *Bacillus*. It has been estimated that natural levels of recombination are probably too low to constrain adaptive divergence of bacterial populations in niche-adapted species.

# Rapid Speciation?

*Yersinia* is the genus of microbes responsible for keeping the human population from rapid growth during the middle ages. The Black Death, or plague, was caused by a species of this microbe. Based on molecular analysis of the extant species, it appears that the molecular progenitor for both *Y. pseudotuberculosis* and *Y. pestis* existed about 1 million years ago. These two species are genetically different from their nearest relative *Y. enterocolitica*. Although 1 million years seems like a long time, it is estimated that *Y. pseudotuberculosis*/*Y. pestis* and *Y. enterocolitica* formed separate species approximately 100 million years ago. The question is when did *Y. pseudotuberculosis* and *Y. pestis* separate into distinct species? There are no fixed genetic differences between *Y. pestis* and *Y. pseudotuberculosis*. *Y. pestis* contains no genetic variation. All isolates are genetically similar. Based on molecular analysis a common ancestor could have existed as little as 2,500 years ago (1,000- to 6,000-year range). It appears that *Y. pestis* arose as a single clone from *Y. pseudotuberculosis* because there is no variation among any isolates and because of the similarity. Although there are large phenotype and ecological differences between the two species there is very little genetic distance—so are they the same or different species?

# Operons

One distinctive feature of bacterial genomes is the *operon*, a cluster of co-transcribed genes that typically provide for a single metabolic function. Bacterial operons have evolved by the assembly of previously unlinked ancestral genes. The selfish operon model is distinct from other models in several ways: It provides a plausible mechanism for the gradual assembly of genes into operons; it provides a selection mechanism for assembly of gene clusters and for their maintenance over evolutionary time; it is consistent with the observation that genes providing for nonessential functions are found in operons; and it does not postulate that gene clusters initially provided any selective benefit to host organisms.

The selfish operon model contends that genes assemble into operons after horizontal transfer into naïve genomes. This model therefore predicts that genes providing for central metabolic functions are least likely to be found in operons, because genomes naïve to these functions are rare. Operons are likely to comprise genes that provide for unusual functions, which can effectively invade naïve genomes (i.e., useful but nonessential metabolic functions). To determine if horizontally inherited selfish operons can have a substantial impact on evolutionary history, the potential for diversification imparted by the gain of introgressed selfish operons must be compared with the potential for diversification generated by mutation and adaptation. To compare these values the rate of horizontal gene transfer among extant genomes must be elucidated.

One estimate of horizontal transfer rate is 31 kb every million years. Using this estimate, *Escherichia coli* has gained and lost nearly 3,000 kb of protein-coding DNA since its divergence from the *Salmonella* lineage. Functions provided by some of these genes would allow *E. coli* to explore novel ecological niches in a rapid and effective manner.

Differences among bacteria may be masked by recombination, which obscures species boundaries. The features that discriminate closely related bacterial taxa probably reflect sets of selective pressures inherent in their individual lifestyles; each species occupies a distinct ecological niche that provides selection for essential niche specific functions. No phenotype distinguishing between *E. coli* and *Salmonella* can be attributed to the differentiation of ancestral genes by point mutations, rather all described differences can be attributed to gain or loss of genes. Although genes providing for distinct functions must ultimately evolve through duplication and divergence, this phenomenon is slow and inefficient and would not allow the competitive exploitation of a novel resource required for bacterial speciation. Such a process would require the absence of strong selection. Novel functions probably evolve when selection is not intense. A function may evolve in a niche where the selection for the function is not critical. However, these functions would not be used to exploit a new niche.

# Genome Economization and Speciation

During microbial starvation, a genome-reducing mechanism (i.e., genome economization) occurs, in which prokaryotic cells in exhausted media can lose a part of their genetic information. If this occurs in nonessential genes, the rate of reproduction may increase. A cell with a smaller genome has a selective advantage over a cell with a larger genome in certain conditions because the cell does not have to replicate or synthesize the extra material. Differences in cell genomes size of 20% have been observed. If we consider that the size of the prokaryote genome is limited to 9.5 Mb, this limit may explain why primitive cells and modern prokaryotes have similar morphological complexity. Nonessential DNA is often located on plasmids or other mobile genetic elements (e.g., antibiotic resistance). Bacteria can be genetically diverse within and among populations, depending on the ecological conditions they are grown under and the amount of DNA they have culled or taken up through selection.

# Hypermutation

Mutator genotypes with increased mutation rates produce rare beneficial mutations more often than wild-type genotype allowing for faster responses to selection. Bacteria may increase their DNA under favorable conditions and this mechanism, which is capable of restoring lost genetic information, provides an advantage for cells in constantly changing environments. Similar DNA often occurs over large phylogenetic distance leading to a widespread horizontal interspecific gene transfer in bacteria (increasing the probability of the selfish operon). Bacteria are the only organisms that have been selected for the ability to take up exogenous DNA actively and recombine it with their genomes. This uptake may be the most important aspect of their evolution. Some barriers to lateral transfer of DNA among taxa do exist and may act as isolating mechanisms and promote speciation.

Prokaryotic genomes can be divided into two parts: exchangeable and non-exchangeable sequences. The latter ones cannot be transferred functionally between species. Change in these genes cause cell death or result in the cell being less competitive and replaced. Prokaryotes can be characterized not by reproductively isolated genomes, but by reproductively isolated sequences. Bacteria diverge when a part of the original exchangeable sequence becomes non-exchangeable due to constantly changing niches. As a result of spatial isolation (from niche changes), the gene flow within the so-called non-exchangeable sequences between two or more populations is interrupted initiating the process of speciation. During this isolation, the exchangeable genes can be transferred between species. This transfer can cause the differences in genomes size within species. This transfer may strengthen sequence isolation through the acquiring of new properties (ecotypes) and contribute to rapid adaptation to novel environments.

It appears that stationary-phase bacteria under stress (e.g., starvation) sometimes produce mutants in response to the stress. Because mutation is random, both deleterious and beneficial mutants arise from multiple molecular mechanisms that may be different from those found in rapidly growing cells. Some studies have shown that *E. coli* collected from many different habitats worldwide increase their mutation rates in response to starvation conditions (Figure 3.4). In contrast, laboratory strains of *E. coli* do not show as elevated a rate of mutation. Given that most bacteria may be living under stressed conditions, especially starvation stress, these mutations may be very important in microbial evolution. Although these increased mutations may result

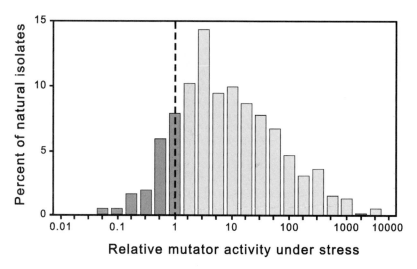

**Figure 3.4** Increase in mutator activity of aging *E. coli* isolates in response to environmental stress (starvation) as determined by measuring the frequency of base substitutions in an RNA polymerase gene in new and old starved colonies. *Dashed line* is the division between new isolates (left of line) and old isolates (right of line). (Reprinted with pemission from Rosenberg SM, Hastings RJ. Modulating mutation rates in the wild. *Science* 300:1382–1383. Copyright 2003 AAAS.)

in increased deleterious mutants, they are probably very important in helping these bacteria escape local extinctions and allowing them to adapt quickly to changing conditions.

# Genome Reduction

*E. coli* and *Haemophilus* both belong to the gamma subdivision of purple bacteria. Inside this subdivision, they appear to be closely related. However, they differ in some very significant ways: genome sizes vary from 4.7 Mb for *E. coli* to 1.8 Mb for *Haemophilus influenzae*. The natural history also is different for the two species. *E. coli*, a normal resident of the gut, can be and often are free living and can adapt to many conditions including changes in salinity and pH to name two. *H. influenzae* is an obligate pathogen requiring specific growth conditions.

Because of the very different preferred habitats of these two species we would expect them to have very different evolutionary ecologies. *Paralous* genes (Fitch, 1970) have been defined as copies issued from a duplication of an ancestral gene, with each copy having diverged before any speciation event. Based on analysis of paralous genes the last common ancestor to *E. coli* and *H. influenzae* was an organism having a genome size and a way of life similar to present-day *E. coli*. The progressive adaptation to parasitic life may have made certain genes dispensable. Alternatively the accidental loss of these genes could have been the stimulus for adopting such a way of life. *H. influenza* has a few genes not found in *E. coli*. Recombination events may be frequent enough to break up large-scale chromosomal arrangements and explain why there is little homology preserved between these two bacteria.

# Ecology of Individuals

# The Individual

## 4

## What Is an Individual?

Natural selection acts on individuals. Individuals reproduce, or they do not. Mutations happen in individuals and if favorable, are passed on to subsequent individuals within their posterity. The ecology of individuals is discussed in this chapter.

What is an individual? Although at the outset it may seem very clear what an individual is, the concept is really not easy to apply to many things in nature. If while walking along a forested stream you bend down and pick up a rock, is the rock an individual? A careful examination of the rock using the principles of geology may give insight into where the rock originated and how the rock came to be found at this particular time and place. The constituent parts of the rock, its mineralogy, can be thoroughly determined. However, these same geologic principles indicate, with few exceptions, that this particular rock did not originate in its current size, form, shape, or location. The rock had been modified by its environment. Is the rock an individual?

We can change the scale of our inquiry and ask whether or not stars, or star systems or galaxies are individuals? At the other extreme, are molecules, genes, or subatomic particles individuals? The answer both philosophically and through common sense is yes, at least from our frame of reference. Individual genes can be sequenced and individual stars can be studied. Are chromosomes or star systems individuals? From a distance, other galaxies look like individuals whereas our own galaxy does not appear so when observed from earth.

David Hull (1992) defined an individual as "spatiotemporally localized material bodies that either remain unchanged through time or else undergo relatively continuous change." The two best examples of an individual are organisms, usually vertebrates, and artifacts such as automobiles, both of which examples have inherent weaknesses that make the designation difficult.

The weaknesses of these two examples can be demonstrated through a series of questions (Hull, 1992):

1. What would happen if you gradually replace the parts of an automobile until all had been replaced? Would this be the same individual or a new one?
2. What if we kept all the old parts and reassembled them. Would this individual be identical to the original or a new one?

3. What if we gradually modified the original car into a van or a truck over time? Would it still be the same individual?

Similar questions could be asked about vertebrate organisms. Hull lists three traditional criteria that need to be met to establish individuality: retention of substance, retention of structure, and continuous existence through time (i.e., genidentity). Unfortunately, organisms do not meet the first two criteria. All organisms replace most, if not all, cellular constituents over their life spans, and many organisms go through significant structural changes. For example, organisms that go through metamorphosis change both form and function. Organisms are very much like cars that have had parts changed over time.

The third criterion, genidentity, has an evolutionary basis. Individuals are allowed to change over time just as long as the changes are continuous and not too abrupt. Evolution is a change in allelic frequency at a single locus. However, our perception of individuals is biased by our own biology. Humans are spatiotemporal entities, which experience continuous, not too abrupt, changes. Other vertebrate species have similar characteristics. When humans direct their observations to nonvertebrate organisms their ability to determine individuals is greatly reduced. This is especially true for clonal and colonial organisms.

For example, the Portuguese man-of-war has been determined to be a floating colony of many different organisms in different life stages. Modified medusae and polyps form specialized structures and provide various functions, including mobility, prey capture, feeding, and reproduction. Each zooid contributes to the survival of the whole through specialized roles. Is this cooperative colony of over a thousand "individuals" an individual also? Earnst Haeckel distinguished between morphologic and physiologic individuals by calling them morphonts and bionts, respectively. However, there is no well-established theory of morphology or physiology such as evolution. We must therefore determine which units function in the evolutionary process, and by doing so, we may determine what an individual is. Individuality is central to selection and evolutionary processes and not incidental to them (see subsequent discussion of units of selection).

The criteria for distinguishing a vertebrate individual seem clear and concise, but these same criteria provide little guidance in defining an individual for many invertebrates, plants, and microbes. Our vertebrate bias has allowed a designation of "typical" for organisms that probably are the most atypical. However, vertebrates may be relatively rare compared with the diversity and abundance of all organisms that do not fit the "typical" designation. The power of evolutionary theory lies in its application to all of life.

The largest single individual in terms of size and biomass is not the great blue whale; it is a fungus. A single individual of *Armillaria ostoyae*, or the honey mushroom, is 5.6 km across and covers an area equivalent to 1,665 football fields. The small mushrooms visible above ground give little indication of the actual size of the whole organism. It has been estimated that this giant mushroom is at least 2,400 years old, but it could be 7,200 years old. This observation strains our reason because it is outside preconceived expectations of what an individual should be like. The fungus grows by extending mycelia and hyphae into the soil and organic matrix. The mycelia and hyphae can fuse, branch, and grow in any direction. There is no symmetry of form and no central corpus. Furthermore, there are no genetically imposed limits on the size of the organism. Size is limited only by the availability of food and space to grow into.

Many higher organisms sequester their germ cells. Primordial sex cells in many higher organisms originate in the endoderm and migrate by amoeboid movement and blood circulation to the gonads, the sequestering location. The gonads, where gametogenesis occurs, differentiate early in development. The somatic cells and germ cells are separate and usually distinct. Genetic change and subsequent alteration of gene frequencies in these organisms follow a pattern of mutation, propagation, and selection (Buss, 1983). Mutations in somatic cells are of little consequence to the organism's offspring and future posterity. However, in clonal organisms mutations can occur in any cell and theoretically be inherited by offspring. Selection can occur after mutation and before propagation.

So far, we have been describing individuals relative to criteria based on vertebrates. Microbial ecology has for the most part avoided the need to determine individuality. Exceptions include medical and industrial microbiology. In these studies, it is imperative to know and recognize characteristics of an individual and to be able to propagate these individuals. In medicine, isolation of pathogens is often required for effective diagnosis and treatment. Many industries that rely on specific microbial functions need to be able to identify the organisms. On the other hand, much of microbial ecology today relies on approaches that do not require isolation or culturing but rather seek evaluation of overall microbial processes. It is much easier to quantify and describe the extent of some process than it is to determine the contribution of individual microbes to the process.

## Study of Individuals

*Autecology* as defined by Alexander (1971) is the study of "a single species and the influence of environmental factors on that species." Autecology of microorganisms requires that individuals be isolated and cultured so that processes controlling growth and function of those individuals can be elucidated. As ecological questions and the tools used to answer these questions become more sophisticated, it has become much easier to forget about individual microbes and address consortia or higher-level processes. *Synecology*, according to Alexander, is the study of "the relationship between the environment and different organisms that make up a biological complex in a single locale." In any given ecosystem, a synecologist would seek to describe the processes and consortia carrying out that process. Descriptions of consortia rely more and more on molecular biological techniques; techniques that do not require culturing. Process description becomes system level description because specific components cannot be adduced without isolation and culturing. Although this is not always the case, it is generally true.

Our understanding of some elegant microbial processes is the result of painstaking autecological studies. These studies sought to determine the impact of individuals on the ecosystem and the reverse, the impact of the ecosystem on individuals.

Given the confusion about what an individual is for many higher organisms it is important that we attempt to define individuality for microbes.

## Study of Individual Microorganisms

What is an individual microorganism? There are at least three different concepts of what an individual is: numerical, genetic, and ecological; each concept can be more or less applied to microorganisms. A numerical individual is a representative of a species that can be counted as a discrete and independent unit. That sounds straightforward. All that is required is some method to visualize each individual and then count the numbers of discrete, independent units. However, the practice is fraught with difficulties and continued debate among microbiologists. For example, plate counts have been used for decades to count culturable bacteria. Colony-forming units (CFUs) are thought to originate from single cells. Is this true? What if instead of one individual starting a colony there were 2, 3, or 10, or more cells? How would this affect our assumption of discrete independent functioning units? By giving the resulting colonies the designation CFU some of the ambiguity of the method is subsumed into the definition. All we are counting is the product of growth of some starting number of cells. Statistically we can through serial dilutions estimate the number of CFUs in a sample.

Our ability to visualize microbial samples has been greatly increased with the advent of fluorescent stains that bind with various cellular components of microbes. Once stained, the samples can be viewed microscopically and the number of fluorescent "individuals" counted. Direct staining has shown that culturing greatly underestimates the number of bacteria in any given sample. However, certain ambiguities are associated with this method. How does an investigator count a chain of cells, such as many coccoid organisms or a cell that has just divided but has not separated yet? Numerical designations of individuals have no evolutionary basis. No information is gleaned about how many of each type of cell or groups of cells makes it into the next generation. All we can know from this approach is cursorily how many cells/individuals there are at one particular moment in time and at one specific place.

Direct and plate counts can provide meaningful information on the ecological state of a particular location. Repeated sampling of a site with suitable replication allows statistical inferences to be made on factors that may be affecting changes if those factors are concurrently measured. Such results are correlative; the microbial count data and one or more other variables increase or decrease together. Cause and effect cannot be determined through such means. For example, we could measure the density of bacteria above and below a sewage treatment plant. At the same time, we could measure levels of phosphorous, nitrogen, and carbon and other factors such as temperature and turbidity. If we saw that the numbers of bacteria were significantly higher below the treatment plant, we might assume that the plant was having an impact on the bacteria. If the concentrations of phosphorous or nitrogen increased over the same sampling stations, we could say that these variables and the bacteria counts were positively correlated. However, we would not know for certain that phosphorous and nitrogen were causing the increase in bacterial counts. The only way to determine actual cause and effect would be to conduct *controlled* experiments (Box 4.1).

**Box 4.1 Experimental Design**

Most students of science have been taught the importance the design of their experiments may have on the outcomes and analysis of data. Failure to carefully design and execute an experiment puts into question the validity of the data generated from the experiment. Researchers often fail to have appropriate controls. Some experiments require both positive and negative controls. Positive controls are experimental units that always show the response the researcher is interested in. These controls are necessary to ensure that the experimental conditions do not block a positive response from an experimental group. If the positive control shows the response, we can assume that the lack of a response is not due to a poorly constructed experiment or to factors the researcher failed to control or understand. In contrast, a negative control is necessary to ensure that spurious results are not obtained because of contamination of other researcher errors. A negative control should not show the response under any conditions. If a negative control does show a response, we can assume that adequate controls and conditions were not present in the experiment.

# Genetic Individuals

A genetic individual or *genet* is a single genetic unit. This concept considers the fate of a genotype through time. Implicit in the use of this concept is the ability to recognize and count unique genets. For many higher unitary organisms the concept has meaning and application. A lion, for instance, would be considered a genet. Each lion produces gametes that have the potential to be passed on to the next generation. Each set of gametes would be representative of the genotype of a specific lion. However, because of recombination, translational errors, and mutations, every gamete is unique. Zygotes formed through fertilization of two unique gametes produce a single, completely unique individual. Each individual (except identical twins) has a unique genotype, and there would be a one-to-one correspondence between numerical individuals and genets. Such a relationship does not occur for microbes, clonal invertebrates, and clonal plants.

# Ramets

By definition, the offspring from clonal organisms are genetically similar to the "parent." In the Rocky Mountains, large stands of quaking aspens (*Populus tremuloides*) can be found. These stands often have numerous "individuals" that appear to be distinct and of different ages. However, genetic analysis of these plants has shown that the individual trees are genetically identical. Each stand is a large clone. Therefore, each stand and not each individual is a genet. The same pattern would be true for all clonal organisms such as poriferas, hydroids, cnidarians, and most microbes, as well as many plant species. For these groups of organisms, the numbers of discrete individuals is not the same as the number of genets. Each colony or clone is called a *ramet*. A ramet is a part of a genet but it is capable of living independently. The buds of a hydra and the original hydra form a genet but each bud is a ramet because it can survive when removed from the parent and continue to reproduce.

Bacteria can occur as single cells or as colonies. In nature, it appears that large bacterial colonies are an exception. Rather, microcolonies, colonies made up of tens to hundreds of cells, seem to be the rule. This is not surprising if we consider the heterogeneity of most environments. Particles available for colonization are not uniform in size, quality, or chemical composition. Large surface areas, such as rocks or plant detritus, are similarly not uniform and are both spatially and temporally variable in terms of quality of available colonization sites. Other environmental variables (see discussion below) change diurnally or seasonally and affect the size of bacterial colonies. Cells disrupted from a microcolony can, after colonization and attachment, begin a new colony.

Similar genets can intermingle. It may be impossible without extensive genetic analysis to determine the number of "individuals" present. Counting of ramets would most likely underestimate the number of individuals. We need to be able to easily determine genotypes. If phenotypes, due to any number of mechanisms (e.g., phenotypic plasticity), are not indicative of the number of genotypes, this approach would be flawed and of little value.

The genet concept implies stability of a genotype through time. How much time is required? Microbes can and do change genetically through mutations, conjugation with other microbes, taking up of naked DNA (i.e., transformation) and by infection by bacterial phage (i.e., transduction). Each of these mechanisms can alter the genotype of a microbe. With the incorporation of whole new genetic sequences or through recombination, the genotypes are changed, and it becomes impossible to tell when a genetic individual begins.

Andrews (1991) in discussing the application of the genet to microbes argues that perhaps we should consider a new model of the genet. In this model we would allow some level of genetic flux caused by conjugation, transformation and transduction to occur. The amount of flux needs to be defined and clarified. How much change is sufficient to declare a new genet? If we consider that species designations are made on 98% DNA sequence similarity or 70% DNA homology (see Chapter 3) then small changes in DNA may have significant phylogenetic and genotype effects.

## Ecological Individual

A third concept is the ecological individual. This concept defines an individual by following it through its life cycle from birth to death. The ecological concept considers the complete organism, including genetic, physiologic, morphologic, and developmental states (Andrews, 1995). Overlap of generations is possible, especially in unitary organisms, because the parental genet undergoes meiosis to produce gametes that fuse through sexual reproduction to form new genets. In clonal or modular organisms, any fragmentation, budding or other disruption of the intact "parent" may result in a similar overlap of generations. This concept is the most readily applicable compared with the genetic or numerical concepts, although it requires following an individual through all phases of the life cycle. This is especially important for organisms that demonstrate major changes in morphology, ploidy, or other genetic changes.

What is the life cycle of a single-celled organism? Bacteria after division hardly go through any perceptible change in size (see Chapter 10 on miniaturization). However, the physiologic state of the cell is subject to change over time. Because bacteria reproduce by binary fission the resulting daughter cells are genetically identical to the original cell. In this case, the life cycle is the physiologic cycle of a single cell and not the life cycle of a genetic individual. The individual continues on in perpetuity; there are just more copies of it. This is an important distinction and relates directly to evolution and fitness. For organisms that reproduce sexually (excluding selfing or hermaphrodites), offspring are genetically different from the parents. For asexual organisms, the offspring are more copies of the parent. Are these offspring a new generation or just more copies of the original generation? Each cell resulting from binary fission is exactly identical, so there is no clear distinction about which was the parent. There is no parent and offspring; there are only offspring.

We have spent some time discussing the concept of the individual. This exercise is important because in both evolution and ecology we need to determine the relative success (fitness) of an organism or set of genes. This requires that we count something. How do we know what to count to get good estimates of fitness? In this sense it is important that we know which entities are interacting. We also need to have fairly clear definitions of what are or are not offspring.

# Niche

The ecology of individuals includes the biotic and abiotic variables that impinge on the survival of an individual. To understand the impact of the environment on the individual and the impact of the individual on the environment we must determine which variables, traits and conditions affect their ecology. Gause (1934) stated: "as a result of competition two species scarcely ever occupy similar niches, but displace each other in such a manner that each takes possession of certain peculiar kinds of food and modes of life in which it has an advantage over its competitor." This statement has been modified over time and is now known as the *competitive exclusion principle*. In essence, this principle states that two organisms that are complete competitors (i.e., organisms that share the same foods, habitat, predators, and impacts of abiotic factors) cannot coexist. The niche is the role of an organism within a community (Whittaker et al., 1973). Niche theory and hypothesis testing are integral in population and community level ecology. We briefly mention the concept as a presage to understanding the impacts of various abiotic and biotic conditions on the ecology of individuals.

## Abiotic Constraints

Although microorganisms can modify their physical environment through feeding, excretion, and respiration, some abiotic variables are independent of these processes and impact the organisms either directly or indirectly. We will consider the following incomplete list as representative of abiotic variables, some of which can be modified by microorganisms: temperature, pH, nutrient source, electron acceptors, redox, and clay particle interactions.

### Temperature
The two most important abiotic variables that limit the distribution of life are temperature and moisture. All living things require water to survive. Even in the most

arid environments moisture is essential for life and organisms have evolved adaptations to conserve and capture what available water is present. Microbes are no exception. Where water is absent, active microbes are absent. Given the immense times that microbes have been on the planet and the global dispersion of microbes there are few surface locations that do not have microbes or microbial spores. However, these organisms cannot reproduce, feed, or respire without water and are therefore metabolically inactive in locations where water is not present.

Environmental temperatures in habitable locations can range from −10°C in Antarctic ice to over 250°C in hydrothermal "black smokers." The earth's temperature is determined by the amount of incoming solar radiation and by the distribution of water and land. The amount of solar radiation striking the earth's surface is affected by the angle of the earth relative to the incoming radiation. For example, solar radiation hitting the earth's surface near the poles (Figure 4.1) covers more area (point A to point B) than a similar amount of solar radiation hitting at the equator (point C to point D). Solar radiation is more diffuse in temperate zones than in the tropics. The radiation in the temperate region must pass through more atmosphere, where the energy can be absorbed, reflected, or scattered. The amount of solar heat energy is also affected by the *albedo* or amount of reflection from the earth's surface. Water has a higher albedo than a forest or grassland.

We have seen that water has certain properties that promote life (see Chapter 2). One important property germane to our discussion about temperature is the *specific heat* of water. The amount of heat a given substance requires to raise the temperature is its specific heat. The specific heat of water is high relative to other substances. Therefore the temperature of water will rise much slower and conversely will drop much more slowly as heat is removed than almost any other material. Because most of the earth is covered with water, this property is especially important in its temperature dynamics. By comparison, land heats much quicker and cools much more rapidly than water. One effect of this difference is that diurnal and seasonal temper-

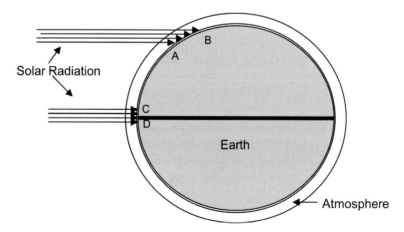

**Figure 4.1** Equivalent amounts of sunlight hitting the earth cover a greater or lesser amount of the earth's surface, depending on where the sunlight is hitting. At the equator, the sunlight is more concentrated than at the poles. (From Krebs CJ. *Ecology: The Experimental Analysis of Distribution and Abundance*, 2nd edition, © 1978. Reprinted by permission of Pearson Education, Inc.)

ature fluctuations of continents are much greater than that of oceans, seas, or large lakes.

Krebs (1978) suggested that an organism could respond to the temperature of a particular habitat in one of two ways: put up with the prevailing temperature or escape the temperature through some evolutionary adaptation. Higher organisms can accomplish the escape through modifications that reduce moisture loss due to heat (e.g., waxy cuticles, conservation of excreta, burrowing, decreased activity, seeking shade) or through adaptations that conserve heat in cooler climates or habitats (e.g., fur, warm-bloodedness, behavior). Microorganisms can modify the local thermal conditions of their habitats but these effects are localized and ultimately subject to the prevailing climate.

What characteristic of temperature is most relevant to the ecology of living things? Is it the maximum or minimum temperature? Is it the variance around the annual or daily mean temperature? Is it the average temperature on an annual or diurnal basis? Even within a specific habitat temperature differentials may exist. Although soils gain and lose heat faster than water there are distinct differences in the temperature dynamics within a soil profile. The temperature profile (Figure 4.2) for both the maximum and the minimum temperature change with depth. Zero depth in this figure represents the air temperature that is always cooler than the surface temperature. Interestingly the average annual temperature at each depth is roughly the same. However, the range between the maximum and minimum temperatures is much greater at the shallower depths than at the lowest depths. With increasing depths, temperature is moderated— high temperatures are lower than at the surface, but minimum temperatures are higher. Organisms living at or near the surface are exposed to much greater extremes and fluctuations in temperature than are organisms living just 2 to 3 m lower.

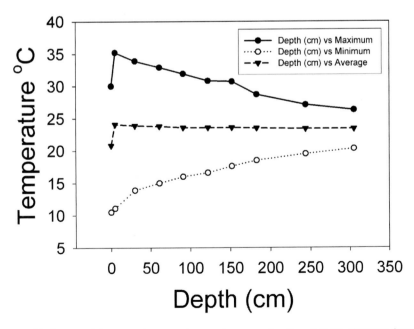

**Figure 4.2** Maximum, minimum, and average temperatures as a function of depth. (Redrawn from Fluker BJ. Soil temperatures. *Soil Sci* 68:43, 1958.)

What aspect of the temperature regime does selection act on? In the soil example, mean temperature at all depths was nearly identical. If selection favors processes that maximize at this average temperature then organisms found nearer the surface would, on average, have more days when they were at less than optimal conditions compared with organisms found deeper below the surface. Conversely, if selection were a function of the variance around the average, we would expect that organisms found near the surface would be able to survive in a wider range of temperature habitats than those found at lower depths. Any organism that was capable of growth or metabolism across the temperature range at the shallowest depths should be able to grow, all other things being equal, at the lower depths. Whereas, organisms adapted to the reduced variance of the lower depths would have a much narrower habitat range.

Temperature affects primarily the enzyme kinetics of a cell. Increases in temperature are usually met, within certain constraints, by an increase in metabolic or enzymatic activity. Increased enzyme activity often results in cell growth or reproduction. However, there are both upper and lower limits to this relationship. If temperatures get too high the enzymes denature resulting in an abrupt slowing of activity followed by death. At the other extreme, as temperatures decrease most enzyme mediated processes slow considerably.

Temperature may act on various aspects of an organism's biology including development, reproduction, survival, growth rate, nutritional requirements, and enzymatic composition and to a limited extent the chemical composition of the cell, as well as the ability to compete with other organisms. These effects are especially observable at the *temperature limits* of the organism. Every organism has upper and lower temperature thresholds. These thresholds are the limits above or below which the organism ceases to function and usually dies. These physiologic limits set by temperature, determine the distribution and abundance of most organisms including microorganisms. One consequence of these differences in temperature dynamics between land and water is that microorganisms living in oceans and freshwater should have smaller temperature ranges than microorganisms living on land.

As a rule of thumb, the temperature limits of many bacteria occur generally over a range of about 30°C. The upper and lower limits vary by species. Optimum growth is usually near the upper limit of the temperature range for a particular species. If we consider all bacteria together, the range of temperature where growth occurs is between −10 to 90°C. Certain life processes such as biosynthesis and bioenergetics have different optimal temperatures. The optimum temperature for biosynthesis is lower than that for bioenergetics. These optimum temperatures, in turn, are affected by other abiotic conditions such as supply of nutrients, pH, osmotic pressure, and salinity. An organism is therefore under different simultaneous selective pressures to maximize various processes. In pure cultures, under optimal laboratory conditions, the effect of temperature on the biological reactions of bacteria is very predictable and usually unambiguous. Attempts to determine similar unambiguous results, under natural settings, have been extremely difficult if not impossible. This difficulty is a result of complex biological and abiotic interactions and because of the different temperature optima of diverse and varied bacteria within a consortium.

Researchers have found that the thermal tolerances of bacteria fall into basically three general groups. Although these designations are for bacteria, they are applicable within certain constraints to other microorganisms. These three groups are

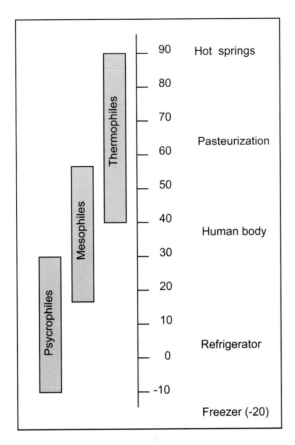

**Figure 4.3** Major classifications of microbes based on optimal temperature ranges. Temperatures are given in degrees celsius. (Redrawn from Nester EW, Roberts CE, McCarthy BJ, Pearsall NN. *Microbiology: Molecules, Microbes, and Man.* New York, Holt, Rinehart & Winston, 1973.)

based on the optimal temperatures for growth as recorded in laboratory experiments (Figure 4.3).

Psychrophilic or cold loving bacteria have temperature optima between 10°C to 15°C. These bacteria die within a few hours if grown at 18°C to 20°C. Many freshwater and most oceanic waters have temperatures that are maintained below 18°C and therefore psychrophilic bacteria evolution has allowed exploitation of much of the world. Within the temperature range for growth, increases in temperature usually promote increased biological reactions. For example, Rheinheimer (1992) shares an example of change in generation time of a bacterium as a function of the incubation temperature (Figure 4.4).

In general, there is a doubling or tripling of the reaction with a 10°C rise in temperature. This would seem to suggest that biologic processes of psychrophilic bacteria should be occurring very slowly. This is not true. Morita and Albright (1965) have shown that *Vibrio marinus* has a generation time of 80.7 minutes at 15°C which only increases to 226 minutes at 3°C. Compared with the bacterium in Figure 4.4, the generation times at 15°C are similar: 1.5 hours for the *Vibrio* and 2.1 hours for the unknown bacterium, respectively. However, at 3°C the *Vibrio*'s generation time had

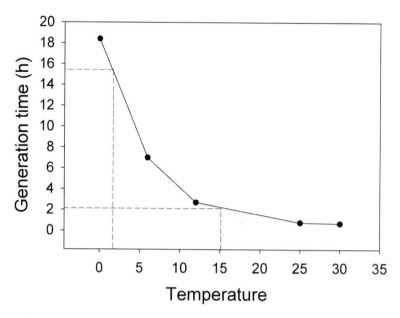

**Figure 4.4** Generation time as a function of temperature. (Data from Rheinheimer G. *Aquatic Microbiology,* 4th ed. West Sussex, UK, John Wiley & Sons, 1992, © John Wiley & Sons Limited. Reproduced with permission. Original data from Christophersen J. Bakterien. In: Precht H, Chrtistophersen J, Hensel H (eds): *Temperatur and Leben*. Heidelberg, Springer, 1955.)

increased to only 3.8 hours, whereas the unknown bacterium had increased to about 15 hours. Clearly, the *Vibrio* is more adapted to growth at lower temperatures.

Adaptation to cold temperatures is extremely important in temperate fresh waters and much of the oceans. In the temperate zone, autumnal leaf-fall in a stream ecosystem is the principle source of carbon and nutrients for growth and metabolism. Leaf fall occurs during the fastest decreases in temperature. Organisms not able to take up and process this allochthonous material while it is available seldom get a second chance. This is true for both particulate and dissolved organic carbon. Microorganisms must be able to take up and assimilate these materials while temperatures are decreasing rapidly or during the period of lowest temperatures. Figure 4.5 shows patterns of leaf decomposition at one site in a Rocky Mountain Stream. The fastest rates were observed for leaves placed in the stream in October. Selection should favor assemblages of microorganisms (both bacteria and fungi) and communities of invertebrates to have life cycles that correspond to the greatest input of carbon and nutrients. It is clear that certain aquatic insects have such life history patterns but it has not been shown that the bacteria and fungi found in the summer months are different from those found during leaf fall.

The Antarctic Ocean according to Rheinheimer (1992) remains longer in the *cryosphere* or cold zones than does the Arctic Ocean, which is influenced by the relatively warmer waters of the Atlantic Ocean. Psychrophilic bacteria with very narrow temperature ranges predominate in the Antarctic Ocean whereas in the Arctic Ocean there are a much higher proportion of mesophilic bacteria.

Temperature near the minimum or maximum of a certain species may cause changes in the morphology. For example, *E. coli* when grown near 7°C produces filaments. In contrast, *Agrobacterium luteum* grows filaments when grown at 30°C.

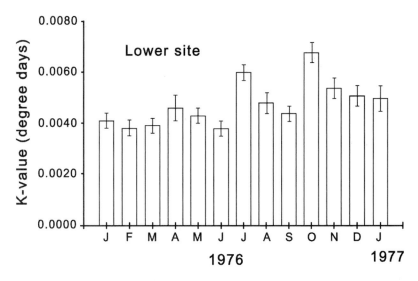

**Figure 4.5** Seasonal rates of leaf litter breakdown. Notice that the maximum rate of decomposition occurs in October. (Adapted from Figure 3 in McArthur JV, Barnes JR, Hansen BJ, Leff LG. Seasonal dynamics of leaf litter breakdown in a Utah alpine stream. *J North Am Benthol Soc* 7:44–50, 1988.)

Optimum temperature for *A. luteum* is around 25°C. Do these filaments provide any selective advantage or are they inconsequential? Although certain biological structures are remnants of past selection and are not necessarily functional in the current environment those structures that develop in response to a specific environmental cue or stress may be adaptive.

Organisms that live at extreme thermal temperatures often have reduced interspecific competition because few other organisms have been selected for these extreme conditions.

Microorganisms for the most part cannot affect the temperature of their habitat. However, there are exceptions. Ice-nucleating bacteria are biological ice nucleators capable of elevating the temperature at which ice crystals form. In essence, these organisms do not change the temperature, but rather the effects of temperature by increasing the temperature for ice formation. In other words ice will form at a higher temperature. This can be a major problem for other organisms especially organisms that have close associations with certain ice-nucleating bacteria. In addition there is metabolic heat produced in the breakdown of organic molecules. This heat can be measured and depending on the environment contributes to the overall temperature.

## pH

Microorganisms, because of their small size, are in direct communication with their environment (see discussion of surface-to-volume relation, Chapter 5). Microorganisms can modify their environment either by making it more suitable for them or less suitable for potential competitors or predators. Although these self-induced changes can be extensive, most microorganisms must cope with physical and chemical changes as imposed by their environment. Many of these environmental changes have direct effects on their fitness. Such changes are often outside the control of the microorganisms.

All materials entering into or out of a bacterium must pass through the cytoplasmic membrane. Factors that affect the membranes are fundamental to the survival and reproductive success of bacteria and other microorganisms. The pH of the environment is one such factor that can directly affect the transport of materials across the cytoplasmic membrane and the fitness of the organism.

The concentration of hydrogen ions or pH is an indication of the intensity of an acid. Water is a weak *electrolyte* and *dissociates* only mildly into the ions that compose water molecules. The primary ions are $H^+$ (hydrogen), $OH^-$ (hydroxyl), and $H_3O^+$ (hydronium). Although pH treats the dissociation of water into primarily $H^+$ and $OH^-$, the hydronium ion ($H_3O^+$) is actually more abundant. The number of hydrogen ions is equal to the number of hydroxyl ions in a neutral solution (pH = 7.0) but their masses would obviously be different. With any change from neutrality, when one ion increases the other ion decreases. A pH lower than 7.0 is an indication of increased hydrogen ions and an acid reaction. A unit change in pH, say from pH 6 to pH 5 is a tenfold increase in hydrogen ions. Bacteria living at pH 3 are experiencing 10,000 more hydrogen ions per unit volume than bacteria living at pH 7.

The pH of a particular microhabitat may be very different from other microhabitats in close proximity and very much different from the pH of the macrohabitat. Our ability to determine environmental effects is constrained by the measuring tools available. Measurement of pH within microhabitats is limited by the size of the probes. Use of microprobes has demonstrated that the pH of microsites can change over very small distances (Figure 4.6). Individual bacteria may be experiencing extreme differences in pH and be only a few micrometers apart.

The pH of a macrosite gives some information that may allow gross generalizations to be made. If the pH of the macrosite is more than 9, we may expect to find only organisms that can tolerate an alkaline environment. However, individual cells may change their local pH through the by-products of metabolism and/or through chemical interactions with various abiotic materials. For example, many bacteria produce

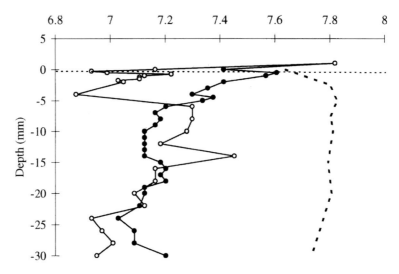

**Figure 4.6** Microelectrode profiles of pH through the surface sediment. Sediment surface shown at 0 mm depth by the *dotted line*. Negative values indicate samples collected below the surface. The *dashed line* is the pH of pore water. (From Woodruff SL, House WA, Callow ME, Leadbetter BSC. The effects of biofilms on chemical processes in surficial sediments. *Freshwater Biol* 41:73–89, 1999.)

organic acids that they release into the environment. The effect of these acids in an alkaline environment would be to reduce the local pH. The effect of these localized changes in pH on a bacterium's neighbors would be dependent on the sensitivity and proximity of the neighbor and the concentration of the metabolic acids. The effect of chemically produced acids would be dependent on the concentration produced and the diffusion rate into the microsites. Inhibition of growth is not necessarily an indicator of the effect of pH on a specific organism but may be caused by the organic acid and not by the pH resulting from the presence of the acid. In other words, the actual molecules of an organic acid may cause an affect on a specific microbe.

Some microorganisms are capable of living at pH values that are fatal to almost all other organisms. The ability to live in extreme environments requires modifications. Given the limited size and morphologies of microorganisms there are not many structural changes possible to allow life under extreme conditions. Changes in pH can affect the morphology of some bacteria. For example, Rheinheimer (1992) describes some microorganisms that tend to produce involutions, with cells that are normally rod shaped becoming enlarged and showing irregular swelling and branching.

For the most part, selection has favored biochemical adaptations over morphological changes in microorganisms. These adaptations usually involve changes to the cytoplasmic membrane, which controls ion transport, and not to the cytoplasm itself. Microorganisms living in these environments need to be able to solve the problem of too many hydrogen ions or not enough. Kushner (1993) pointed out that the internal pH of bacteria that live in extreme pH environments might be near the optimum pH. For example, Hsung and Haug (1975) determined that lysates derived from *Thermoplasma acidophilum* were at pH 6.5, even though the bacteria were grown at pH 2. Other researchers have determined that the maintenance of the internal pH is a passive process and maintenance of the $\Delta$pH does not help the cell with viability. Death of *acidophiles* (acid-loving organisms) is due to starvation (lack of ATP) and not to acidification of the cytoplasm.

Alterations to the cytoplasmic membrane affect the movement of materials into and out of the cell. The energy source for the movement of materials by active transport across cell membranes is the *protonmotive force*. The protonmotive force can be generated either by electron transport or by ATP hydrolysis (Figure 4.7). Certain carrier molecules with binding sites for protons ($H^+$) and specific substrates use this force to transport materials into or out of a cell. During the process both the protons and the substrate are transported into the cell. Eating for microorganisms is energy dependent. Both electron transport and ATP provide energy for active transport for aerobes, phototrophs, and anaerobes, indicating the ancient origin of these systems.

According to Kushner (1993), the existence of microorganisms living in extreme pH environments has created fascinating problems in bioenergetics. The protonmotive force has two components, $\Delta$pH with acid outside the cell and $\Delta\Psi$ or membrane potential, with (+) outside the cell. If we consider microorganisms under extreme pH challenge, we might expect some differences in these values and their orientation with respect to the cell membrane. $\Delta\Psi$ of many acidophilic microorganisms is (+) within the cell, whereas alkalophiles have very high $\Delta\Psi$ (+) outside.

Considering the general model of the protonmotive force (see Figure 4.7), it should be apparent that organisms living in an alkaline environment need to conserve the $H^+$ gradient by not allowing the protons to interact with the bulk medium outside the

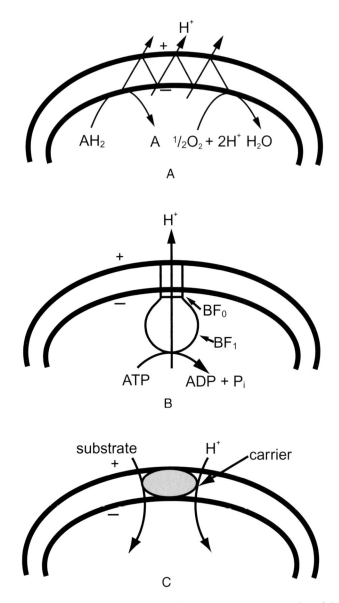

**Figure 4.7** Protonmotive force as the energy source for active transport: generation of the protonmotive force by respiration (**A**); generation of the protonmotive force by ATP hydrolysis (**B**); and active transport by a carrier molecule with binding sites for a particular substrate and for proteins (**C**). (From Gottschalk G. *Bacterial Metabolism.* New York, Springer-Verlag, 1979, 2nd ed., 1985.)

cell. For certain *alkalophiles* (alkaline-loving organisms), it has been suggested that they have special channels for conserving the $H^+$ gradient developed during respiration. Any modification that allows an organism to survive in a habitat that is generally inhospitable would be selected because exploitation of this habitat would be in the absence of competitors. Although the diversity of organisms living in extreme environments is amazing, this diversity is much lower than other more "normal" environments suggesting that adaptations to live in extreme environments carries certain evolutionary costs.

Other adaptations or structural changes to microorganisms living in extreme pH environments are unusual surface components. Kushner (1993) identifies ester-linked lipids in both acidophilic and alkalophilic archaebacteria that may make them resistant to *hydrolysis* under acid, alkaline or high temperature conditions. Some acidophilic and alkalophilic organisms have unusual *glycoproteins*. The exoenzymes of these organisms are of interest because they are stable under extreme pH conditions and heat conditions and may be important in certain industrial processes.

Most bacteria can grow only within a narrow range of pH 4 to 9 with the optimum ranging from 6.5 to 8.5 (Table 4.1). However, *Thiobacillus thioxidans* and *Thiobacillus ferrooxidans* can survive at pH 1. These organisms through the oxidation of sulfur or its reduced compounds produce metabolic useful energy and sulfuric acid. In effect, these organisms, through their feeding, create an inhospitable environment for almost all other organisms. Both species are obligate acidophiles that are found in hot acid springs and mine residues. Bacteria that are able to grow at pH greater than 7.0 (range, 7.3 to 10.6) are designated alkalophilic. Fungi are capable of growth across a much wider range of pH.

**Table 4.1**  Range of pH for Various Organisms

| Organisms | Range of pH for Growth |
|---|---|
| **Bacteria** | |
| *Thiobacillus thiooxidans* | 0.9–4.5 |
| *Thiobacillus ferrooxidans* | 1.5–4.0 |
| *Thermoplasma acidophilum* | 1.0–4.0 |
| *Bacillus acidocaldarius* | 2.0–6.0 |
| *Sulfolobus acidocaldarius* | 0.9–5.8 |
| *Bacillus alcalophilus* | 8.5–11.6 |
| *Actinomycete* sp. | 8.0–11.5 |
| *Flavobacterium* sp. | 7.0–11.2 |
| *Exiguobacterium* sp. | 7.0–11.5 |
| *Plectonema nostocorum* | to 13 |
| *Spirulina* sp. | 8.0–11.0 |
| *Synechococcus* sp. | 6.5–10.0 |
| *Ectothiorhodospira* sp. | 8.0–10.0 |
| *Natronobacterium* and *Natroncoccus* | Opt. 9.0–10.0 |
| **Fungi** | |
| *Aspergillus, Penicillium*, and *Fusarium* | 2–10 |
| *Phycomyces blakesleanus* | 2.0–7.0 (opt. 3) |
| *Aconitum velatium* and | Grow in 2.5 mol/L |
| *Cephalosporium* | H₂SO₄ and 4% CuSO₄ |
| **Algae** | |
| *Chlorella pyrenoidosum* | 2–10 |
| *Cyanidium caldarium* | <2–5 |
| **Protozoa** | |
| *Polytomella caeca* | 1.4–9.6 |
| Flagellates and amoeba | Grow at pH 2 |

Copyright 1993 from Kushner DJ. Growth and nutrition of halophilic bacteria. In: Vreeland RH, Hochstein LI (eds): *The Biology of Halophilic Bacteria*. Boca Raton, FL, CRC Press, 1993, pp. 87–103. Reproduced by permission of Routledge/Taylor & Francis Group, LLC.

## Redox Potential

*Oxidative-reduction* or *redox potential* of water controls the abundance and forms of various ions. Reduction and oxidation occur simultaneously. An ion can be reduced only as it gains electrons that come from the oxidation or loss of electrons from another ion. Oxidation is the loss of electrons and reduction is the gain of electrons. An electrical current will flow between solutions that differ in the concentrations of reduced and oxidized substance. Redox is the measurement of the current flow across a neutral salt bridge and conducting filament to nonreactive electrodes of a noble metal and is measured and expressed in mV. The intensity of the current flow is related to the states of oxidation and reduction in the two media. This electromotive force is designated as $E_h$.

Ecologically the $E_h$ or redox potential is very important as it controls the availability of certain ions and thereby directly determines the physiology of the microorganisms present in the medium (Figure 4.8).

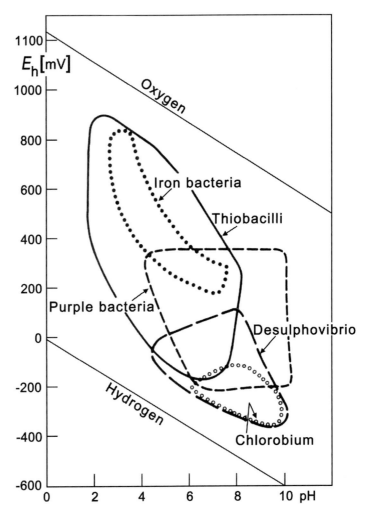

**Figure 4.8**  The $E_h$ values and physiology of various groups of bacteria. (From Rheinheimer G. *Aquatic Microbiology*, 4th ed. West Sussex, UK, John Wiley & Sons, 1992, © John Wiley & Sons Limited. Reproduced with permission. Original figure in Baas Becking LMG, Wood EJF. Biological processes in the estuarine environment. *Kon Ned Akad Weten Proc B* 58:160–181, 1955.)

**Table 4.2** Redox Values at Which Important Reductions and Oxidations Occur and Associated Oxygen Concentrations

| Redox Couples | $E_h$ | Dissolved $O_2$ (mg/liter) |
|---|---|---|
| $NO_3^-$ to $NO_2^-$ | 0.45 to 0.40 | 4.0 |
| $NO_2^-$ to $NH_3$ | 0.40 to 0.35 | 0.4 |
| $Fe^{+++}$ to $Fe^{++}$ | 0.30 to 0.20 | 0.1 |
| $SO_4^-$ to $S^-$ | 0.10 to 0.06 | 0.0 |

From Table 13-1, Cole GA. *Textbook of Limnology*, 1975, C.V. Mosby Company, St. Louis.

The ability of some microorganisms to exploit many diverse and seemingly harsh environments is due in part to their nonreliance on oxygen based respiration. Higher organisms respire using oxygen as the terminal electron acceptor but microorganisms are not so constrained and have evolved numerous ways to capture energy. In general, anaerobic bacteria require $E_h$ values much lower than those required by aerobic bacteria (Table 4.2). Most lakes have water column $E_h$ values in the range of 400 to 500 mV indicating well oxygenated systems. Microorganisms can actively alter the redox potential of their environment through the consumption of $O_2$ or other electron acceptors. Consumption of $O_2$ results in a lowering of $E_h$, especially when $H_2S$ is produced. $E_h$ values for lake sediments and overlying water drop rapidly to <0 within millimeters as the oxygen is consumed (Figure 4.9). The production of $O_2$ through photosynthesis causes a rise in $E_h$.

## Salinity

Ions of specific elements are needed for metabolism and growth of all organisms. Inorganic ions are needed in the activation of various enzymes, in the transport of materials across membranes, and in induction mechanisms. Ions also aid in the prevention of lysis and the maintenance of intracellular solute concentrations (Morita, 1997). At salinities above 5%, certain morphological changes occur in some bacteria. Rod shaped bacteria become longer and filamentous. Furthermore, increases in salinity have been shown to interfere with the mechanisms of reproduction. The cells are capable of growth but are unable to divide.

All marine and freshwater organisms are affected by the salinity of the environment they live in. Microorganisms that live in freshwater are usually *halophobic*, or salt intolerant, whereas 95% of microorganisms in marine systems are *halophilic*, or salt loving. Most marine microorganisms cannot live in freshwater and most freshwater microorganisms die rapidly when introduced into marine systems. Many marine bacteria lyse quickly when moved from seawater to freshwater. It has been shown that a lowering of the salt concentration destroys the mucopeptide layer of the cell wall in some bacteria (MacLeod, 1968), which results in a change in osmotic pressure in the cell caused, by the weakening cell wall that is sufficient to rupture the remaining wall layers.

All organisms, including plants and animals, originated from marine systems, although the line of descent may be well removed in space and time from salt water. However, the marine environment 3.8 billion years ago and the marine environment today are very different. Regardless of the difference between the ancient and modern seas, tolerance of freshwater is a secondary adaptation. As with all other

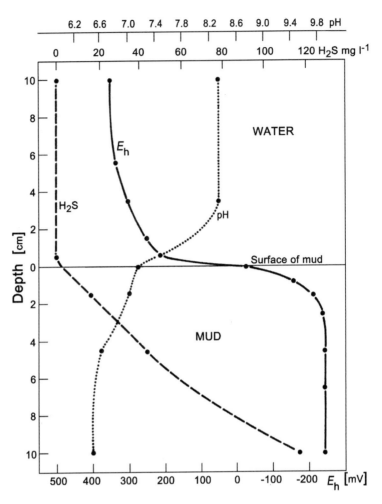

**Figure 4.9** Change in $E_h$ with depth. Notice the rapid change from overlying water to the mud surface. (From Figure 5.7 in Rheinheimer G. *Aquatic Microbiology*, 4th ed. West Sussex, UK, John Wiley & Sons, 1992, © John Wiley & Sons Limited. Reproduced with permission.)

environmental conditions, microorganisms have succeeded in exploiting a wide range of salinities.

Selection has resulted in halophobic species and most of these species are found in freshwater and soils. These organisms are adapted to much lower salinities than those found in the seas. Given the continuous import of freshwater and soil bacteria into estuaries and coastal areas what is the fate of these organisms? Do they become more salt tolerant or are they immediately killed?

Salinity in the open ocean remains fairly constant with an average of 3.5% and ranges from 3.3% to 3.7%. Higher values can be found in sea waters that are subject to evaporation, such as the Red Sea, where the salt content can approach 4.0%. With little or no variation in the salt content of the oceans, salinity has little impact or limitation on marine microorganisms. However, coastal and estuarine systems may have very high variation (0.1% to 4.4%) in salinity because of run-off, rain, and drought or through the constant inflow of freshwater from rivers. Some bacteria that grow optimally in freshwater can tolerate salinities of 2% to 4%. For example, bacteria that live in the sand of beaches experience extreme fluctuations in salinities on a diurnal basis. These strong fluctuations have given a selective advantage to microorganisms capable of growth and metabolism within the salinity range set by the tide and inputs of freshwater. In other words these organisms have a wider range of tolerance.

The communities of microorganisms have been determined, in part, by the degree of salinity. The physiology of organisms found in seawater is very different from that found in freshwater organisms. In extreme saline environments, only a very few taxa (mainly bacteria and flagellates) have been able survive, although these organisms can be at very high densities because there are few predators and competitors.

Salinity alters some of the properties of water. Seawater has a higher density, specific gravity, and osmotic value and a lower freezing point than does freshwater. Temperature affects the level of salinity tolerance. The range of salt tolerance for bacteria is at its maximum at the optimum growth temperature. An increase in the temperature optimum of a bacterium comes with an increase in salt requirements and a decrease in the temperature optimum causes a reduction in the salt requirement.

In culture, some freshwater bacteria are able to adapt to higher concentrations of salt and some halophilic bacteria can adapt to a lower salt concentration. The most adaptable bacteria are those that have the highest natural range of salt tolerance. Bacteria that are obligate halophilic or halophobic have narrow salt concentration ranges. We would expect that bacteria found along coastal margins, in estuarine sediments, or beach sands to be the most likely to adapt to changes in salt concentration or to demonstrate the widest range of tolerance for salinity.

Freshwater systems have lower salt concentrations, and the salts may differ from lake to lake or stream to stream, depending on the geology or the basin. Lakes and rivers may have NaCl or $MgCl_2$, $MgSO_4$, or $NaHCO_3$, but NaCl predominates, and the highest salt concentration in salt lakes is 28%, compared with only 3.5% in seawater. Most microorganisms in freshwater are halophobic and cannot, under natural conditions, grow in water with more than 1% salt concentration.

Figure 4.10 demonstrates the response of halophobic and halophilic bacteria growth to changes in salt content. The freshwater bacterium has a steady decrease in growth with increasing salt content. This particular bacterium is halotolerant because although growth decreases with salt content, some growth occurs at all levels. Growth in an obligate halophobic bacterium would have decreased to zero after the salt content maximum had been passed. Bacteria living in brackish water experience increases and decreases in salt content. The curves for the brackish and the marine bacteria both show distinct peaks of growth associated with narrow salt content ranges. The shape of these curves indicates that these organisms are obligate halophiles, but the optimum salt content differs significantly between the two species. All of these bacteria were collected from the Baltic Sea (Rheinheimer, 1992).

## Light

Radiant energy from the sun is an important ecological and evolutionary factor in both terrestrial and aquatic ecosystems for at least three very different reasons. The first important biological reason is that light is essential for *photosynthesis*. Organisms are usually adapted to the light regime of their habitats, and any ecological limitations on where an organism can live and reproduce are attributed to these adaptations. As a process, photosynthesis is incredibly inefficient, with less than 1% of light hitting the earth being converted into chemical energy. Fortunately, the light

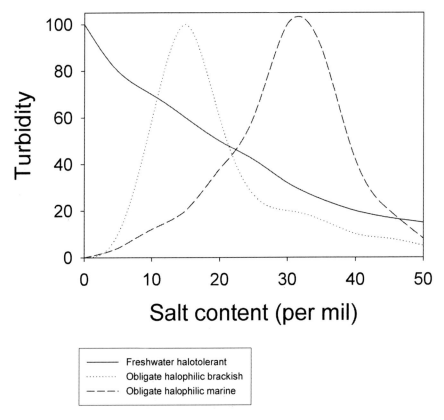

**Figure 4.10** Growth curves as a function of salt content for halotolerant freshwater bacteria, obligate halophilic bacteria from brackish water, and obligate halophilic bacteria from the marine environment. (From Figure 5.8 in Rheinheimer G. *Aquatic Microbiology*, 4th ed. West Sussex, UK, John Wiley & Sons, 1992, © John Wiley & Sons Limited. Reproduced with permission.)

just keeps on coming. The constant input of light from the sun is sufficient to overcome the low capture efficiency of *phototrophic* or light eating organisms. A fundamental premise of evolutionary biology is that individuals of a species cannot do everything in the best possible way (Krebs, 1978). Selection for the most efficient means of capturing light in one habitat limits the ability of the same organisms to capture light in a very different habitat. Said another way, selection cannot favor a "superbug" capable of surviving in all habitats because there are trade-offs between the requirements of one habitat and those of another habitat.

Second, the timing of many *diurnal* and seasonal biological rhythms is stimulated by light. *Photoperiodism* or predictable repeating behaviors that are stimulated by light include for example, breeding/reproduction, feeding, and pollination in plants. The daily vertical migration in aquatic systems of planktonic organisms including microorganisms is generally due to light. Phototaxis, or directed movement toward light, has been observed in the purple nonsulfur bacteria. Photosynthesis by these organisms is anoxygenic, and instead of water hydrogen, reduced sulfur compounds, or organic compounds serve as the electron donors.

Diurnal and seasonal patterns in breeding and feeding in higher organisms have been studied quite extensively. Flowering plants can be categorized based on their response to photoperiod: *short-day plant* flowers, *long-day plant* flowers, and *day-*

*neutral plant* flowers. Careful research has determined that flowering plants measure the length of the dark period, not the light. This response to photoperiod in plants has clear adaptive significance. Short-day plants that flower in the spring are able to put energy into seedling growth during the growing season, whereas short-day plants that flower in the fall store nutrients during the growing season and overwinter as seeds. Long-day plants often have seeds that require extended ripening times. Day-neutral plants are adapted to uncertainty. These plants flower when conditions such as water are right. Many microbes respond to changes in both wavelength and intensity of solar radiation. These characteristics of light change seasonally and diurnally, and we might expect adaptive responses by microorganisms to these changes.

The third important affect of light is that of light damage. Light, particularly ultraviolet (UV) light, can cause significant damage to bacterial and fungal cells. However, UV is not the only wavelength of light that can cause a negative affect on microorganisms. Blue light (366 to 436 nm) can inhibit some processes (e.g., nitrite oxidation). Red light does not seem to have any measurable affect. Sensitivity is species specific and can vary significantly even between closely related species.

Some bacteria are inactivated and finally killed by exposure to light. This is particularly true of the nonpigmented species. Pigmented bacteria that contain carotenoids are generally not inhibited by light. Bacteria found in the air are usually pigmented, and they are usually more resistant to UV damage. In water, UV light is quickly filtered out. Blue light can penetrate deeper (about 1 m), but any microorganisms deeper than 1 m probably have little light damage.

Although the atmosphere filters many harmful wavelengths, some of these wavelengths make it through to the earth's surface. Ultraviolet radiation is particularly harmful to organisms. Ultraviolet rays damage many different types of organic molecules, including nucleic acids and are potentially lethal to most forms of life. Some microorganisms (e.g., *Dienococcus*) are capable of repairing the damage caused by UV radiation at rates that allow them to survive in environments with increased levels of natural radiation. The ability of oxygen and ozone to filter UV suggests that early life was subjected to very high levels of solar radiation before photosynthesis developed and oxygen began to accumulate in the atmosphere.

It is important to understand the basic physics of light to recognize the potential impacts, both positive and negative, on microbial organisms. Light behaves as both waves and as particles (photons). All wavelengths of light travel at the same speed in a vacuum ($300,000 \, km \, sec^{-1}$). However, the shorter the wavelength, the more energy there is in the photons emitted at that wavelength. The energy of a photon is inversely proportional to the wavelength. Photons of violet light (430 nm) have almost twice the energy as photons from red light (650 nm). Gamma rays have the highest energy and radio waves the lowest. Visible light is only a small portion of a continuous spectrum of electromagnetic radiation and the differences between visible and nonvisible are measured in nanometers. Why is only this small band of wavelengths utilized by living organisms? Such varied functions as sight or vision, photosynthesis, diurnal and seasonal rhythms, the bending of plants toward light all depend on this little band of electromagnetic radiation. Is this simply coincidence or are there larger forces at work?

All living things are made up of similar building blocks, various molecules of life such as proteins, nucleic acids, and fats. The configuration of these molecules is held together or in association with other molecules by hydrogen and other weak bonds.

It is the specific structures of these molecules that give them their function. Wavelengths with energy greater than that of violet light are capable of disrupting or breaking these weak bonds and affecting the function of the molecules. Wavelengths in the infrared region cannot cause changes in electron configuration but will increase the amount of heat in the cell because water absorbs these wavelengths.

It has been pointed out that the visible light spectrum is important to biology because it is the only spectrum that is available. Our modern atmosphere filters out higher-energy wavelengths with $O_2$ and $O_3$ and infrared radiation through water vapor and carbon dioxide; essentially, visible light is all that makes it through to the earth's surface. Organisms that evolved to use other wavelengths would be at a selective disadvantage relative to those organisms using the visible spectrum.

Most of the earth is covered by water. Water affects the amount of light through several processes. A significant portion of light that hits the surface of a body of water is reflected and is lost unless backscattered from the atmosphere. Light intensity quickly diminishes in aquatic systems as a function of water and various materials either suspended or dissolved in the water. As discussed elsewhere, the angle of incidence greatly affects the amount of radiant energy available. Bodies of water in the temperate regions would reflect more light than similar size bodies of water in tropical regions. Wetzel points out that wave action can greatly increase light loss due to reflection. Between 10%–20% of the light is reflected depending on the angle of incidence. In addition to reflection, light can be lost from water by scattering. Scattering is caused by light hitting suspended particulates and by the materials dissolved in the water. Scattered light can go in all directions and some of this radiant energy is available for water dwelling photosynthetic organisms.

Gases found in the atmosphere are transparent to the wavelengths of light in the visible spectrum. That is most of the visible light is not filtered but reaches the earth's surface. However, carbon dioxide and water are not transparent to infrared wavelengths. Light energy absorbed by the earth is radiated back toward space as heat or infrared radiation. Because $CO_2$ and $H_2O$ are not transparent to infrared the heat is captured by the atmosphere and continues to warm the earth's surface. This is the well-known greenhouse effect. Fortunately for life, the input and loss of heat are in a state of equilibrium.

A primary effect of light on the ecology of microorganisms is directly related to the processes of photosynthesis. The ability to capture and harness solar radiation altered the potential diversity and complexity of life on earth. Photosynthesis by microorganisms is much more diverse than that found in higher plants and will be discussed in detail in the chapter on the ecology of feeding.

**Pressure**

Organisms that live at extreme depths in the oceans and some lakes experience very high hydrostatic pressures. For every 10 m change in depth the hydrostatic pressure increases by approximately 1 atm (= 101.325 kPa). In the deep Pacific trenches the pressure can reach more than 1,100 atm. It has been estimated that over 90% of the oceans are over 1,000 m deep. Pressure at that depth is 100 atm. Almost 25% of the oceans experience pressure of more than 500 atm. These pressures are a significant ecological factor that affects the distribution and survival of living organisms. These pressures can crush most man-made devices.

Counts of viable bacteria obtained from surveys of sediments collected from more than 10,000 m indicated densities between $10^4$ to $10^6$ cells per g wet weight of sediment. Not all microorganisms can live at these extreme pressures. Some of these microorganisms that grow at >500 atm are unable to survive or grow at sea level. Any organism that requires high pressures to grow and survive is termed *barophilic*. However, some microorganisms can survive at great depths and at also at sea level pressures and these organisms are called *barotolerant*. Barotolerant organisms have much wider ranges of tolerance for pressure. Any organism that cannot grow at high pressures is called *barophobic*. Most soil, freshwater, and many marine bacteria are barophobic.

The temperature of the deep waters is quite cold (about 4°C) and so barophilic organisms are also psychrophilic. These organisms can grow only at cold temperatures and high pressures. As shown in Figure 4.11, some bacteria show a pressure dependent response in growth. At low pressures no growth occurs but with increasing pressure more cells are able to divide reaching an optimum growth rate at around 300 atm. We might expect that growth at high pressures and low temperatures to be slow relative to growth at the surface. This is basically true, but true barophilic organisms cannot grow at all at the surface, and the comparison is therefore relative and almost meaningless. Barophobic bacteria cannot grow at the cold temperatures and high pressures and barophilic bacteria have just the opposite pattern. Clearly, natural selection has favored adaptation to these extreme environments.

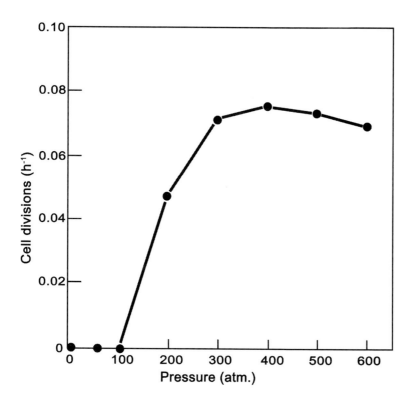

**Figure 4.11** Effect of pressure on bacterial growth. (From Figure 5.3 in Rheinheimer G. *Aquatic Microbiology*, 4th ed. West Sussex, UK, John Wiley & Sons, 1992, © John Wiley & Sons Limited. Reproduced with permission.)

Some processes, such as sulfate reduction, do occur at extreme depths but at fairly slow rates. However, like the growth rates discussed previously, deep-sea sulfate reducers cannot reduce sulfate at surface pressures or temperatures.

Although many species of bacteria perish at high pressures or at quick decompression intervals, some bacteria seem unaffected. For example, cells of *E. coli* that were exposed to 1,000 atm and then brought quickly back to surface pressures did not show any decrease in viable cells while other species died.

Certain enzyme systems are affected by pressure, including those involved in methane and hydrogen production, phosphatase activity, and sulfate and nitrate reduction. Some systems in barophobic bacteria are permanently inactivated by pressure. For example, succinic acid dehydrogenase is shut down in *E. coli* cells exposed to high pressures, but this same system was unaffected in barophilic bacteria. DNA synthesis is affected by pressure in all but barophilic organisms. Many barophilic bacteria are flagellated and motile at high pressures but lose the flagella at atmospheric pressure.

Adaptation to extreme pressures has been demonstrated for a much more diverse group of organisms than previously thought. Both fungi and bacteria have species that can survive at high pressures.

# Growth and Feeding

5

## Growth and Surface-to-Volume Ratios

Organisms can grow larger or smaller. Most people think of growth as an increase in size. We speak of the growth of the economy, and every mind's eye sees an ever-increasing line of profits. However, the stock market can grow smaller; it probably has to. During growth in higher organisms there is usually an increase in size and mass. Starvation, however, may cause an organism to grow smaller, at least in mass.

Bacteria can also grow bigger or smaller. The cue is environmental, and the results may ensure the survival of the organism. Although all bacteria are small, there are differences in the size of cells. Some cells grow only as rods. When grown in rich media, media that contains everything necessary for growth, some bacteria get big and fat, at least at the bacterial scale of things. These "big" bacteria can become 0.5 μm long. A micrometer needs some reference point for people to appreciate. The period at the end of this sentence is about 1 millimeter (mm) long. A millimeter is 1,000 μm. About 2,000 of these "big" bacteria can line up end to end, and they would just fit inside the period. The smaller bacteria are less than 0.1 μm long.

Normally during cell division, bacteria increase in size so that the resulting daughter cells are about as big as the starting cell. The daughters are usually a little smaller and increase in size after feeding. These observations of growth are made under laboratory conditions (i.e., perfect conditions). Samples brought in from nature and observed under a microscope seldom have the large bacteria, but rather only little, small bacteria. If you grow these small environmental bacteria under the laboratory conditions, they sometimes increase in size into the big bacteria. When environmental conditions are right or wrong, the bacteria continue to divide but this time without the increase in size. The result is smaller and smaller cells. We call this *miniaturization*. Why would bacteria want to grow even smaller in size than they currently are?

To answer that question, we need to pause and consider cereal boxes. Why is it cheaper to buy a large economy or family-sized box of cereal than to buy the same volume or weight of cereal in several small boxes? The answer has nothing to do with the amount of product, in this case cereal, but rather with how much material it takes to wrap the cereal. To make things simple let us say that our family size cereal box is a cube 20 cm wide, 20 cm tall, and 20 cm deep. To determine the surface area of the box, we need to calculate the area of each side and then multiply by the number of sides of the cube (= 6). The area of one side would be $20 \times 20 = 400 \text{ cm}^2$. The surface

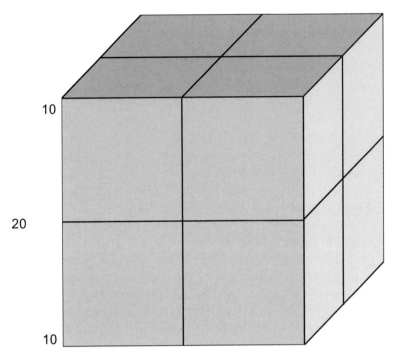

**Figure 5.1** Surface area of a cube 20 × 20 × 20 cm. The cube has been divided into 8 cubes that are each 10 × 10 × 10 cm.

area of the box would be 400 × 6 = 2,400 cm², which is the amount of cardboard needed. The volume of our box is calculated by multiplying the length × width × depth. In our case, all of the sides are 20 cm long, and the volume is 20 × 20 × 20 = 8,000 cm³ (Figure 5.1).

Let us say we want to make our smaller boxes cubes with 10-cm sides. As shown in the illustration, we would have eight 10 × 10 × 10 cm boxes that would have the same volume as our original 20 × 20 × 20 cm box. Each of these smaller boxes would have a surface area of 10 × 10 × 6 = 600 cm² and a volume of 10 × 10 × 10 = 1,000 cm³. The volume of all eight smaller boxes would be 1,000 × 8 (i.e., the number of smaller boxes) = 8,000 cm³ which is exactly the same volume of our large box. Now let us look at the surface area of the smaller boxes. Each box is only 600 cm², but there are eight boxes, and the total surface area needed to cover the cereal in smaller boxes is 600 × 8 = 4,800 cm². The surface area of the big box was 2,400 cm², or one half the area of the smaller boxes. It would take the manufacturer twice as much cardboard to cover the same amount of cereal hence the increase in price.

Another important measure used in considering the change in surface area and volume is called the surface-to-volume ratio. The surface-to-volume ratio is calculated by dividing the surface area of an object by its volume. In our previous example, the surface-to-volume ratio of the large package would be 2,400 ÷ 8,000 = 0.30. The same calculation for the smaller boxes would be 600 ÷ 1,000 = 0.60. This value is two times bigger than that of the larger box, which means there is an increasing amount of surface area per unit volume. If we were to make the boxes even smaller, this ratio would continue to increase.

This same phenomenon occurs with all shapes of bacteria and is not restricted to boxes. The surface area and volume of individual bacteria have been estimated. Remember bacteria are very small to begin with, and when they divide without growth, their surface-to-volume ratio gets larger and larger. If we estimate bacterial surface area and volume based on a $0.1$-$\mu m$ cube (i.e., approximate size of some small bacteria), the surface-to-volume ratio is very large. $1 cm = 10$ millimeters $= 10,000 \mu m$. Our box bacteria would have a surface area of $0.00001 \times 0.00001 \times 6 = 0.00000000006 cm^2$ and a volume of $0.00001 \times 0.00001 \times 0.00001 = 0.0000000000000001 cm^3$. These measurements would give us a surface-to-volume ratio of $0.00000000006 \div 0.0000000000000001 = 600,000$! This is a huge number and clearly shows that the smaller bacteria get the larger their surface-to-volume ratio gets.

By increasing the surface-to-volume ratio a microorganism has more surface area per unit volume. It is through its surface that a bacterium gets its food and nutrients. By getting smaller during tough times of nutrient depletion or scarcity an individual bacterial cell increases the amount of surface area that may come in contact with food or nutrients. In short because of their very small size microbes are the most environmentally sensitive organisms. When times are good they grow up to their maximum size like other organisms but when things get tough they get smaller making themselves more responsive to food when it is available.

# Ecology of Feeding

All living things must obtain energy to perform the basic metabolic functions of life and they must assimilate materials necessary to synthesize the basic constituents of their being. Much of the diversity of life that we can see and catalog is a function of the many and diverse modes whereby organisms capture energy and assimilate materials. Many people see feeding as teeth and chewing. Lions feed, cows feed, ants feed, and humans feed, but do plants feed? Do fungi and bacteria feed? They do, but the methods employed in the gathering of energy and materials are anything but intuitive.

*Trophism* means feeding and numerous terms have been coined that are indicative of various feeding modes, such as *phototrophy, autotrophy, heterotrophy, chemotrophy,* and *lithotrophy.* These and other trophisms are discussed in detail later in the chapter.

To feed an organism must have food. What constitutes food? Andrews (1991) defines food as anything that provides energy, or an electron donor, or a source of carbon and other *bioelements. Resources* are any quantity that can be reduced in amount by the activity of an organism (Begon et al., 1986) and include both food and space. Food resources for microbes can include $O_2$ and $NO_3^-$ (i.e., electron acceptors), $NH_4^+$ as a nitrogen source, $H_2$ acting as an electron donor or glucose or acetate, which would be used as both organic carbon and energy and as electron donors. Depending on the microorganism these resources could and probably would change.

Microbiologists use the term *substrate* to mean the nutrient component of a resource. Substrates have well defined meaning in laboratories where the media is exactly formulated and the constituents known both chemically and physiologically.

However, in nature substrates are complex and often poorly defined. Attempts to describe the resource base of free-living microorganisms pose significant challenges. Chemical characterization of potential resources is limited by the detection capabilities of the technologies employed. In the complex milieu surrounding free-living microorganisms concentrations of available resources may be adequate for an individual microorganism to feed but below the ability of a researcher to detect.

Regardless of whether a resource or food source can be measured by humans, microorganisms face at least three "decisions" relative to their food. Andrews (1991) identifies these questions as

1. What do I eat?
2. How do I get my food and keep it?
3. How do I metabolize and allocate it efficiently so as to propagate my genes into the next generations?

The first question essentially deals with resource categories. What are the available food items, and how much of these items are there? For microorganisms, we must adjust our thinking about resource categories and allow both inorganic and organic materials. Some microorganisms are capable of eating light, and we must understand the similarities and differences between higher plants and bacteria in their modes of photosynthesis.

The second question relates to the acquisition of resources. For higher animals acquisition is easily visualized with predators chasing down some fleeing prey. For other higher organisms the presence of a mouth gives clues about acquisition. However, there are animals that lack mouths and all aspects of a gastrointestinal tract (e.g., tube worms *Pogonophora*, many parasites). Acquisition of food in these organisms is similar to that of bacteria and involves the movement (active or passive transport) across membranes of resources. Fungi can seek out food items. Some bacteria have taxis and can move either toward or away from resources (discussed later).

The final decision microorganisms need to make refers to the optimal allocation of their resources. Of the three decisions, this one has the greatest impact on the evolution of the organism. Failure to allocate energy toward the appropriate activity may result in death of the organism and the failure to pass genes on to offspring. There are no superbugs or bacteria capable of maximizing all processes in all habitats and under all conditions. Rather selection acts to achieve compromise across a limited range of options. These options are those available to the organism within the constraints of a specific habitat.

# Metabolic Energy

The energy needed for all aspects of an organism's life including metabolism (both anabolic and catabolic), feeding, reproduction, movement, dispersal, inter and intraspecific interactions, and growth comes from the generation of ATP. There are only two basic mechanisms for generating ATP. The first mechanism is *oxidative phosphorylation* or *respiratory chain phosphorylation*. In this mechanism ATP is generated by

**Table 5.1** Classification of Organisms Based on Energy Source, Electron Donor, and Carbon Source

| Energy Source | Substrates by Electron Donors | | Carbon Sources | |
|---|---|---|---|---|
| | **Inorganic** | **Organic** | **Carbon Dioxide** | **Organic Compounds** |
| Light | Photolithotrophs Use $H_2O$, $H_2S$, S, $H_2$ | Photoorganotrophs Use succinate, acetate | Photolithoautotrophs plants, algae, cyanobacteria, some purple and green bacteria | Photoorganoheterotrophs heterotrophs, some bacteria |
| Chemicals | Chemolithotroph Use $H_2$, $H_2S$, $NH_3$, $Fe^{2+}$, $NO_2$ | Chemoorganotrophs Use many organic substrates | Chemolithoautotroph hydrogen bacteria, colorless sulfur bacteria, nitrifying bacteria, iron bacteria methanogenic, and methylotroph bacteria | Chemoorganoheterotroph animals, most bacteria fungi, many protists |

Data from Andrews JH. *Comparative Ecology of Microorganisms and Macroorganisms.* New York, Springer-Verlag, 1991, with kind permission from Springer Science and Business Media.

the flow of electrons from some inorganic or organic donor to a terminal electron acceptor like $O_2$ or $NO_3^-$. In the process electrons move from a relatively negative redox potential to a relatively positive potential or from higher energy levels to lower energy levels. ATP is synthesized from ADP and inorganic phosphate. This can be cyclic or noncyclic photophosphorylation as found in photoautotrophs or heterotrophic respiratory chains.

The second mechanism is *substrate-level phosphorylation*. This mechanism does not require the flow of electrons and occurs in the degradation of high-energy phosphoryl bonds in certain organic substrates.

Organisms can obtain energy derived from the flow of electrons from light (i.e., phototrophs) or from chemicals (i.e., chemotrophs). As can be seen in Table 5.1, the source of the electrons (i.e., donors) can be inorganic or organic molecules for both energy sources. *Organotrophs* obtain electrons from organic molecules whereas *lithotrophs* get electrons from inorganic sources. There are many organic and inorganic molecules that can be used as electron donors. The use of specific donors is dependent on the redox conditions, as shown in Figure 5.2, and the electron acceptors because of thermodynamic constraints (Leigh, 2002). Electron-donating and electron-accepting reactions are always coupled, and energy-yielding reactions always involve electron flow from a more negative to a more positive redox potential (downward from left to right in Figure 5.2). In general, organisms that use relatively more negative electron acceptors are restricted to electron donors that are even more negative. Microorganisms that can use more positive electron acceptors are able to catabolize a greater variety of substrates.

# Role of Carbon

All living things are carbon based. Carbon forms the structural basis of most important biological molecules, including amino acids, nucleic acids, fatty acids, lipids, cellulose, lignin, hormones, and steroids. Organisms can get the carbon used to

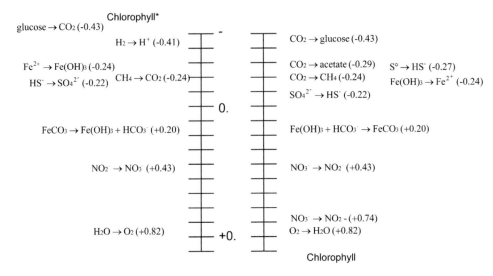

**Figure 5.2** Redox potentials for electron-donating and electron-accepting reactions. Chlorophyll* indicates light-activated, reduced chlorophyll. (Redrawn from Leigh JA. Evolution of energy metabolism. In: Staley JT, Reysenbach A (eds): *Biodiversity of Microbial Life*. New York, Wiley-Liss, 2002. This material is used by permission of Wiley-Liss, Inc., a subsidiary of John Wiley & Sons, Inc.)

synthesize these many and diverse molecules from two basic sources: carbon dioxide or other organic compounds. Plants and many algae use $CO_2$ as the carbon source and most students' understanding of photosynthesis is based on this model. In this model light energy is captured and used to fix $CO_2$ into organic molecules. The cyanobacteria and some purple and green bacteria use a similar mechanism. However, there are major differences in where these organisms live. The purple and green bacteria are obligate anaerobes and as such are found in specific environmental conditions.

## Microbial Feeding Strategies

Bacteria have representative species in every one of the categories listed in Table 5.1 however; complex macroorganisms are restricted to the photoautotrophs and the chemoheterotrophs. There are no eukaryotic chemoautotrophs or photoheterotrophs. What are the ecological implications of these observations? Why are complex multi-cellular organisms restricted to only a few nutritional modes? The answers to these questions are partially embedded in the origin of feeding and energy acquisition from the earliest of times.

If we consider that most metabolic processes do not require oxygen, certain logical deductions become apparent. First, it appears that metabolism evolved in the absence of oxygen. Early life experts predict that the atmosphere of earth before photosynthesis was made up of reducing and not oxidizing gases (see Chapter 2). Life evolved the necessary machinery to grow and reproduce in the absence of oxygen. Bacteria have been around for more than 3 billion years. During most of this time, there was not any free oxygen. The expansion into new niches was accomplished by the evolution of light energy capturing mechanisms (e.g., photolithotrophy, photo-organotrophy) and by the ability to capture electrons from many different sources. Many other microorganisms that were unable to use light energy made use of the energy stored in chemicals (both inorganic and organic) to allow them to fix carbon and synthesize organic molecules. These processes also evolved in the absence of oxygen.

If we accept the symbiotic origin of mitochondria and chloroplasts then it is still interesting why only a specific subset of nutritional modes were selected by higher organisms. The DNA associated with chloroplasts is strikingly similar to that of cyanobacteria. Cyanobacteria are photolithoautotrophs like higher plants, or rather we might say higher plants are like the cyanobacteria from which they derived their chloroplasts. Why were the other bacterial chlorophylls not incorporated into higher plants, or if they were, why were they selected against in favor of the cyanobacteria model?

Perhaps evolution and ecology can help us understand these patterns found in nature today. Based on 16S rRNA there are five major groups of photosynthesizing bacterial groups within the domain Bacteria. These include the cyanobacteria, the purple bacteria, the green sulfur bacteria, the heliobacteria, and the green filamentous bacteria (Stackebrandt et al., 1996; Pierson, 2002). Only the cyanobacteria are oxygenic, use chlorophyll a, and have two photosystems in which they are able to oxidize water and produce oxygen. The cyanobacteria are basically indistinguishable from higher plants in the way they do photosynthesis. All of the remaining phototrophic bacteria are anoxygenic, are unable to use water as an electron source, do not use chlorophyll a, and have only one photosystem. All non-cyanobacterial phototrophic bacteria live within relatively narrow environmental ranges. The only constraints are the availability of electron acceptors and donors and the ability of light to penetrate to where these bacteria are found. Because of the use of different chlorophylls bacterial phototrophs are able to use almost the complete visible light spectrum and are able to exploit habitats that are not available to higher plants.

Light competition among higher plants is common with plants in the upper canopy getting more of the available light energy than plants in the understory. Because all higher plants use the same wavelengths of light, plants living in the understory are energy limited. These plants have adapted to the low light levels, but at the cost of slower growth and reproduction. Weedy plant species that colonize open fields are able to grow rapidly and reproduce with high numbers of offspring because there is essentially no competition for the light. After plants establish that grow over these weedy plants, the weedy plants are unable to compete for light and they die out. The structure (i.e., species diversity and vertical distribution) of plant communities is in essence driven by competition for the same wavelengths of light.

The diversity of bacterial light-harvesting pigments has allowed phototrophic bacteria to survive in complex, often dense communities, with different species able to use different wavelengths of light. The density of one type of light-harvesting phototroph may be so great that they effectively absorb all of the available light of a particular wavelength. Below these organisms, no other phototrophs that use the same wavelengths of light would be able to survive. However, most of the nonabsorbed light would be transmitted through the layer, and phototrophs that can use this light energy can survive. Pierson (2002) suggests that the high diversity of light-harvesting mechanisms in bacterial phototrophs has had profound effects on the ecological distribution of phototrophic bacteria. Any environment exposed to light and suitable to life will have phototrophs no matter what the restriction of wavelengths due to environmental conditions including other phototrophs. Microbial phototrophs can be in such high numbers that layered communities of conspicuously colored groups can be readily observed.

In water, shorter wavelengths of light (blues) penetrate deeper than longer wavelengths. Pigments such as carotenoids and phycobilins may have evolved under these conditions (Pierson, 2002), whereas pigments that harvest near-infrared and reds may have evolved in shallow mat communities (Pierson et al., 1990).

Whereas all phototrophs make biologically available energy from light by photosynthesis, only microbial phototrophs do not have to grow as autotrophs. Photoheterotrophy is the form of carbon metabolism used by many of the purple bacteria and filamentous green bacteria. For the heliobacteria, it is the only form of carbon metabolism. Photosynthesis in bacteria is not equivalent to autotrophy. Photosynthesis and autotrophic metabolism probably evolved under different scenarios. Ecologically, photoheterotrophy provides increased versatility and allows growth and survival in the absence of light.

A phototroph needs to be vertically distributed relative to the wavelengths of light. Even if entire wavelengths are absorbed by a specific group, the remaining light is still available to other groups. Layered photosynthetic communities can be found in temperature or chemically stratified lakes or in microbial mats. Stratified lakes have oxic surface layers and steep gradients of decreasing oxygen with depth. An anoxic zone is found associated with hydrogen sulfide derived from the bottom sediments. Light penetration is affected by absorbance, scattering, and suspended and dissolved materials and by the organisms found in the water column.

Organisms that feed on light have advantage over strict heterotrophs in that they do not need to actively forage. The food comes to them day in and day out. Light is distributed to the earth at a fairly consistent rate. The amount of light changes with latitude and season, but generally speaking, light at the earth's surface is not limiting and is obviously renewable. Individuals can increase their light foraging capacity by increasing the light-harvesting pigments or proteins. There are limits on the number of pigment molecules that can be stored in a single cell. Some phototrophs migrate vertically in the water column on a diel basis. Such movement should be considered a form of foraging, because these organisms are increasing the probability of getting the right wavelengths and increasing their potential to grow and reproduce.

Non-energized chlorophyll has a large positive redox potential (lower right side of Figure 5.2) and can accept electrons from a wide variety of sources, including hydrogen, sulfide, ferrous iron, and water. After chlorophyll becomes energized from light, it has a large negative redox potential that allows the fixation of inorganic carbon to occur.

There are certain costs from being exposed to light. Certain wavelengths of light can be harmful to living organisms and cause intracellular damage. In particular UV-B, UV-A, and photosynthetically active radiation (PAR) may affect the physiological state of microbes exposed to these wavelengths in both terrestrial and aquatic systems. In one study, short-term exposure of *E. coli* populations to simulated solar radiation had a sublethal effect. Populations showed a loss of culturability and the formation of viable but nonculturable cells (Muela et al., 2000). When these same populations were exposed to solar radiation for longer periods, increased levels of cell damage were observed with little loss of cellular integrity. The greatest effect on these populations was observed with UV-B radiation, for which there was loss of culturability similar to exposure to the simulated solar radiation.

Why would bacteria and other microorganisms live in habitats where exposure to solar radiation may affect their ability to survive? Light is important to many organisms directly through photosynthesis but it may also benefit heterotrophic organisms indirectly by increasing the availability of various organic compounds. Photochemically transformed leachates from aquatic and terrestrial primary producers were provided to bacteria to determine whether UV light increased the availability of these substrates (Anesio et al., 2000). The results were dependent on the source of the organic matter. Leachates derived from terrestrial vegetation became more available whereas leachates from aquatic macrophytes became less so after exposure to UV light. From these data we might expect that bacteria in wetlands and rivers to behave differently. Rivers with extensive flood plain forests or riparian zones would input large amounts of organic matter in the fall. With the loss of leaves there should be an increase in UV light reaching the water and the dissolved leachates coming from the leaves. These data suggest that organic matter may become more available during these times.

# Costs of Feeding

In the process of survival, there are many possible solutions to obtaining and allocating energy and food, but each solution has associated costs. The following discussion is largely based on Andrews's work (1991). An *optimal solution* is the one that has the lowest associated evolutionary costs. Evolutionary costs are usually measured in energy units. Natural selection should favor the best overall efficiency in energy capture and utilization. Overall efficiency would give an organism, on average, a competitive advantage over others within the constraints of a particular habitat and allow these organisms to leave more copies of their genes.

Because of energy constraints an organism cannot be selected to have huge reproductive outputs and have numerous anti-predatory secondary structures. Some balance is achieved between competing attributes. Phylogeny and the gene pool set boundaries for all attributes. For example, a weedy species of plant that reproduces by producing many small seeds cannot be selected to produce large seeds with heavy seed coats. Selection has favored one aspect of this plant (high reproductive output) over seed defense. If we consider that an organism does not have a single "choice" or a single activity to optimize but rather there are often many different functions and activities, some of which are competing for the same energy, then it is easy to see that every activity or function cannot be maximized. An optimal solution is the best compromise from the limited options available to a particular organism. Consider this problem for microbes. A single microbe, say a bacterium, cannot be selected to be able to use hundreds of different substrates and be an obligate parasite of some higher organism. The parasitic strategy and the strategy to use many organic compounds are at odds with each other. More than likely the parasite would never be exposed to all of those compounds.

Optimal foraging theory predicts strategies that maximize net energy intake per unit time spent foraging. *Net energy* is the difference between the amount of energy taken in and the costs of acquisition. *Energy efficiency* as used in ecology measures the rate of productivity (Andrews, 1991) or the amount of energy input to the organism less the energy output (acquisition) per unit time. Energy efficiency may or may not be directly related to the number of gene copies in the next generation but operationally it is a good currency and allows hypothesis testing. The use of fitness, an evolutionary concept, as the currency is fraught with difficulties in higher organisms because of the complexities in following an organism throughout their life. It would be difficult to definitively determine whether changes in resource acquisition or any number of other factors that impinge on an organism resulted in reproductive differences. However, in microorganisms many of these difficulties would be removed. Microorganisms make few "decisions" other than to feed and to reproduce. Although this is not entirely true (whether to make antibiotics or extracellular polysaccharide), it is generally so.

Optimal foraging theory has two major constraints: First energy is the common denominator. Energy is used because of its role in metabolism and the acquisition and use of all other resources. However, most resources are probably *limiting* (a limiting resource is one that affects the growth and competitive ability of an organism; a non-limiting resource does not constrain growth) because of the efforts necessary to obtain them is often incomplete and requires energy. Selection favors a compromise that optimizes foraging for all limiting resources. Because all resources are not equally important, the compromise will be biased toward the following conditions: first, for resources that are most extensively used in metabolism and for which there are no substitutes; second, those that are most limiting; and third, those that give the greatest return per investment in foraging; and fourth, can be most easily selected for, given the existing gene pool.

The second constraint on optimal foraging theory is that it needs to take into consideration the conversion or digestion of the resource. In higher organisms there are various ways to maximize the net rate of obtaining energy including modifications of the gut and the retention time of the food. Together foraging and digestion should be selected to maximize net energy return. Optimal foraging consists of searching for the food item, capture and handling of the food item and conversion of the food item. Let us consider each of these functions as they relate to microorganisms.

At any level, resources can be limiting. The availability of nutrient resources is determined in part by the proximity and abundance of competitors and by the source and input rate of the nutrient. For example, if the source of a particular limiting nutrient is the decomposition of a specific organic compound and this compound is only produced by a certain primary producing organism, the delivery rate of the compound is limited by the numbers of the producing organism and the rate of decomposition after release from the producer. If the input rate of a limiting resource is constant, availability will be determined by the number of organisms competing for the resource. In nature, there is rarely a single limiting resource. Because organisms cannot be ultimately efficient in all aspects of their lives, compromises have to be made with respect to acquisition and metabolism of resources.

# Generalists and Specialists

Organisms that can use many different resources at somewhat similar efficiencies are called *generalists* and those that are restricted to a few resources are known as *specialists*. These designations are not restricted to trophic interactions but include diversity of habitat use. Theoretically generalists should not be as efficient in using any specific resource as well as a specialist otherwise specialists should be selected out of existence. In certain situations specialists, because of their ability to use the resource very efficiently, will out-compete any generalists. Outside those situations generalists will out-compete specialists because of their ability to use more than one resource.

In microbiology organisms that can feed on a wide spectrum of molecules are referred to as versatile. For example, *Pseudomonas putida* can use over 77 carbon compounds as their sole carbon and energy source. *P. putida* is said to have high *versatility* and would be considered an ecological generalist. In contrast, *Methylococcus* feeds only on methane, has low versatility and is an ecological specialist.

*Versatility* has two components. First is the ability to do many things. Versatility under this definition is increased by extending the range of activities but it is not altered by the degree of complexity of the activity or by the competency with which the activity is executed. Second, an organism that masters many different situations quickly and easily is said to be versatile. This definition has a time component to it. The speed of response is emphasized, not the degree or breath of the accomplishments. Versatility under this definition includes those individuals who adjust quickly to new situations, with time being of the essence. Versatility is a relative term and must be used in reference to some other organism, place or time.

Knowing what to compare is critical in understanding relative terms such as generalist, specialist, and versatility. Should we compare microbes with plants or higher animals or microbes with microbes? If only intramicrobial comparisons are valid, do we compare heterotrophs with autotrophs or only heterotrophs with heterotrophs and autotrophs with autotrophs? Before we can address these and other questions relative to microbial activities, it is important to understand the differences between microbial and higher organismal trophic relationships.

# Optimal Foraging and Microbes

How applicable is the concept of optimal foraging to microbes? Do bacteria or fungi selectively pick food items? What constitutes food-handling time for bacteria and fungi? When should microbes switch food resources and under what kinds of conditions? Are there differences in microbial feeding with and without predation? Answers to each of these questions will provide insight into whether a concept such as optimal foraging can be applied to microbial systems. Let us look at each question separately.

Can microbes assess food quality in such a way that they are able to select preferred food items over non-preferred items? Studies have been conducted with bacteria in which different food resources were added. These studies are discussed below as they relate to the differential uptake of the food items.

Do microbes handle their food? If so what behaviors are in support? For higher organisms food handling involves manipulating the food in such a way that the food item is suitable to eat. For instance, birds have to remove seed husks, thrushes have to remove snail shells, and all aspects of prey capture are in effect food-handling behaviors. Predators can be sit-and-wait or active pursuers. Sit-and-wait predators take up a location that has a high probability of having prey come by. This strategy includes web-building spiders, aquatic filter-feeding invertebrates, anglerfish, snapping turtles, frogs, and many others. The predator expends little energy in hunting for prey as the prey come to them. Predators that pursue or seek their prey include many species of birds, large cats, wolves, and most of the animals that are seen on nature programs running down or attacking prey. This strategy requires sometimes large expenditures of energy. Microbes use both strategies.

When considering bacteria as predators, it is necessary to rethink predators and prey. Most bacteria are not predacious in the sense of higher organisms; they do not consume other living organisms. This is not exclusively so, and there are examples of predacious bacteria. Fungi can and do attack living tissues as well as feeding on dead organic matter. Other fungi are able to trap nematodes. However, most microbial "prey" items are various molecules or compounds derived from living or dead organisms. Complexity of the molecules determines the energy that may be harvested from eating all or part. The number and types of chemical bonds and the presence or absence of certain elements further determines the quality of the food. The potential diversity of organic molecules is immense. All classes of organic compounds can be considered potential food items for microbes. Even toxic or inhibitory compounds may eventually be degraded or eaten by some microbe given enough time.

Aquatic bacteria that are attached to surfaces (*biofilms*) are essentially sit-and-wait predators. The flow of water brings dissolved organic matter (DOM) and particulate organic matter (POM) to the attached bacteria. If we consider the effective distance of diffusion and the relative flow rate of the overlying water, it is clear that most DOM and POM is transported through the system without being "captured" by microbes. Water flowing over a solid object creates a boundary layer in which there is little or no flow. The thickness of this boundary layer is primarily determined by the flow rate.

Lock has developed a conceptual model of biofilm structure and function (Lock, 1993). In this model the spatial distribution of microbes is determined by the nature of the substrate and the flow rate of the water. Biofilm organisms exude an exopolysaccharide matrix (EPM) in which they are embedded. The EPM performs several functions including protection from grazing organisms, holding exoenzymes in place and the capture of organic matter. Davey and O'Toole (2000) monitored temporal changes in a biofilm that include species diversity, resource diversity, and the competition among biofilm inhabitants (Figure 5.3).

Part of the cost of obtaining food by some microbes is the production of EMP and exoenzymes, especially in aquatic environments. If a bacterium produces and releases enzymes needed to begin the degradation of some organic substances and those enzymes are washed away before or during the process, the bacterium loses both food and the energy and resources used to make the enzyme. Having a mechanism that keeps the enzymes in place is essential. However, after an enzyme is released from a microbe, it will perform its function independent of the producer and the products produced by the enzyme are available for any and all microbes in the near vicinity.

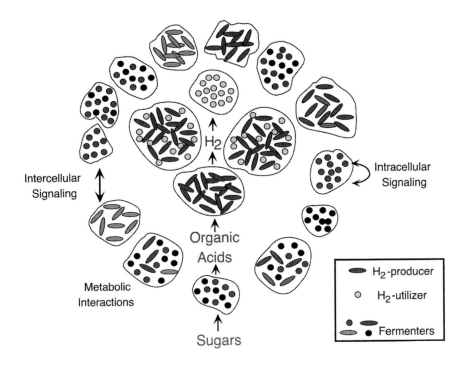

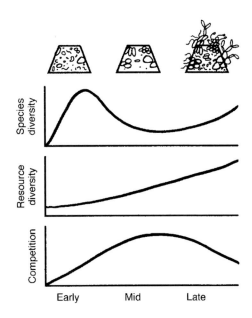

**Figure 5.3** Model of biofilm interactions among and within various microbial groups. (Top figure from Figure 7 in Jackson CR, Churchill PF, Roden EE. Successional changes in bacterial assemblage structure during epilithic biofilm development. *Ecology* 82:555–566, 2001.) Changes in species diversity, resource diversity, and levels of competition during biofilm formation. (Bottom figure from Figure 1 in Davey ME, O'Toole GA. Microbial biofilms: from ecology to molecular genetics. *Microbiol Mol Biol Rev* 64:847–867, 2000.)

The earliest stages of decomposition or mineralization of a compound often entail considerable modification of the compound. For example, the insertion of oxygen, cleavage of an aromatic ring, net reduction or oxidation, and decarboxylation have little or no net return of energy or materials to the bacterium. Enzymes that perform these early modifications of a substrate are often carried on plasmids and are not necessarily part of the organisms "normal" suite of metabolic activities. Enzymes that perform central metabolic functions are usually encoded on chromosomes.

Microbes, if they behave like higher organisms, should take up and process molecules with the highest payback first. In many experiments where different substrates are presented in the growth media bacteria seem to select substrates based on the notion of energy payback. However, in nature, bacteria or microbes in general seldom have a constant delivery of the "right" food. Rather DOM is a complex mixture of new and old carbon compounds in various stages of mineralization. Having the ability to capture this material outside the cell allows microbes the time to perform the early modifications.

The release of exoenzymes is risky if the probability of obtaining useable energy and carbon are unlikely either from low delivery rates of the substrate that the exoenzymes can work on or by competition from either intraspecific or interspecific competitors. The distribution of similar cells may in fact aid in the initial modifications because exoenzymes produced by neighbors will break down trapped substrate within a volume that often includes other cells. Because little net energy is obtained by exometabolism and exoenzymes have a metabolic cost, two different strategies can emerge. All cells produce and excrete exoenzymes and benefit from this apparent assemblage level cooperation, or some cells cheat and take advantage of the exoenzymes of other cells but do not themselves produce exoenzymes. As Lock (1993) points out, we do not know the half-life of exoenzymes. How long do these enzymes persist in a biofilm? If the enzymes persist for extended lengths of time then some bacteria can sit back and enjoy the benefits of other bacterial cells' enzyme production. Depending on the high life of exoenzymes, newly colonizing individuals may be able to benefit from exoenzymes produced by aging or dead cells while they are trying to get established. Each of the latter two cases is an example of what is known as cheating.

## Cheating

Cheating has evolutionary consequences. Cheating or selfish traits should increase in frequency if the traits are heritable relative to traits that benefit the group or assemblage. Microbes that take advantage of enzymes produced by other cells will have more energy and intracellular materials available for reproduction than the enzyme producing cells. If no cells produce enzyme, no substrate will be available for energy or carbon. An alternative strategy is to make enzyme only when the substrate is present. This would require some mechanism to detect the substrate extracellularly. Detection, production of enzyme, and uptake of modified substrate have to occur faster than the residence time of the bacterium in the biofilm. Cheating is discussed in more detail in the section on living in communities.

# Free-Living Microorganisms

We have been discussing the handling time of microbes associated with a biofilm, but what of free-living cells? Suspended cells do not have the luxury of an external capture mechanism such as EPM. However, sea and lake rain (i.e., cellular debris and detritus settling through the water column) are quasi-biofilms and may facilitate handling of food by pelagic bacteria. The ability to take up or break down specific molecules by free-living microorganisms is dependent on the concentration of the substrate and the membrane associated enzymes or transport molecules. Under these conditions we might expect that lower molecular weight compounds and substrates are preferentially taken-up by these cells. Experimental results are difficult to obtain from nature, however, laboratory results are supportive.

To test preferential uptakes of different mixed substrates, researchers have used batch and continuous culture techniques (Box 5.1). Smith and Bull (1976) examined the growth of *Saccharomyces fragilis* on a mixture of four compounds: glucose, fructose, sucrose, and sorbitol. These four compounds are the major constituents of coconut water. The utilization of the different substrates varied depending on the experimental approach. When grown in batch culture glucose and fructose were used simultaneously with concurrent "inversion" of sucrose into glucose plus fructose, and sorbitol was used as a substrate only after the other sugars had been consumed. Under continuous culture different patterns of utilization were observed dependent on the dilution and subsequent growth rates of the cultures. When the cultures grew at rates below 0.05 hour$^{-1}$, all four substrates were consumed completely. At rates between 0.05 to 0.2 hour$^{-1}$, sorbitol appeared in increasing concentrations in the media; between 0.18 and 0.35 hour$^{-1}$, sucrose remained un-metabolized and above this dilution rate both fructose and glucose began to accumulate. These results are representative of numerous experiments conducted with several other species of microbes.

Batch and continuous culture experiments mimic different aspects of nature. Some microbes experience batch-like conditions and other microbes experience continuous throughput of substrates. Therefore the results of these types of experiments give insights into the ecology of microbes in different habitats.

---

**Box 5.1   Batch Versus Continuous Cultures**

Various aspects of microbial growth can be captured by two different procedures. Microbes can be grown in batch or continuous culture. Both procedures provide insight into microbial growth, but they differ in what those insights are. In batch cultures, the microorganisms are inoculated into some media and allowed to grow. Samples can be collected over time to show the population dynamics under the constraints of continuous depletion of substrates and continuous accumulation of waste products. At some point, sufficient nutrients are not available for growth, and the population begins to decline or crash.

In continuous culture, two separate aspects of microbial growth—cell turnover and exponential growth at a slow rate—can be mimicked and controlled under laboratory conditions by growing the cells in a *chemostat*. This device (Figure 5.4) consists of a growth chamber that is connected to a reservoir of media and to an outlet. The volume of media remains constant because inflows match outflows molecule for molecule. Under these conditions, there is a continual turnover of the populations. Turnover in nature may result from delivery of sufficient nutrients and from predation, dying, and elimination through various environmental conditions. In the chemostat, the rate of turnover can be determined by the investigator.

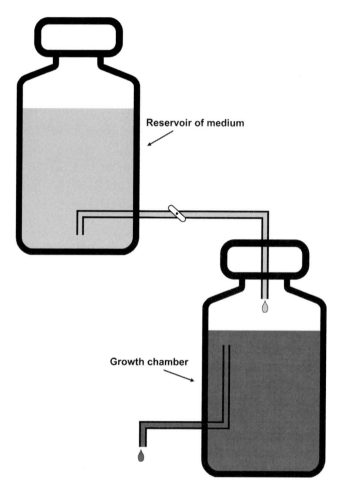

**Figure 5.4** Basic components of a simple chemostat including reservoir of media, a delivery device that can control the rate of input, and a growth chamber with an outlet that allows the removal of bacteria and spent media.

The sequential utilization of substrates in batch cultures supports optimal foraging predictions. Microbes do preferentially take up or "eat" food items. From the example given above sorbitol was not consumed until all of the sugars had been consumed. The mechanisms involved in repressing the uptake of other available substrates has been elucidated for *E. coli* and other bacteria and involves *catabolite repression*, in which degradation products of a readily metabolized carbon source interferes with the synthesis of enzymes that would be needed in the uptake and metabolism of other carbon sources.

Handling time includes the secretion of the polysaccharide matrix, the secretion of exoenzymes, and the subsequent uptake by passive or active transport of the substrate. As we will discuss in detail below, consortia of microbes often facilitate the complete mineralization of a substrate. However, some versatile microbes are capable of completely mineralizing a substrate to $CO_2$ and $H_2O$. The ability to perform all sequential degradations of a resource would be exceedingly important in extreme environments where metabolic and species diversity is low relative to more tempered habitats.

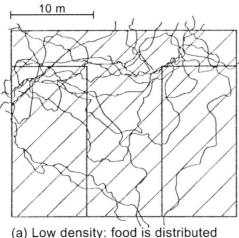

(a) Low density: food is distributed
throughout all areas

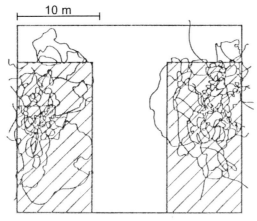

(b) High density: food is distributed
in specific areas only

**Figure 5.5** Foraging patterns of thrushes as observed in patches with high and low levels of forage. (From Smith JNM. The food searching behaviour of two European thrushes. I. Description and analysis of search paths. *Behaviour* 48:276–302, 1974.)

We have been considering the acquisition of a resource by microbes that sit and wait or that are suspended in their growth medium such as pelagic organisms. Let us consider microbes that actually search or seek for food items. Most people have observed Robins as they forage for insects and worms on a lawn. The birds run rapidly in a direction, halt, listen, observe, and then run toward potential food items. Smith (1974) mapped the movement patterns of European thrushes as they foraged (Figure 5.5) during an experiment where he manipulated the prey density. In low-density situations, the birds traveled longer distances, turned less sharply, and turned less frequently than birds foraging in the high-density treatments. In the high-density treatments, birds sometimes foraged into nearby areas without food but quickly returned to the high-density patches, whereas in the low-density treatment, the birds foraged across the entire study area.

Some bacteria are motile. Why are they motile? If we consider the speed of bacterial movement, it clearly is of little value in escaping predation. Anyone who has observed pond water through a microscope cannot help but be amazed at the intensity of movement. Organisms, usually rotifers and microcrustaceans, seem to be rushing here and there. Rotifers, nematodes, copepods, and other predacious organisms feed regularly on bacteria. Bacteria are ineffective in avoiding predation through movement. Why then do bacteria exhibit movement?

If you add ink particles to a drop of water and then observe the drop under magnification, the particles are moving rapidly. This movement of particles by abiotic forces is called *brownian motion*. Brownian motion can result in a nearly uniform distribution of particles over time within the liquid medium. Immotile bacteria suspended in a drop of water would in time be distributed throughout a liquid medium. Bacteria can increase the probability of finding food through directed movement.

The movement of bacteria has been described and mapped (Figure 5.6). These maps crudely resemble the maps of the thrush foraging (see Figure 5.5). Movement consists of "runs" that maintain a particular direction for a brief time followed by "twiddles" during which the bacterium reorients. Bacteria are able to move toward or away from various stimuli. This ability is truly remarkable considering the simplicity

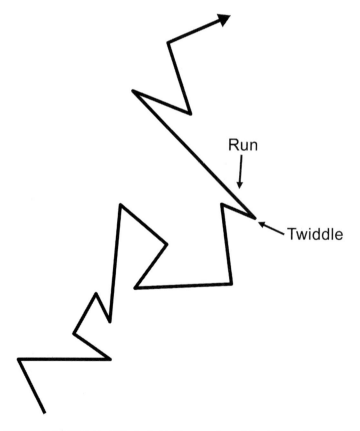

**Figure 5.6** Directed movement of motile bacteria. Movement consists of directed runs and reorienting twiddles.

of bacterial cells compared with the complex structure and physiology of higher organisms that display similar responses. Bacteria are single cells with a very limited genome and yet some bacteria actively forage for food. The twiddles or reorientation of the bacteria has the trappings of environmental evaluation and directed response.

# Food Chains and Webs

In this chapter, we are considering aspects of an individual's ecology. In most environments, complex assemblages and communities exist where interactions and associations affect the behaviors and abilities of organisms to obtain food. We address some of these interactions in later chapters.

Trophic relationships of higher organisms are visually expressed in *food chains*, *food webs*, *pyramids of numbers/biomass*, and *pyramids of energy*. A food chain shows the linkages of energy and carbon from a primary producer through some top predator. Food webs attempt to show linkages among all organisms within a location or ecosystem. Food webs are usually extremely complex and difficult to construct because all possible linkages and interactions are not known. Figure 5.6 demonstrates typical food chains and food webs. For higher organisms, the numbers of primary producers and their biomass usually far exceeds the numbers and biomass of herbivores that feed on the primary producers. Similar relationships exist between the numbers and biomass of herbivores and their predators. Large predators are usually fairly rare. Which begs the question: why is there this pattern of decreasing number and biomass with increasing feeding chain length?

The usual answer is based on the transfer of energy and carbon between trophic levels. Not all the energy consumed by an organism can be used or extracted by that organism and not all of the carbon is converted into consumer biomass. Some carbon is respired as $CO_2$ and some carbon is eliminated as waste. A consumer needs energy for growth, catabolism, reproduction, feeding, movement, intra and interspecific competition and other aspects of survival so that not all energy is used to make consumer biomass. Much energy is lost as heat. The rule of thumb applied to the transfer of energy and carbon is 10%. That is about 10% of the energy and carbon available at one trophic level is used by the next higher level. This rule of thumb is based considerably on intuition and less on solid empirical evidence. However, the principle makes sense and whether the actual transfer is greater or less than 10% is not important to the discussion at hand. Whatever the transfer efficiency, it does not take many trophic linkages before there is not much carbon or energy available. Large predators are rare because there is not enough prey carbon and energy available to make very many of them. Large predators are forced to have very large feeding ranges in order to capture enough energy to reproduce.

Exceptions to these general patterns have been found in nature. In fact inverted pyramids have been observed where the numbers and biomass of consumers is higher than the standing crop of primary producers. One such example is that of the plankton. Planktonic algae and diatoms have very high production rates but they are consumed/grazed almost as fast as they reproduce. Therefore at any one moment in time the numbers and biomass of producers is lower than the numbers and biomass of

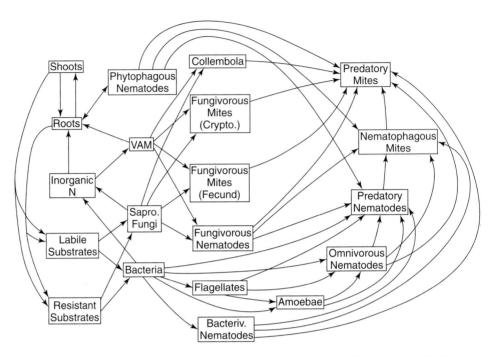

**Figure 5.7**  Ecological food web. This example is from a shortgrass prairie. (From Coleman DC. Through a ped darkly: an ecological assessment of root-soil-microbial-faunal interactions. In: Fitter AH, Atkinson D, Read DJ, Usher MB (eds): *Ecological Interactions in Soil: Plants, Microbes, and Animals*. Boston, Blackwell Scientific Publications, 1985.)

their grazers. However, if one could measure the total amount of producer biomass produced annually it would far exceed the biomass of grazers produced annually.

Conspicuously absent from or de-emphasized in most discussions of food webs and pyramids are microbes. Decomposition, if acknowledged, is difficult to incorporate into the webs and chains. Some authors include microbes as part of a parallel saprovore web, which also includes some invertebrates (Figure 5.7). In these schemes, the saprovore linkages seldom exceed three links whereas the herbivore chains can have four or more links. The saprovore system is limited to three links because certain processes are lumped together in the saprovore component of the chain. In essence, all decomposers are assumed to be doing the same things! Bacteria and fungi are lumped together because they reduce organic detrital biomass. Within the saprovore link exists a complex food chain/web that is lost by the generalization of saprovore.

Trophic relationships within the saprovore box can be collectively called the *dissimilatory food web* (Figure 5.9). Trophic links in dissimilatory food webs can be numerous and varied and are based on the starting resource substrates. Some versatile bacteria can break down a complex substrate completely without assistance from any other organisms. These organisms would form the simplest dissimilatory chain with every link on the chain being the same organism.

In many dissimilatory webs, the roles of various microbes are unknown. However, entire consortia are required to completely mineralize many substrates. In some instances, roles are known, and not all microbes within a consortium actually depolymerize a molecule. Some microbes provide growth factors for those microbes directly involved in the primary attack on a substrate or they may remove inhibitory products.

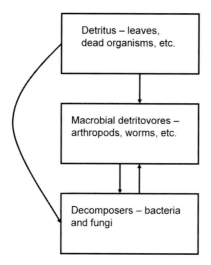

**Figure 5.8** Saprovore food chain consisting of naturally occurring organic matter, microbial and microbial detritovores and decomposers.

Dolfing (2001), in a mini-review, described the "microbial logic" behind incomplete oxidation of various organic compounds. Dolfing examined the microbial degradation of organic matter in methanogenic ecosystems. This process is a multistep process in which subsequent groups of bacteria use the products of the first groups in a chain. One particular group, the acetogenic bacteria, produce $H_2$ and acetate. Using a thermodynamic approach, it can be shown that in methanogenic ecosystems, the acetogenic oxidation of substrates such as propionate, butyrate, and benzoate yields more energy than their complete oxidation to hydrogen and carbon dioxide. In other words, a multistep process with several groups of bacteria is better able to harvest the energy in compounds than a single organism completely degrading some substrate.

Hegeman (1985) identifies three stages of mineralization:

1. Initiation of the process by surface bacteria and sediment or soil bacteria. This primary attack is characterized by the excretion of proteases and other extracellular hydrolytic enzymes. Metabolism is often inefficient (fermentations) and the bacteria often require one or more growth factors found in fresh decomposing plant matter.
2. The second stage when the fermentation products and materials unused by the initial decomposers become the focus of attack. This phase is usually aerobic because most of the fermentation products cannot be degraded anaerobically. The bulk of the carbon is respired as $CO_2$.
3. A final stage is characterized by the slow release of $CO_2$ aerobically from the most refractory compounds that remain. The one-carbon compound—oxidizing bacteria and other specialists participate.

Although we can quibble over the details in Hegeman's scheme, the important point is that there are numerous bacteria and fungi involved in the breakdown of organic

# Dissimilatory Food Webs

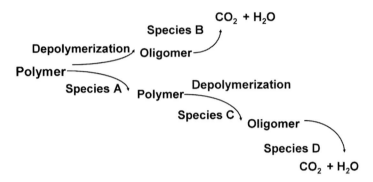

Figure 5.9  A possible dissimilatory food web where some large complex organic molecule is converted into carbon dioxide and water through the combined activity of several species of microorganisms.

# Single Location

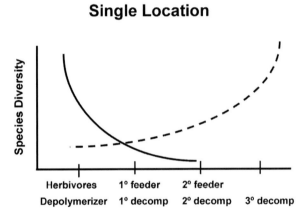

Figure 5.10  Species diversity of various trophic levels in a dissimilatory and classic food chains.

material. There is a succession of microbial species or types over time, each with different mineralization roles. In stage three of the scheme, Hegeman suggests that during the final stage of mineralization, specialist bacteria participate. There are bacteria that are able to break down specific carbon compounds. In dissimilatory food webs, the number of potential specialists increases with trophic linkage (Figure 5.9), whereas the number of organisms in higher linkages of classic food webs decreases. Many more organisms can be classified as herbivores than as primary predators and as we have discussed above very few organisms are top predators. The reverse pattern is true in dissimilatory webs. There are relatively fewer microbes that are capable of the initial attack on some complex molecule. Sometimes, there may be only one type that can begin the mineralization process. However, as depolymerization proceeds smaller and smaller molecules are formed and more and more microbes are able to use these smaller molecules. The species diversity of tertiary and higher linkages in dissimilatory webs increases and may be very high. The actual number of species within a specific habitat or location that are involved in the complete mineralization of some complex molecule is seldom determined. We know very little about the make-up of various consortia found in nature. Because the nature and timing of organic inputs can be very different both spatially and temporally we might expect that the consortia responsible for the breakdown of these inputs to be nearly as spatially and temporally variable. Rarely is there a single organic resource, and in many terrestrial and aquatic environments, the diversity of organic inputs can be very high and exceed the diversity of the plant species within the habitat because of the production of various specific plant secondary compounds. Whereas sugars, amino acids, and nucleic acids and other building blocks of life are similar among plant species, many plants produce novel compounds that aid in defense. These compounds are imported into dissimilatory food webs on abscission and leaching of leaves and twigs. How interconnected are the trophic levels of dissimilatory webs that begin with unique compounds? Do the consortia overlap? Are there groups of specialist microbes, microbes that feed on one-carbon compounds, present in many consortia?

Bagwell and Lovell (2000) have shown that certain diazotrophs found in salt marsh displayed extensive physiologic diversity. In this study, seven physiologically dissimilar strains showed utilization patterns for at least one class of ecologically relevant compounds. These particular organisms appear to be physiologically adapted to use specific substrates or classes of substrates and demonstrate a level of functional redundancy. Functional redundancy in nitrogen fixation across extensive habitat heterogeneity ensures that an important environmental function is maintained.

# Fermentations

Under anaerobic conditions the terminal electron acceptor has to be something other than oxygen. If this terminal hydrogen acceptor is an organic compound the reaction is called fermentation. Pyruvic acid can be converted to many different products, including ethyl alcohol, acetic acid, lactic acid, succinic acid, butyric acid, butyl alcohol, isopropyl alcohol, propionic acid, formic acid, and other compounds. In all fermentations, most of the energy contained in a glucose molecule is not realized by the fermenting organism. Many of these metabolic products have commercial value and form the basis of several industries. Under controlled laboratory conditions, these products can be captured. In nature, these products can accumulate and either be inhibitory to other organisms or provide the beginning substrates for metabolism for other organisms. In many instances, these substances are inhibitory to the organism producing the substance.

# Ecology of Sex

6

## Reproductive Ecology

At first cut, it seems that an entire section devoted to the ecology of microbial reproduction is somewhat overkill. Most microbes reproduce by binary fission. End of
story. However, the story behind microbial reproduction is much deeper, gives clues
to the ecology of reproduction for all other organisms, and involves strategies that are
found only among the microbes. Given their immense success at survival a careful
examination of what we know and a lifting of the curtain on what we do not know
is justified.

Ecology in simplest terms is the interaction between the abiotic and biotic. Reproductive ecology is the study of abiotic and biotic factors that affect the reproduction
of an organism. Natural selection favors traits that increase the spread of genes. As
Emlen (1973) points out, natural selection should favor selection of appropriate
breeding sites and breeding times. In the reproductive ecology of higher organisms,
opposing selection factors result in a compromise between optimal reproduction and
survival to reproduce. For example, the best breeding areas may not be the best place
to obtain food, or the best time to reproduce may be the time when predators are
feeding. As such, the reproductive ecology of higher organisms is a compromise. But
what of microbes? Are there favored breeding sites or times for microbial species? Are
there constraints on the age of reproduction (i.e., are new daughter cells as likely to
reproduce as cells that have existed for longer periods)?

The study of the reproductive ecology of higher organisms involves observations
of the organism in both space and time. Where are the preferred sites for reproduction, and can these sites be used for obtaining food, or are they exclusively used for
reproduction? When do organisms first reproduce? Are there seasonal or diurnal patterns to reproduction? For many organisms, much of their reproductive ecology has
been described. It would seem that for microbes, the simplest of organisms, their
reproductive ecology would be straightforward and of little interest. However, the
questions posed for higher organisms are just as valid for microbes, and the answers
will provide insight into microbial processes.

As with most things microbial, we must begin with an appreciation for spatial scales.
Birds can travel thousands of miles to reach ideal breeding locations. Male frogs and
toads have elaborate choruses that bring females large distances to them. Many
mammals and insects release chemical cues (i.e., *pheromones*) that waft long distances

on the breeze, alerting males of female receptivity. Many plants release pollen on the wind and by so doing have some probability of fertilizing another like plant some distance away. In each of these examples, the organisms or their reproductive cells travel long distances to find mates and to reproduce. Most prokaryote microbes do not have to travel any distance to reproduce. They simply divide in situ. However, are all locations equally optimal for microbial reproduction? The short answer is no. The long answer is still no, but it demonstrates the effects of billions of years of natural selection acting on many different microorganisms and demonstrates some of the most novel and exciting behaviors and mechanisms known in biology.

# Microbial Reproduction

Considering the success of asexual organisms to survive over billions of years and their remarkable ability to exploit new, unusual, and even extremely harsh environments, sexual reproduction certainly gets a lot of attention. Certain aspects of sexual reproduction have been touted to provide apparent advantages for the spreading of genes and gene combinations. However, there are evolutionary costs to sexual reproduction. Regardless, most of the organisms that people experience reproduce sexually. Based on these experiences, some have argued that sex is obviously better than no sex, because sex is such a pervasive characteristic in biology. However, this is not true. Most reproduction is asexual (without meiosis) in eukaryotes and prokaryotes. Higher organisms have restricted meiosis to a few select cells, the germ cells. All other cells in the bodies of higher organisms are the products of asexual reproduction, cells giving rise to identical or nearly identical daughter cells without the fusion of nuclei. The apparent disconnect is the equating of reproduction with sex.

Let us define sex as the formation of individual organisms that contain genes from more than a single source or parent (Margulis and Sagan, 1986). Sexuality is the coalescence of DNA from previously separate sources into a single cell. Reproduction is evolutionarily essential, but sex is not. Many scientists subscribe to a classification of all living things into five major kingdoms. These kingdoms include plants, higher animals, fungi, protists, and bacteria. Margulis and Sagan point out that all five kingdoms have multicellular members and contain species that lack sexuality throughout their life cycle. However, bacterial sex is a very old phenomenon and probably originated 3,500 to 2,500 million years ago in the Archea. Meiotic sex appears only 2,500 to 580 million years ago. Bacterial sex is different from meiotic sex found in higher organisms.

Microorganisms have developed novel mechanisms that provide some of the advantages of sex. Bacterial sex differs from higher organisms in that no fixed numbers of genes are transferred between recombinants, nor is there a limit on the number of sources. DNA from virus particles and plasmids has evolved ways of getting into and out of bacteria. These pieces of DNA can recombine with DNA within the bacterium or they may not and remain separate entities within the cell.

These novel bacterial mechanisms are probably very ancient. Margulis and Sagan suggest that bacterial sex, as defined here, probably arose as a means of chemical defense. Based on observations and experimental evidence, it appears that threatening conditions such as exposure to toxic or inhibitory substance, marked changes in food abundance or other environmental change may have promoted sexual exchange by extrachromosomal entities. Not all bacteria engage in sharing plasmids or other extrachromosomal DNA. This is in and of itself very interesting. A number of interesting questions arise from the observation of differential uptake and incorporation of extrachromosomal DNA:

1. Can we predict which types of microbes should be obligate asexual and which microbes should take advantage of sex, at least at certain times?
2. What are the conditions that should favor bacterial sex?
3. Can we predict when bacteria should attempt to cull extra DNA from their cells?
4. Should bacteria census the available extra DNA as a hedge against future environmental challenge?

Each native bacterial species has some *range of tolerance* for environmental conditions in which they are found. Within this range of tolerance, there is a presumed much narrower optimal condition. If the environment changes beyond the range of tolerance, the organism either dies or goes dormant. Sudden changes to the environment may result in elimination of many individuals and species that are not capable of withstanding the change. Individuals who carry extra DNA that allows them to survive under the new conditions would then multiply. Under these new conditions, the extra DNA is an advantage and not a cost. However, if the environment swings back to the original conditions, the extra DNA may be an evolutionary burden.

Each of the sexual mechanisms used by bacteria to obtain new DNA is discussed thoroughly in most microbiology and molecular biology text books. In this chapter, we briefly review the mechanisms and then discuss them in regard to their potential evolutionary ecological impacts. The three mechanisms are *conjugation*, *transformation*, and *transduction*.

## Conjugation

Probably the major route of sexual transmission of extra DNA to bacteria is by plasmid conjugation. In conjugation, one bacterial cell acts as a donor, and the other as the recipient. There is only a one-way transfer of DNA. Plasmids that can be transferred between bacterial cells are called conjugative plasmids. Broad host-range conjugative plasmids can be transferred to a large number of Gram-positive and Gram-negative bacterial genera (Davison, 1999). Conjugative plasmids are typically large and can carry many types of bacterial genes, including those for substrate metabolism, DNA repair, and resistances to heavy metals and antibiotics (R plasmids). They also encode functions that mediate the transfer of plasmid DNA to any of a wide variety of recipient cells. Not only can conjugative plasmids cause their own transfer; some can mobilize the transfer of other plasmids residing in the same donor cell. Conjugative plasmids therefore have a high capacity for disseminating plasmid-encoded traits throughout the environment.

Conjugative transfer of plasmids in the environment is well documented (Davison, 1999). Plasmids can be transferred in polluted and nonpolluted soil and water environments, on plant roots or leaves, in marine ecosystems, and in polluted aquatic environments. A large number of bacterial genera have been shown to participate in the transfer of plasmid DNA in the environment, including *Bacillus*, *Rhizobium*, *Pseudomonas*, *Ralstonia*, *Escherichia*, *Enterococcus*, *Vibrio*, *Alcaligenes*, *Aeromonas*, and *Burkholderia*. The ability to transmit and receive genetic traits from heterologous bacteria appears to be widespread in nature.

Conjugative transfer of plasmid DNA also occurs in the guts of humans and animals, and involves pathogenic and commensal bacterial genera such as *Escherichia*, *Shigella*, *Salmonella*, *Enterococcus*, *Klebsiella*, and *Erwinia* (Davison, 1999). Transfer of antibiotic resistance to pathogenic bacteria is of critical importance to human health; this is evidenced by the increasingly difficult task of treating patients infected with bacteria resistant to multiple antibiotics. Because many of these bacteria also have environmental niches or reservoirs, it is likely that they have both the capability and the opportunity to swap DNA with environmental bacteria.

## Transposons

The capacity of conjugative plasmids to carry a variety of *exogenous* genes is large. Exogenous genes are genes that come from outside the organism. The most common means by which plasmids pick up these genes is *transposition*. Transposition is the movement of certain pieces of DNA from one location to another location on the chromosome or to some other plasmid. *Transposons*, the mobile genetic elements that contain individual or multiple resistance factors can and do move onto plasmids and thereby are mobilized into other cells. Afterward, even if the conjugative or conjugation-mobilized plasmid is unable to replicate in its new host, transposons contained thereon may move into host chromosomal or plasmid DNA and pass the resistance phenotype to the recipient bacterium.

One of the best studied transposons is Tn21 (Liebert et al., 1999). Tn21 derivatives carry genes for mercury resistance and one or more genes specifying resistance to antibiotics. Antibiotic resistance genes are usually contained in *integrons*, mobile genetic elements that are common in nature. Another transposon that is capable of transmitting antibiotic resistance among a wide variety of Gram-positive and Gram-negative bacterial species is the conjugative transposon Tn916 (Rice, 1998).

Resistance to mercury is accomplished through a complex gene system collectively known as *mer*. These genes are often carried on plasmids and transposons. A priori, if we were to predict the distribution of *mer* along an environmental gradient of mercury contamination, we would expect that the haplotypes of *mer* to be more similar at spatially closer locations. However, what was found was that sites above and below a source of mercury were indistinguishable, but haplotypes at the contamination source were very different from the other two (Figure 6.1). How is this possible? Bacteria have extremely limited capabilities to move and are incapable of movement upstream of distances exceeding 500 m. The downstream movement of water is the predominant vector of transport. Selection was favoring certain gene combinations at the confluence site over other genotypes. We can only presume that bacteria from upstream are passing through this reach, but it is clear that they are not colonizing.

## Transformation

Another possible mechanism for the spread of extrachromosomal DNA is natural *transformation*. Transformation is the taking up of naked DNA from the environ-

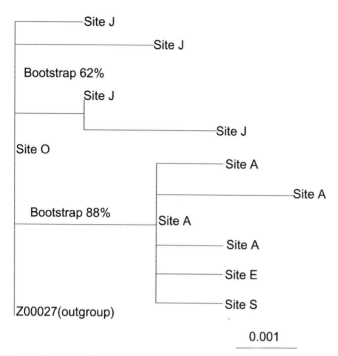

**Figure 6.1** Phylogenetic relationship of various *mer* haplotypes found along one stream course. Notice that haplotypes found at site *J* cluster together. Site *J* is found between the other sites.

ment. Many bacteria are *competent* for the uptake and incorporation of *exogenous* DNA and the transfer of DNA among bacterial pathogens is well established. Environmental DNA is relatively abundant and can be actively secreted from viable bacteria or released nonspecifically after bacterial cell death. Free DNA can be quite stable, and it can survive in the environment for months adsorbed to sand or clay particles. The transfer of antibiotic resistance by transformation of *Pseudomonas* and *Acinetobacter* with free DNA (in soil) has been demonstrated (Nielsen et al., 1997; Stewart and Sinigalliano, 1991). Studies with boiled cell lysates showed the ability of *Acinetobacter* in soil to be transformed by released DNA encoding antibiotic resistance from *Acinetobacter*, *Pseudomonas*, or *Burkholderia* species (Nielsen et al., 1997). Transformation can also occur in aquatic environments, such as with *Acinetobacter* grown in biofilms on sterile stones and dipped into a river, where free DNA concentrations would be expected to be relatively low (Williams et al., 1996). Natural transformation therefore appears to be a relatively efficient process under certain conditions and has the potential to allow the spread of genetic determinants such as metal and antibiotic resistance. Conjugation and transduction enable pieces of bacterial genomes to be transferred from one bacterial cell to another. Conjugation and transduction can provide novel DNA which can confer a selective advantage under a variety of environmental conditions, but the primary role of these processes is as a vector for the transmission of parasitic or infectious material. In contrast, transformation evolved as a unique process to transfer chromosomal genes between individual cells (Mongold, 1993). Natural transformation follows a highly evolved pathway that allows the uptake, binding, and recombination of DNA from the environment.

Transformation provides the benefits of recombination. Bacterial mutants that are deficient for recombination are much more sensitive to DNA damage, the very process that initiates the enzymes needed for recombination. Recombination is usually genetically conservative. In other words, recombination requires genetic sequences to be quite similar. Put these two benefits together, and it appears that recombination is a mechanism that leads to DNA repair. Because DNA damage is potentially possible to all cells, those cells that can repair the damage will have a selective advantage.

## Transduction

Although it may have a lesser capability than conjugation or natural transformation for the widespread dissemination of various genes, a third possible means for transmitting genetic material among bacteria is by bacteriophages (phages) through the process of *transduction*. Phages are extremely abundant in nature and outnumber bacteria in aquatic systems (Wommack and Colwell, 2000). Lysogenic infection by phage is increasingly being implicated in the transfer of bacterial virulence factors, and has been shown to encode such phenotypes as serum resistance, toxin expression, and host cell adherence or modification (Miao and Miller, 1999). Phages from *Pseudomonas aeruginosa* (Blahova et al., 1994a,b), *Staphylococcus aureus* (Fouace, 1981), *Streptococcus pneumoniae* (Buu-Hoi and Horodniceanu, 1980), and *Streptococcus pyogenes* (Hyder and Streitfeld, 1978) are able to transduce antibiotic resistance to other bacteria. The production of shiga toxin-producing progeny phage from *E. coli* can be induced by the DNA-damaging action of quinolone antibiotics (Zhang et al., 2000), resulting in an increased potential rapid spread among bacterial populations. Because many heavy metals cause DNA damage (Hartwig, 1998), a similar increased spread of phage-encoded metal or antibiotic resistance may be possible in environments contaminated with heavy metals.

# Advantages and Disadvantages of Sex

It is important to compare and contrast asexual and sexual reproduction. Asexual reproduction can occur through binary fission and vegetative propagation, through which a single cell or groups of cells develop into a new organism. Binary fission is not a simple process; it has several stages and set points that affect the rate of division. Division of the cell into two separate cells requires a duplication of intercellular components and especially the nuclear material. Division without duplication of cellular constituents other than DNA results in the *miniaturization* of the cells. When environmental conditions (i.e., temperature, pH, oxygen, and carbon source) are favorable for cell growth, microbes are able to duplicate their cellular constituents. Elongation of the cell occurs, the nuclear material divides, and the cell envelopes begin to grow inward. The cytoplasmic membrane grows into the center of the cell forming a septum; the cell wall grows inward forming a thickened wall which subsequently divides forming two cells. These simple statements mask the intricacies of the processes required to regulate and coordinate cell division. The apparent simplicity of the process can become extremely complex as we introduce unique microbial evolutionary mechanisms.

Reproduction by binary fission generally results in two identical daughter cells. If a microorganism is reproducing, even if the rate of reproduction is not at the optimum, environmental conditions can be considered favorable. Each successful replication results in additional copies of genes that are able to survive under the prevailing environmental conditions. When environmental conditions are constant or the range in conditions does not exceed the thresholds for survival of the organism, asexual reproduction is a favorable mode of reproduction. However, if the environmental conditions change frequently, randomly, or significantly such that new mutations are required for survival, asexual organisms may not be able to survive. This is especially true for traits that require more than one mutation to confer a selective advantage.

Sexually reproducing individuals must form gametes that have one-half the number of chromosomes as the parent. The process of halving the chromosomes is called meiosis. Meiosis occurs in most eukaryotes and offspring are formed by the fusion of two gametes to form a zygote. This process is called *syngamy*. Higher plants and animals have gametes that differ in size between males and females, whereas in lower eukaryotes, there is no difference in the size of the gametes but there may be different mating types. These mating types restrict gamete fusion to a narrow subset of mates. Sex provides one distinct advantage over asexual reproduction and that is the advantage of sexual recombination. Sexual recombination presumably increases the rate of adaptive evolution especially in changing environments but also in constant environments. There are at least two advantages of sexual recombination: spreading of favorable combinations of genes and elimination of deleterious mutations.

Each mutation in an asexually reproducing organism must arise independently and sequentially. If gene combination $aB$ is favorable but the genetic background at time $T_0$ is $Ab$, there are two separate pathways of mutations that would result in the same adaptive genotype. Single mutations that change $Ab$ to $AB$ or $ab$ must be followed by mutations of $AB$ or $ab$ to $aB$. Either pathway requires two sequential mutations. The intermediate genotypes ($AB$ or $ab$) must provide some "intermediate" selective advantage over the original genotype to increase the genotype in the population. If the intermediate genotypes are less fit than the original genotype, the final genotype cannot be achieved.

Given that the intermediate has some selective advantage over the original, it would take several generations for the intermediate to increase in proportion in the population to the extent that the probability of the final mutation occurring would be possible and not just probable. Under this regime (i.e., asexual reproduction), complex favorable mutations have a very low probability of becoming fixed in a population. Even if the intermediate genotype has higher evolutionary fitness than the original, long time frames are needed to change the gene frequencies. Bacteria have had very long time frames. However, the environment must remain fairly constant to allow the new favorable mutants to accumulate. Any change in the environment would change the selection regime, which might favor different combinations of genes.

Crow and Kimura (1965) contrasted the rate that favorable mutations could arise in either sexual or asexual populations. As with most evolutionary models, population size was very important in their model (Figure 6.2). The advantage of sexual recombination was only apparent in very large populations. In small populations any one mutation could become fixed before a second mutation arises and recombination

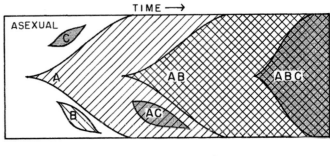

LARGE POPULATION

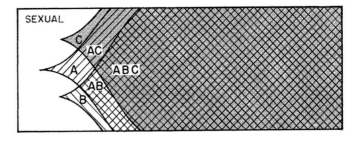

Figure 6.2 Evolution in sexual and asexual populations. (From Crow JF, Kimura M. Evolution in sexual and asexual populations. *American Naturalist* 99:493–450, 1965, published by the University of Chicago.)

would provide no advantage. In their example three new mutations A, B, and C provide an advantage when all three occur together. As discussed above, in an asexual population these mutations must occur sequentially (i.e., A→B→C, A→C→B, B→C→A, B→A→C, C→A→B, or C→B→A) within a single lineage. In sexually reproducing populations, each mutation can occur in a different lineage and then be spread throughout the population through sexual recombination. The time necessary for the spread of the favorable combination is significantly reduced in sexually reproducing populations compared with asexual populations.

The second advantage of sexual recombination is the removal of *deleterious mutations*. Most mutations are deleterious; they provide no advantage to the organism and often are harmful in that the mutation negatively affects the function of the gene. However, mutations occur at some rate in all populations, and deleterious mutations that do not result in immediate death accumulate. In asexual populations, the number of individuals with no deleterious mutations (i.e., the superior genotype) will decrease over time through genetic drift or through mutations such that given enough time, no individuals will be found without some deleterious mutations. The frequency distribution of the number of deleterious mutations will shift toward increasing deleterious mutations. Because back mutation to the wild type is highly unlikely, over time, the fitness of a small asexual population is predicted to decrease. This change in the frequency distribution of deleterious mutations has been termed *Muller's ratchet*. A ratchet is a tool that moves an object, such as a bolt, in only one direction: on or off. Muller's ratchet moves asexual populations toward decreasing fitness. In sexual populations, recombination can result in offspring that have none or few deleterious mutations. The effect of Muller's ratchet is most pronounced in small populations.

Population size directly affects the accumulation of favorable mutations and the elimination of deleterious mutations. Large sexually reproducing populations accumulate favorable combinations of mutations more rapidly than asexual populations, whereas small sexually reproducing populations can eliminate deleterious mutations faster than asexual populations, which cannot eliminate these mutations.

This section is focused on individuals and yet the advantages of sex that we have been discussing are not advantages to individuals but rather to populations. Mutations generally have negative impacts on individuals. In asexual populations the individuals will pass these negative traits on to their offspring. The negative aspects of asexual populations seem great. Yet, microbial populations have survived longer than all other organisms combined. Why and how? To address these questions we must continue a discussion of populations for the moment.

The numbers and types of microbial species found in any one single location are sometimes mind-boggling. Counts often exceed $10^6$ to $10^9$ bacterial cells per gram of soil or per milliliter. Even if we consider the diversity of microbes present, the potential population sizes of each type remains very large. Based on such an assumption, Bell (1988) suggested that bacteria might be immune to the effects of Muller's ratchet because of their small genomes and enormous populations. It seems likely that this is the case. However, it is not necessarily the total number of individuals in a sample that determine the population size, but rather the number of individuals that contribute genes into the next generation. In higher organisms, this is called the *effective population size*, which is defined as the actual number of adults that reproduce. There are at least four reasons why the effective population may be less than the census population: variation in the number of progeny produced by females, males, or both; unequal numbers of males and females; overlapping generations; and fluctuations in population size. Of these four reasons, only two seem to apply to microbial populations.

Overlapping generations lowers the effective population size because offspring may actually reproduce with their parents. Parents and their offspring carry identical copies of the same genes, and the effective number of genes would be reduced. In asexual populations, the daughter cells produced are identical to the original cell, and although there are more copies of that cell, there is no increase in the number of unique genes in the population. Consider two populations of bacteria that began as a single cell or as a colony of 3,000 cells. After many divisions, you count the number of cells per milliliter and find approximately $10^6$. What is the effective population size of both flasks? The answer is the same, and it is 1 if we assume no spontaneous mutations. All cells in both cultures would be genetically identical. It did not matter how many cells we inoculated the culture with, because there was no genetic variation in the inoculant. If we were to redesign our experiment and take 1 cell from 50 different colonies or 600 cells from 50 different colonies, the effective population size, excluding mutations, would be 50 regardless of how large the census population was.

In nature, although we may have many individual microbial cells in a census count, the actual number of unique genotypes may be relatively small. In essence, we can have complete overlap of generations. The parents exist through time as copies of themselves. Random mutations would change the effective population size.

The second reason that the effective population size may be smaller than the actual is the problem of fluctuating population sizes. The rate of genetic drift increases when population sizes are small. If we have a population of bacteria with 100 different genotypes present and the population size decreases because of some change in the environment, by chance alone, we may lose numbers of the unique genotypes. When the remaining individuals reproduce, the effective population size will have been reduced because of the loss of those other genotypes.

Although Bell (1984) thought that bacteria were immune to the effects of Muller's ratchet because of their large population sizes, perhaps we should carefully reexamine the assumption of large population sizes. Some bacteria may be very susceptible to Muller's ratchet. Some of the novel evolutionary mechanisms of bacteria may be in response to trying to overcome the effects of having essentially small effective population sizes.

# Rate of Reproduction

All microbes do not reproduce at the same rates in the same environments. Rate of reproduction is controlled in part by the availability of suitable energy and carbon sources, by the physical environment (e.g., temperature), but also by the genetic make-up of the organism. Not all genotypes can reproduce in all types of environments. Some microbes are obligate for a particular system or condition. *Obligate* means that certain limiting conditions must be met in order for the microbe to survive and reproduce. For example, some bacteria are obligate pathogens. These bacteria cannot survive and reproduce outside of a host. Other bacteria are obligate anaerobes. When grown in the presence of oxygen, these cells are killed or severely impaired. Still others require certain growth conditions such as elemental sulfur or other minerals. Obligate bacteria must have the minimum threshold of growth requirements met to reproduce. Conditions outside this threshold prevent growth. Some bacteria have extremely flexible and plastic growth conditions. These bacteria are able to reproduce albeit at reduced or enhanced rates under a wide variety of environmental conditions.

If we consider that many bacteria are in a dormant state in the environment, it would seem that the prevailing condition for most bacteria is lack of suitable conditions. Even those bacteria with wide growth tolerances may not be actively growing in all of the environments they can be isolated from because conditions are suboptimal for growth.

All things being equal, some bacteria still grow at faster rates than others. Some genotypes of the same species of bacteria grow at faster rates than others. This should not be surprising, given that evolution will select for various combinations of genes that, under the appropriate conditions, produce more offspring. If one particular genotype has a selective advantage in environment A and another genotype has a slight advantage in environment $A_0$, the genotype from environment A should reproduce slower in $A_0$ than in A and vice versa for the other genotype. Based on laboratory experiments these general predictions seem to hold. Researchers have conducted competition experiments with *E. coli* strains that have various nutritional constraints. From these studies, it has been shown that some differences are apparently neutral in that there is no competitive advantage for the strains with or without the genetic difference. However, other studies have shown that one strain does have an advantage over other strains under certain environmental conditions. In nature, where the environment varies over both spatial and temporal gradients, differences between strains may result in changes in species make-up and functional capacity of the habitat. Competition among strains of related and unrelated microbes will increase or decrease in intensity, depending on the degree of overlap for critical limiting resources.

There are very few habitats in nature where the environmental conditions are constant either spatially or temporally. Variation in space and time in one or more important variables results in differing selection regimes often across very narrow spatial or temporal scales. For example, detritus particles are far from being homogenous. Concentrations of organic substrates within the detritus particle change constantly as the material is decomposed or mineralized by microbes and physical processes. Waste products of degradation can accumulate near hot spots of decomposition. There is an infinite set of possible detrital particles in which selection should vary across subsets. Temporally the detritus is changing. The complexity of the detritus rain and the accumulating sediment detritus vary based on timing and amount of inputs, currents, and other characteristics such as geology. In short, the range of conditions that an individual bacterium or fungi may experience is significant. Bacteria adapted to a certain range of conditions should have decreased reproductive rates when the bounds of those conditions have been breached.

Very little is known about the temporal stability of bacterial strains found in nature. For instance, a researcher samples a specific habitat at a specific time and determines that a certain species of bacteria dominates. He further characterizes the isolates obtained from the sample using various techniques to determine what strains of this species are found. If the researcher samples the same location the next day, what would you predict would be the dominant strain (i.e., the same strain found the previous day)? What if the samples were collected a week or month or 6 months later—would the same strains be found, and if they were, would they be in the same relative abundances?

In some ways, the answers to these questions are unknowable. Sampling in nature is destructive. That is some material is removed. It is impossible to sample the exact same location twice because the location has changed from sampling. To overcome this problem, researchers must sample multiple locations within the habitat and determine the distribution of strains and species based on the average of all sites. Spatially, we expect that species and strains will have variable abundances because of the heterogeneity found in nature (as in the detritus example). We assume that species and strain distributions will be more similar across narrow spatial scales as opposed to larger scales, but is this true? We discuss microbial distributions in space and time more fully in the next section. However, the fact that microbes are not homogenous with respect to their environment relates directly to reproduction. Because all habitats are not optimal, reproduction should vary both spatially and temporally. It does.

# Plasmids and Extrachromosomal DNA

Extra genetic material can be isolated from a wide variety of bacteria. This material can provide genes necessary for living in variously harsh or novel environments. Although for the most part, microbial ecology is a fledgling science, plasmid ecology is still in the egg. Plasmids have some interesting properties that have important evolutionary consequences. These pieces of double-stranded, self-replicating, circular DNA replicate independently of the bacterial chromosome. In this discussion, we are restricted to comparisons between bacteria of the same species with and without plasmids.

Bacterial genomes are small but nevertheless range up to 9.5 Mb. The best-studied organism, *E. coli* has a genome size of approximately 4,700 kb. The genome of *E. coli* has been completely sequenced, and from this, it is apparent that there is conservation in DNA. There is not much extraneous material. How true this is for other bacteria is the focus of intense research. Most of the genes in *E. coli* are functional genes or genes involved in central metabolism or cell division. Any changes in these genes would presumably result in cell death or greatly reduced abilities. In higher organisms, there is a phenomenon known as *gene duplication*. This is an unfortunate choice of terms because all genes must duplicate for an organism or a cell to reproduce, whether through meiosis or through asexual means. As used, a gene is duplicated singly or in concert, and cell division does not take place. The resulting cell has another copy of the gene or genes. Any gene in the vicinity of a replicating fork of DNA will probably be duplicated, and any incomplete replication that is followed by a crossover can result in a duplication of this region. As long as one set of the genes maintains its function, the other set is free to change. Some plants have taken this to the extreme, and multiple copies of the entire genome are maintained. Although there may be different ways that a bacterium can duplicate one set of genes, the three mechanisms (i.e., conjugation, transformation, and transduction) described previously increase the probability of a cell getting novel DNA.

Genes located on the chromosome must be conserved to maintain metabolic and structural genes, which if compromised through any mutation such as insertions, deletions, or substitutions may cease to function and result in cell death. Plasmids are a major evolutionary innovation that uncouples conservatism and the need for change. Because plasmids are semiautonomous from chromosomal DNA, they are free to evolve rapidly without affecting the essential biochemical pathways directed by the genome. There are certain evolutionary benefits that come from extrachromosomal DNA to bacteria (Reanny, 1976, 1978).

Mutations, recombination, and translocation generate novel genetic combinations, but they differ in the rate that these changes occur or accumulate (Table 6.1). Mutation rate is the number of changes in DNA bases per unit time. These changes include substitutions, deletions, and insertions. The rate of recombination is how frequently sections of DNA from one piece are transferred to another section of DNA. The translocation rate measures the frequency that pieces of DNA move to a new location on the same chromosome or extrachromosomal DNA.

**Table 6.1  Comparisons of Different Rates**

| | |
|---|---|
| Average mutation rate | $10^{-6}$ |
| Average recombination rate | $10^{-4}$ |
| Average translocation rate | $10^{-2}$ |

Millions of years may be needed to develop new traits through random accumulations of spontaneous mutations. Recombination occurs at two orders of magnitude faster rates than mutations and can generate new combinations of genes in a single cell division. In terms of rate, translocation has several advantages over the other mechanisms, and it is probably the most efficient mechanism. Several advantages are described by Reanny (1978): increased transfer rates are as high as 1%; although recombination is usually confined to closely related organisms, promiscuous plasmids are able to transfer DNA among widely related taxa; and by relegating "temporary" genes to extrachromosomal loci, there can be an economization of the chromosomal genes.

DNA from plasmids and viruses allow duplicate genes to be maintained, which otherwise would destabilize the genome if they were tandemly located in the chromosomal DNA. Chromosomal DNA in microbes is highly *conservative* (i.e., not very much extraneous material). In bacteria, there is usually a single circular piece of DNA. The length of this chromosome varies but is ultimately restricted by the size of the prokaryotic cell. This need for conservatism is directly opposed to the need to be able to change concurrently with environmental change.

Plasmids are classified based on compatibility groups, which means that a single cell is incapable of accepting identical or similar extrachromosomal elements. The effect of these incompatibilities is that a single cell is much more likely to get heterologous plasmids than homologous plasmids and new or novel DNA. The most important consequence is increased genetic flexibility.

Reanny recognized a second key aspect of extrachromosomal elements: *transmissibility*. He states that this is "perhaps the most effective adaptive strategy devised by selection." Transmissibility is the ability of extrachromosomal elements to disseminate genes in an infectious manner and greatly accelerate an adaptive response. The worldwide spread of antibiotic resistance among diverse taxa is the prime example of this infectious adaptation.

Evolutionary theory, especially some species concepts, has an axiom of genetic isolation. Higher organisms are defined as species based on the assumption of no interbreeding. The ability of phage and plasmids to interact with genomic DNA across multiple species barriers certainly puts into question the concept of microbial species, as well as the idea of genetic isolation. Cells in ecological proximity to each other may be more important in microbial adaptations and gene exchange than is genetic relatedness. Microbes that are able to draw from the collective genetic experience of the whole microbial ecosystem should have higher chances of surviving than those individuals that lack such ability. However, some bacteria do not have plasmids.

# When Would Plasmids Be Favorable?

Let us compare the need for plasmids in two very different *open systems*. An open system is one in which energy flows and materials cycle. Streams, rivers, grasslands, and all of earth's ecosystems are open systems. The primary energy source is the sun but that is not always the case (see earlier discussion on alternate electron donors). The human gut, blood, and lymphatic system can be infected by various microorganisms at times and under various conditions. Humans and many other higher organisms are in a state of *homeostasis*. Temperature, dissolved nutrient concentrations, ionic strength, pH, and dissolved gases are all maintained within fairly tight constraints and limits. Deviations from the norm for any of these physiological conditions have significant consequences. In contrast, diurnal and seasonal variation is the norm in most environmental systems, such as lakes, streams, and terrestrial habitats. Except for some hot springs, caves, springs, and thermal vents, natural ecosystems have fairly large spatial and temporal variations across most measurable parameters. Which habitats are most conducive to the maintenance and dispersal of plasmids?

## Genes on Plasmids

Before answering that question let us review some of the genes carried on plasmids. Metal and antibiotic resistance or tolerance genes, ability to degrade complex molecules, and other catabolic functions are some. Table 6.2 is not a complete listing of plasmids and their functions nor does it give a thorough picture of the diversity of bacteria that harbor plasmids. However, there are some things fairly apparent from this list. There are no clinical isolates listed, and these bacteria are commonly isolated species from a wide array of habitats. Bacteria that in the course of their life have some probability of being exposed to metals, complex organics, or xenobiotics may carry plasmids. Based on this list, would it be possible to predict which natural (i.e., undisturbed) environments should have bacteria that harbor the most plasmids?

Our use of natural or undisturbed as a descriptor of the environment is self-serving and relates to our ability to detect disturbance. At the microbial level, feeding and excavation by an earthworm may be an extreme disturbance. Ecologists often speak of pristine or nearly pristine conditions. In this context, pristine refers to habitats that have not had significant human intervention or disturbance. If we consider only habitats that might be pristine or nearly so, can we make a prediction about the occurrence of plasmids?

Before we make our predictions, let us examine Table 6.2 again. Based on the abilities provided by the genetic information on these plasmids we might make certain predictions according to ecological understanding. Table 6.2 lists resistance or tolerance to various heavy metals. Are there naturally occurring locations where metal concentrations are unusually high? For example, might we consider hydrothermal vents that are rich in sulfur compounds or geologic formations with exposed metals.

Metals have been leaching or eroding from geologic formations for eons and these metals tend to move down the slope or stream. Naturally occurring higher incidences of plasmids that confer metal resistance may be found at places with exposed veins of metals or in metal-rich rocks.

Table 6.2 also lists plasmids that allow the breakdown of various naturally occurring and man-made or *xenobiotic* organic compounds. Complex organic compounds that are not easily mineralized by microbes are said to be *recalcitrant* or *refractory*. However, there are few naturally occurring organic molecules that cannot be broken down, given enough time and the right mix of microorganisms. Let us consider mineralization of organic matter in one type of ecosystem.

Science Library

Due Date:        18/03/2010  23:59

Title:           Microbial ecology : an
                 evolutionary approach / J.
                 Vaun McArthur.
Author:          McArthur, J. Vaun.
Classmark:       579.17 M


Item number:     1613882153

++++++++++++++++++++++++++++++++++++++++
* Please return this item to the library *
* by closing time on the specified due   *
* date. Details of our opening hours can  *
* be found on our website.                *
*                                         *

   www.cardiff.ac.uk/insrv/opening
++++++++++++++++++++++++++++++++++++++++

**Table 6.2** Properties of Plasmids Conferring Resistance to Metals and Oxyanions, Encoding for the Degradation of Some Naturally Occurring Organic Compounds and for the Degradation of Some Predominantly Manmade Xenobiotics Organic Compounds

| Strain | Plasmid | Markers | Size (kb) |
|---|---|---|---|
| *Staphylococcus aureus* | pII147 | Cd, As, Hg, Bi, Pb, penicillin | 30 |
| *A. eutrophus* CH34 | pMOL28 | Ni, Co, (Zn), $CrO_4^{2-}$, Hg, Tl | 165 |
| *A. eutrophus* CH34 | pMOL30 | Cd, Co, Zn, Cu, Pb, Hg, Tl | 240 |
| *A. eutrophus* DS185 | pMOL85 | Cd, Co, Zn, Cu, Pb | 250 |
| *Alcaligenes* sp. | pMER610 (IncHI-2) | $TeO_3^{2-}$, Hg | >250 |
| *Klebsiella aerogenes* | pHH1508a (IncHII) | Sm, Tm, $TeO_3^{2-}$ | 208 |
| *Salmonella typhimurium* | pMG101 | Ag | — |
| *Escherichia coli* | R773 (IncFI) | $AsO_4^{3-}$, $AsO_2^{2-}$ | — |
| *Pseudomonas fluorescens* | pQM1 | Hg, UV | 254 |
| *Pseudomonas aeruginosa* | pUM505 | $CrO_4^{2-}$ | 100 |
| *Nocardia opaca* | pHG 33 | Tl | 110 |
| *Rhodococcus erythropolis* | — | As compounds, Cd | |
| *Rhodococcus fascians* | pD188 | Cd | 138 |
| *Xanthomonas capestris* pv vesicatoria | — | Cu | 186 |
| *Proteus* sp. | Rts 1 (Inc T) | Cu | 210 |
| *E. coli* (pig isolate) | pRJ1004 (IncII,K.) | Cu | 116 |
| *Pseudomonas syringae* var tomato | pPT23a | Cu | 101 |
| *Pseudomonas syringae* var tomato | pPT23c | Cu | 67 |
| *Pseudomonas putida* PpG1 | CAM | Camphor | 500 |
| *Pseudomonas oleovorans* PpG6 | OCT | Octane, decane | 500 |
| *P. putida* R1 | SAL1 | Salicylate | 85 |
| *P. putida* PpG7 | NAH | Naphthalene | 83 |
| *P. putida* PaW1 | TOL | Xylene, toluene | 115 |
| *Acinetobacter calcoaceticus* RJE74 | pWW174 | Benzene | 200 |
| *Pseudomonas convexa* PcI | NIC | Nicotine, nicotinic acid | — |
| *P. putida* NCIB9869 | pRA500 | 3,5-Xylenol | 500 |
| *Pseudomonas* sp. CIT1 | pCIT1 | Aniline | 100 |
| *P. putida* CINNP | pCINNP | Cinnamic acid | 75 |
| *Pseudomonas* sp. CF600 | pVI150 | Phenol | — |
| *Acinetobacter* sp. A8 | pSS50 | PCBs | 53 |
| *Arthrobacter* sp. M5 | pKF1 | PCBs | 80 |
| *Pseudomonas putida* AC858 | pAC25 | 3CBA | 117 |
| *Pseudomonas* sp. B13 | pB13 | 3CBA | 104 |
| *Alcaligenes* sp. BR60 | pBR60 | 3CBA | — |
| *Alicaligenes eutrophus* JMP134 | pJP4 | 2,4-D | 75 |
| *Pseudomonas* sp. E4 | pUU204 | 2-Chloropropionic acid | 293 |
| *Pseudomonas diminuta* | pCS1 | Parathion | 68 |
| *Arthrobacter* sp. | — | S-Ethyl-N,N'-di-propylthiocarbamate | |
| *Pseudomonas* sp. DBT | — | Dibenzothiophene | 80 |
| *Flavobacterium* sp. 50001 | pRC10 | 2,4-D | 45 |
| *Pseudomonas putida* ST | pEG | Styrene | 37 |
| *Pseudomonas putida* RE204 | pRE4 | Isopropylbenzene | 105 |

PCBs, polychlorinated biphenyls; 3CBA, 3-chlorobenzoic acid; 2,4-D, dichlorophenoxyacetic acid.
Data from Mergeay M, Springael D, Top E. Gene transfer in polluted soils. In: Fry JV, Day MJ (eds): *Bacterial Genetics in Natural Environments*. New York, Chapman & Hall, 1990, pp. 152–171.

## Plasmids in Streams

In temperate streams, trees and plants that line the stream banks and flood plain (*riparian zone*) drop their leaves, usually in the fall in a single large pulse of material. On entering the stream, soluble compounds are leached quickly from the leave matrix. These compounds include sugars, amino acids, and various organic acids (e.g., tannins), as well as certain plant secondary compounds such as syringic acid, ferulic acid, sinapic acid, and many other compounds. Most of these compounds are quickly removed from the water by bacteria or adsorption. The remaining leaf tissue begins to decompose through the action of aquatic *hyphomycete* fungi and bacteria. These organisms break down much of the remaining structural tissue and in the process release or produce as byproducts compounds that are considered *refractory* or *recalcitrant* (i.e., not easily broken down). Such compounds and molecules are washed downstream. In some stream systems, the diversity of riparian vegetation changes along the watercourse. Microorganisms in lower reaches of a stream may experience more sources of organic matter and presumably higher diversity of organic compounds than microorganisms located in the upper reaches. Acclimation to novel sources of organic matter can occur.

Two studies conducted in different physiographic regions of the United States determined that the bacterial assemblages are acclimated or adapted to the primary sources of dissolved organic matter that they have been exposed to (McArthur et al., 1985; Kaplan and Bott, 1985). Kaplan and Bott demonstrated that stream-bed bacteria are acclimated to the sources of dissolved organic matter that have been generated by the surrounding land use. The use of foreign dissolved organic matter (DOM), as they called it, required an acclimation period. They showed that upstream-downstream bacterial assemblages preferentially use different molecular sizes of DOM (Figure 6.3).

In the other study (McArthur et al., 1985), bacterial response to leachates made from riparian sources along a prairie stream differed depending on which reach of stream the leachates were added to. Leachate prepared from grasses found along the headwaters of the stream was degraded and bacterial densities increased at both grassland and forest reaches (Figure 6.4). However, leachates prepared from leaves collected from riparian forest trees could only be used by bacteria found in the reach of stream flowing through the gallery forest. The forest leachate had an inhibitory affect on the grassland bacteria. In many prairie streams, the headwaters are in grassland, and the transition from grassland to forest is abrupt. Lower reaches of the stream have inputs of organic compounds originating from grasslands and from trees in the gallery forest. In contrast, grassland reaches of stream can only have organic matter from grassland origin. The grassland assemblage could grow better on the ambient water (i.e., water collected from the grassland site not nutrient additions) than they could on oak leachates (Figure 6.5). In contrast, the forest assemblage grew better on the oak leachate than the ambient water.

No attempt was made in either of these studies to determine whether the bacteria had naturally occurring plasmids. We might expect that the number and diversity of plasmids that aid in breaking down naturally occurring organic matter to be higher in the lower reaches as opposed to the upper reaches of a stream. No data is available to support or reject this prediction because few if any studies have screened for plasmid numbers or diversity along any such continuum.

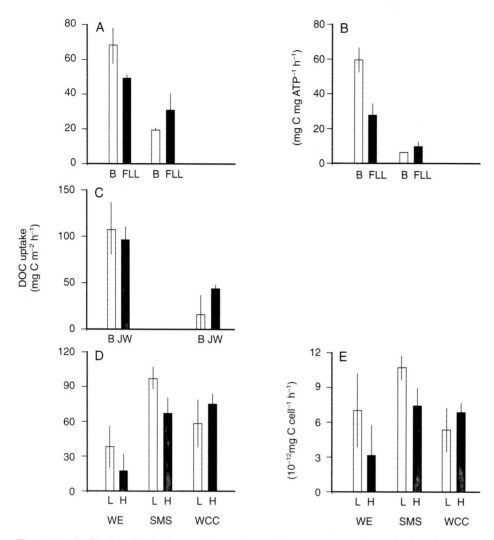

**Figure 6.3** Acclimation of bacterial assemblages along a stream in response to introduction of novel organic matter. B, bovine manure; FLL, forest leaf litter; H, high molecular weight forest litter; JW, jewel weed; L, low molecular weight forest litter; SMS, Saw Mill Creek; WCC, White Clay Creek; WE, West Creek. (From Kaplan LA, Bott TL. Acclimation of stream-bed heterotrophic microflora: metabolic responses to dissolved organic matter. *Freshwater Biol* 15:479–492, 1985.)

Given sufficient time stream microbial communities can acclimate to novel sources of organic matter and quite quickly. It may be interesting to see if acclimation is accompanied with acquisition of plasmids particularly if the new source of organic material contains recalcitrant or complex compounds. Do other microbial communities show similar acclimation responses?

## Plasmids in Lakes

Consider the following example from freshwater lakes. What would we expect the response of a bacterial community to be when it is exposed to some novel organic materials? Would we expect bacteria collected from the same basic environment, say a freshwater lake that had been exposed to algal exudates during a summer bloom to

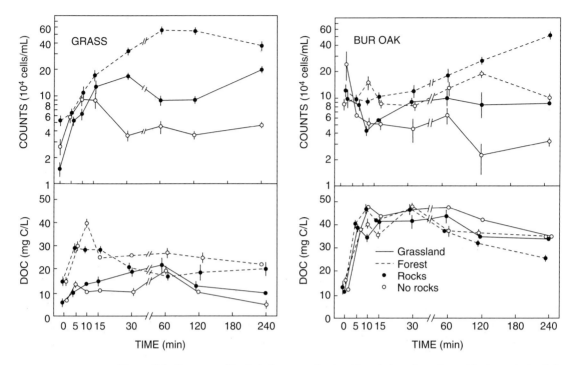

**Figure 6.4** Response of bacteria from grassland and forested reaches of a prairie stream to leachates derived from senescent oak leaves or grass. With both leachates, the forest assemblage of bacteria demonstrated growth, whereas the grassland assemblage grew only with grass leachate and showed a decrease in population when exposed to the oak leachates. (From McArthur JV, Marzolf GR. Interactions of the bacterial assemblages of a prairie stream with dissolved organic carbon from riparian vegetation. *Hydrobiologia* 134:193–199, 1986, with kind permission from Springer Science and Business Media.)

respond to these exudates the same as bacteria collected from a lake without such a bloom? Adaptation or acclimation to these substrates does occur. Casamatta and Wickstrom (2000) took bacterial isolates from a lake where there was a large algal bloom and other isolates from a lake without such a bloom and experimentally tested the response in terms of positive or negative chemotaxis, antibiotic response, and growth yields. Those bacteria collected from the lake with the algal bloom showed increased positive chemotaxis toward the algal exudates, decreased antibiotic response, and higher yields than did the bacteria from the other lake.

## Hot Spots for Plasmid Transfer

Given that the transfer of plasmids by conjugation is perhaps one of the most important means of adaptation and evolution in microbial communities, we should ask under what conditions we might expect plasmid transfer to occur. Plasmid transfer has been observed in a wide variety of habitats and under differing conditions. For example, conjugal transfer has been detected in freshwater and terrestrial systems. However, in soils, conjugative transfer may be at very low frequencies, except where nutrient levels are high (Thimm et al., 2001). Soils, like most habitats, are extremely heterogeneous with respect to nutrients, temperature, pH and other physical charac-

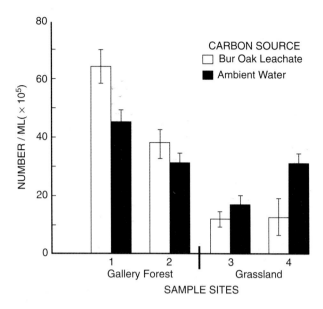

**Figure 6.5** The response of grassland and forest bacterial assemblages to the addition of oak leaf leachates. Sites are spatially separated along the same stream system. Ambient water is water collected at the site the bacteria were collected from. No additional nutrients or carbon sources were added to this ambient water. (From Figure 2 in McArthur JV, Marzolf GR, Urban JE. Response of bacteria isolated from a pristine prairie stream to concentration and source of soluble organic carbon. *Appl Environ Microbiol* 49:238–241, 1985.)

teristics. In areas of the soil where there are high levels of nutrients and other suitable conditions increased transfer frequencies should be observable. These hot spots for plasmid transfer have been detected in the *rhizosphere*, on detritus, and after the addition of manure (Thimm et al., 2001).

One potential hot spot for plasmid transfer may be soil-associated invertebrates. Intraspecies conjugative plasmid transfer between introduced donor and recipient bacteria have been observed in lepidopteran larvae, nematodes, collembolas, and earthworms. *Transconjugants*, bacteria that have obtained a plasmid from some donor, were found in earthworm casts (deposited material that has passed through an earthworm gut) but not in the bulk soil surrounding the casts (Thimm et al., 2001). Although the transfer rates were barely above detection, it is clear that passage through the earthworm gut facilitated conjugation between donor and recipient cells. The density of earthworms can be very high and the potential for plasmid transfer much higher in these areas. Thimm et al. (2001) give four reasons why earthworms may be important sites of the spreading of plasmids in soil bacteria:

1. Earthworms can be highly abundant in soils.
2. Because of burrowing and feeding mode, they interact closely with the soil microbial community.
3. Their guts are continuously inoculated with a high diversity of bacteria and organic substrates (nutrients) through feeding.
4. Microbial activities are general stimulated compared with bulk soil within earthworm guts because of nutrient concentrations.

These reasons generally apply to other soil invertebrates. Not all bacteria are consumed within invertebrate guts, and those that pass through appear to have a higher incidence of plasmids.

# Transformation in Nature

In freshwater and seawater, dissolved DNA (dDNA), DNA that can pass through a 0.2-$\mu$m filter can be found at concentrations between 0.2 to 88 $\mu$g L$^{-1}$ (Matsui et al., 2003). Naturally, *competent* (able to transform naked DNA) bacteria have been found which are able to take up this DNA. dDNA is composed of bacterial DNA, algal DNA, and DNA from other sources. Bacterial DNA may come from the lysis of cells through the action of bacterial phage or through the grazing activities of bacteriovores.

Bacteria may also actively excrete DNA into the environment, but the mechanisms responsible for this action are not clear. *Bacillus subtilis* released plasmids and transformable chromosomal DNA when grown on minimal media. Algae may be another stimulus that causes the release of plasmid DNA from bacteria (Matsui et al., 2003), at least from *E. coli*, and may facilitate horizontal gene transfer in aquatic environments. In this case, the algal stimulus may be for the increased phosphorus concentrations accompanying the dDNA release. The algae are not stimulating the bacteria to release DNA to aid the bacteria but rather to increase their own survival but in the process DNA is released into the environment and subsequently taken up by other bacteria.

Plasmid ecology is still poorly understood. However, it is clear that plasmid transfer is a highly effective means of spreading genes within and between bacterial populations and species. In an evolutionary context, all other things being equal, two individuals of the same species that differ only in whether they carry plasmids may or may not have equal survivorships or growth rates. It all depends on the environment in which the individuals are found.

Bacteria that carry extragenetic material (i.e., plasmids and transposons) may have an advantage when the conditions are such that the traits they carry allow increased feeding or uptake of nutrients or allow them to survive under suboptimal or harsh conditions. However, there is still that evolutionary cost of carrying extra information when it is not needed. In higher organisms, other traits working together may ameliorate this cost. For example, they can move to richer feeding areas easily.

In bacteria, there is an economy of genomic DNA. Any extra DNA, whether chromosomal or extrachromosomal, depending on the size, could be a significant proportion of the total DNA in the cell. For example, the total genomic DNA content of *E. coli* K12 strain EMG2 was found to be around 4,700 kb (Smith and Cantor, 1987) and *Pseudomonas aeruginosa* strains ranged from 4,400 to 5,400 kb (Bautsch et al., 1988). If these organisms contained only one copy of a plasmid that was 500 kb, it would represent almost 10% of the total DNA in the cell. Replication of a cell carrying that plasmid would need 10% more phosphorus, 10% more nucleotides, and 10% more of everything needed to double that DNA than a similar cell without the plasmid. The extra nucleotides would have to be synthesized from extracellular reserves and the phosphorus taken up from the environment. Under conditions in which the plasmid's function is not needed, cells carrying that plasmid would be expected to grow at a much slower rate than those without the plasmid because of these added costs of replication.

Most comparisons of growth rates and generation times of specific groups of bacteria with and without plasmids or other extrachromosomal DNA have been done in continuous culture systems known as chemostats. Chemostats have nice properties that allow tight controls and meaningful interpretations. However, they are fairly simple systems that mimic only aspects of nature. From these types of studies, it has been shown that the presence of many but not all plasmids increased the host cells generation times (Dykhuizen and Hartl, 1983). In one study, the effect of 101 *R factors* on the growth of *E. coli* showed that one fourth of these R factors increased generation time by more than 15%, and most R factors that were greater than 80 kb increased generation times (Zund and Lebek, 1980).

When bacteria are grown in association with algae, the production and release of extracellular plasmid DNA increases relative to situations where no algae are present (Matsui et al., 2003). In other words, algae may be involved in bacterial gene transfer by causing the bacteria to release transformable plasmid DNA into the environment. Why would algae stimulate bacteria to release plasmid DNA? Is the response advantageous to the algae? Are the bacteria the only ones that benefit? Under what natural conditions might we expect such stimulation? The answers to these and other questions need to be answered.

The ecology of individual microbes is interesting and seldom explored. An understanding of the evolution and autecology of these organisms requires that we understand concepts that are difficult to apply to asexual organisms. Species concepts and speciation need to be developed and applied to prokaryote and asexual microbes. Much of evolutionary theory is based on well-defined species concepts. Good ecology can be conducted without knowing what species are present or whether a species concept can even be applied. However, the application of many ecological principles requires some understanding of species.

Much of microbial ecology is essentially gene ecology because of the method-based approach to the science; therefore, it is important that students understand the limitations and potentials of these methods. A working knowledge of basic molecular ecology and the origins of life are important to understanding various microbial responses in the environment including their distributions and abundances. As addressed in the following chapters, the collective responses of individuals determine the population and community level interactions and responses.

# Living Together in Populations

# Fundamentals of Microbial Population Ecology

7

## Populations in Ecology

Populations in ecology are often defined as groups of interbreeding individuals. This definition is not satisfactory for most of the world's biota. All asexual and parthenogenic organisms, regardless of their densities, would not be populations based on that definition. A more general definition of an ecological population might be a group of like organisms living in a defined area in a specific period of time. Whether the number of organisms is 2 or 10 million does not matter. The boundaries of populations vary both in space and time. They are difficult to determine and are usually defined by the specific investigator.

For example, a researcher may define the population of interest to be the enteric bacteria found in the gut of a single human host. The spatial bounds of this population may appear to be set. However, enteric bacteria from other humans and animals may enter the specific human host, and the real population includes all potential colonizers and the resident members. The strains of enteric bacteria of a single human host may change over time. The structure of a population at time $T_0$ may be very different from the population found at time $T_1$. The physical boundaries of some populations are ridiculous to consider but still may have important ecological impacts. One such example might be considering the population of a specific species of bacteria found in the oceans.

## Properties of Populations

Populations have properties (primarily statistical) that cannot be applied to individuals. They have meaning only in populations. These include measures such as *density* or *size* of the population. Density refers to the number of individuals per unit area. We do not speak of the density of an individual. There are four characteristics that affect the size of a population: births, deaths, immigration, and emigration. *Secondary characteristics* of populations include age distribution, genetic composition, and the distributional pattern of the organisms of interest. Each of these secondary characteristics is a function of individual characteristics summed across all individuals.

## Density

The most fundamental parameter of a population is the density. Density like most population parameters is usually estimated based on a random sub-sampling of the habitat of interest. It is usually impossible to count, measure, and record every individual in a population. This is especially true for microbes where the population densities can exceed $10^8$ cells per gram of soil or per milliliter of water. It is also nearly impossible to even know exactly the physical bounds of the population (habitat) and the absolute density of the species of interest. In the process of measuring density, other population parameters can be estimated. These parameters include births and immigration, as well as death and emigration.

## Natality and Fecundity

The size of a population increases because of birth rates and rates of immigration. *Natality* means essentially the same thing as birth except that it covers all forms of population increase due to reproduction including binary fission. *Fecundity* is a measure of the potential reproductive output of a population and it is inversely related to the level of parental care provided to the offspring. In other words, the highest levels of fecundity occur with the lowest level of parental care. We discuss parental care as it relates to microbes in another section. Natality rate is the number of organisms produced per female per unit time. This definition has little meaning in microbial populations because there are no females per se. In binary fission, there is no mother and resulting daughters rather there are only the daughters.

## Mortality, Longevity, and Senescence

*Mortality* is the measure of the death of organisms. However, mortality is not just the observation of organisms dying, but rather determining why they are dying and at what age. Implicit in the notion of mortality is the concept of *longevity*. Longevity is a measure of how long a single organism lives. Krebs (1978) discusses both physiologic and ecological longevity. Physiologic longevity is the length of time an individual can possibly survive under optimal conditions. Ecological longevity is the average length of life in a given habitat and under given environmental conditions. Both of these concepts are difficult to apply to microbes. Rarely does a single microbial organism survive without some reproduction and once reproduction occurs, it is impossible to determine which of the two cells is the oldest. They are exactly the same age. Microbial populations do *senesce* (die of old age), so how do we estimate longevity?

## Immigration and Emigration

The other two population parameters of immigration and emigration are in many ecological studies considered to be equal and are not measured. The number of individuals moving into a habitat equals the number leaving. Immigration and emigration together constitute dispersal. Immigration is very important in colonization of new populations or where "new" habitat has formed or been exposed. The measurement of microbial movement out of a particular habitat is fraught with difficulties. It is not known if the microbes that are being transported out are "native" or autochthonous to the habitat or whether the emigrating microbes are just passing through.

Emigration assumes that an organism was established within a habitat and has subsequently "chosen" to leave because of degradation of the habitat, biological interactions (especially intraspecific and interspecific competition), or other factors assessed by the emigrating organism. Emigration may also occur as the excess reproductive output from a resident group of organisms that were unable to find or secure suitable habitat.

One additional caveat in regard to populations is that populations are not restricted to individual organisms. We can have a population of aggregates or populations of colonies. The concept of what is an individual is very important in defining what a population is or is not.

# Microbial Population Ecology

In this section, we discuss various aspects of populations especially as they relate to microbes. In our discussion, we sometimes discuss phenotypic traits that promote the survival of one population over another. We assume at times that all individuals within a population are equivalent, and we sometimes highlight their differences and the underlying variation found in natural populations. We discuss ways in which microbes communicate within groups and discuss the evolutionary consequences of such communication strategies. Fundamental to our discussion are estimates of population growth, expansion, and senescence. Let us first consider the growth of populations using simple but established population models.

## Population Growth

The growth of a population is a function of those factors that increase population density and those factors that decrease population density. We have already identified these factors as birth rate and immigration that increase density and death rate and emigration that decrease density. For simplicity, we assume that immigration rate and emigration rates are equal, although this assumption has not been verified for microbial populations. However, it is not known whether birth rates exceed immigration rates and similarly whether death rates exceed emigration rates. If we consider a population that is continuously reproducing, the net in change in population size over time can be expressed as

$$\frac{dn}{dt} = bn - dn = n(b - d)$$

where $b$ is the birth or natality rate, $d$ is the death rate, and $n$ is the population size. The quantity $(b - d)$ is usually written as $r$. Substituting, we get

$$\frac{dn}{dt} = rn$$

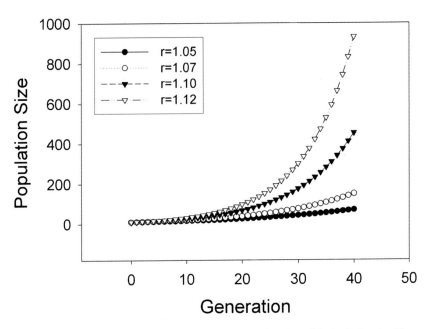

**Figure 7.1** Change (increase) in population size over time as a function of the intrinsic rate of increase (*r*). Notice that small positive changes in *r* greatly affect the population size after only 40 generations.

This can be read as the change in population size (*dn*) over the change in time (*dt*) is equal to the net reproductive rate (*r*) times the population size. How fast the population size changes is clearly a function of *r* or the net reproductive rate. If *r* is constant and *r* > 1, the population will increase geometrically without any upper limit; if *r* is constant and *r* < 1, the population will decrease until it goes extinct. For example, if we consider four populations with an initial population size of 10 individuals but that differ in *r* then it becomes readily observable from Figure 7.1 that small changes in *r* can result in extremely large differences in the population size even after just a few generations. Similarly if *r* < 1 then the rate to extinction can be very fast (i.e., few generations) or prolonged. However, if *r* stays below 1, the population will go extinct (Figure 7.2).

For most populations, *r* is seldom constant over all generations. Natality or death rates change with increases in population size due to factors such as the availability of food, fewer suitable habitat locations, and toxic waste products. The value of *r* is usually a function of the population size.

The actual relationship between population size and *r* can take many possible shapes, but the simplest and easiest to explain is a straight line showing an inverse pattern. The slope of this line is variable and can change for different species or for different environmental conditions (Figure 7.3).

The point where the line intersects *r* = 1.0 is the equilibrium point. At this point, the population size will remain fairly constant. It may oscillate around this population density, but over time (all other things being equal), it will converge toward the equilibrium value. As can be seen when the population is small, *r* is at its maximum, and the population will increase in size rapidly, but as the population increases, the value of *r* begins to decrease, causing the rate of increase to decelerate. If the popu-

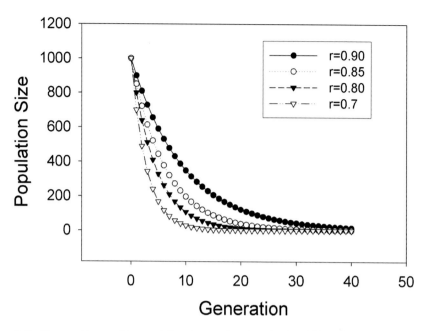

**Figure 7.2** Changes (decrease) in population size as a function of *r*, for which the value of *r* is less than 1. Notice that all populations go extinct eventually.

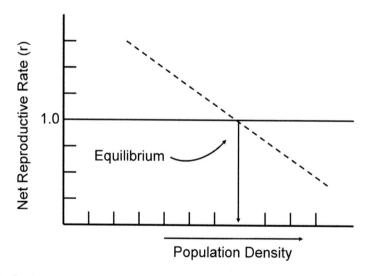

**Figure 7.3** Graphic representation showing that the equilibrium density of a population is where *r* is equal to 1.0. The slope of the dotted line is variable for different populations and species. Only one possible relationship is shown.

lation is larger than the equilibrium value, *r* will be less than 1, and the population size will decrease to the equilibrium point.

The pattern of growth seen in Figure 7.1 is typically that given for microbial and human populations. This is the classic J-shaped curve of exponential or geometric growth. Exponentially growing populations cannot persist indefinitely for at least three reasons. First, there is not an infinite supply of food to support unlimited continuous growth of a population. Second, there is not enough space for the population to expand into. Third, wastes generated by such population growth would increase and become toxic or inhibitory to the population. However, some populations do experience exponential growth with a subsequent crash of the population when available resources are consumed. This is especially true of microbial populations grown in batch culture with no new addition of resources.

In nature, it is possible to sample any habitat and find thriving populations of microbes. Repeated sampling over time indicates that the population/community densities remain fairly constant although there may be diurnal or annual cycles that fluctuate; however, there does not appear to be "boom or bust" type of growth. Microbial populations in nature do not appear to exceed their available resources. The intrinsic rate of natural increase ($r$) must be changing with the population size in order for these sample estimates of microbial density to remain fairly constant in a pattern similar to that shown in Figure 7.2. When populations have limited resources the population will increase until competition for the limiting resource affects the fertility and longevity of the population in which case the population's growth rate reduces until the population ceases to grow. This type of growth (Figure 7.4) is called sigmoid, or S-shaped growth. Sigmoid growth curves have an upper boundary designated $K$, or the *carrying capacity* above which the population cannot be maintained, and the population density usually approaches this upper bound smoothly (i.e., the growth curve does not exceed $K$ by much). Oscillations around the carrying capacity occur, but the oscillations are generally not large; otherwise, the population has the potential to crash and the population go extinct.

The *carrying capacity* of an environment is the maximum number of individuals the environment can support. Carrying capacity can be determined by the availability of food, other macro or micronutrients, the amount of space for growth, the number of sites suitable for reproduction, the temperature regime, the amount of water, and any number of other factors. In nature, the carrying capacity may be set by more than one factor and it is the interactions between factors that determine $K$.

Sigmoid or *logistic* growth curves are based on the equation $\frac{dn}{dt} = rn$ with an additional term added that reduces the rate of increase as a function of the size of

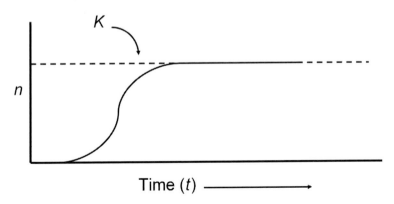

**Figure 7.4** Sigmoid curve of population increase, showing the population obtaining an asymptote. This asymptote is $K$, or the carrying capacity of the population.

the population. The new equation can be written as $\dfrac{dn}{dt} = rn\left(\dfrac{K-n}{K}\right)$, where $K$ is the upper asymptote or bound of the population. The term $\left(\dfrac{K-n}{K}\right)$ is a representation of the unused capacity of the environment. If there are $K$ sites for growth and reproduction then $K - n$ would be the number of available sites and we would expect the population to grow until those sites were filled. $K$ is not restricted only to physical sites but includes the other factors mentioned previously. However, this term does decrease the overall increase of populations and results in a smooth S-shaped curve of growth. Consider a population for which $r = 1.05$, the starting population size is 50, and the carrying capacity of the environment is 10,000. The population growth for this population would be that shown in Figure 7.5. Note that the population slightly exceeds the carrying capacity at time 11, but this produces a negative value for $\left(\dfrac{K-n}{K}\right)$ and the population quickly returns to $K$.

This model of population growth has been applied with varying success to vertebrates including human populations as well as invertebrates and bacteria. Workers examining population growth of *E. coli* (Figure 7.6) did one of the first applications in 1911. Hutchinson (1978) states, "In general, organisms with simple life histories and reproduction by division . . . give smoother curves than do bisexual animals with complicated life histories."

The logistic equation is an oversimplification of natural populations; however, it does describe many observed situations. One of the most serious oversimplifications is that the maximum rate of growth occurs when the population is at its lowest density. For most sexually breeding organisms, this creates an almost insurmountable

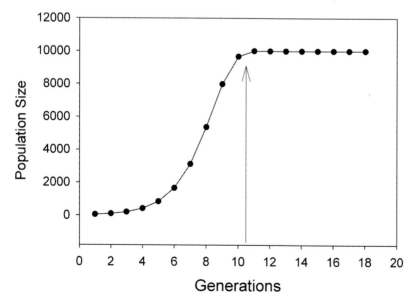

**Figure 7.5** Growth of a population for which $r = 1.05$, the starting population size is 50, and the carrying capacity of the environment is 10,000. The population size exceeds the carrying capacity after 10 generations but then quickly oscillates down to the carrying capacity.

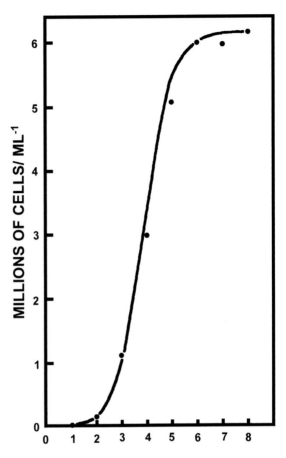

**Figure 7.6** Growth of *Escherichia coli* at 37°C in peptone. (From Figure 8 in Hutchinson GE. *An Introduction to Population Ecology.* New Haven, CT, Yale University Press, 1978.)

problem. Sexes must find each other to reproduce. The smaller the population size, the greater chances of not being able to find a mate, which results in a decrease in the rate of increase. However, for all microbial populations that reproduce by fission the maximum rate of increase can be when the population is at its lowest density. For these organisms a single cell is all that is needed and at such low densities competition between like organisms would be greatly reduced. The result would be accelerated growth.

Various oversimplified assumptions of the logistic equation have been pointed out but the point is seldom made that the logistic equation is descriptive and not explanatory and that the mechanistic basis of the decline in the rate of increase as a function of increasing population density is not explicitly recognized. What is the difference between something that is descriptive and something that is explanatory? A descriptive equation describes the phenomenon by accurately modeling the variables. The logistic is fairly good at describing the effect population size has on the rate of increase of a population. However, the equation does little to explain why such a response happens. Explanatory models get at mechanism or proximate reasons behind a process. Explanatory models can even be used to gain insights into ultimate or evolutionary processes. The logistic equation does not provide explanations.

## Density Dependence and Independence

Considerable research has demonstrated for many organisms that as the population size increase the birth or natality rate decreases and the death rate increases. When the natality rate and death rate equal each other, or $b - d = 0$, the growth rate by definition is also equal to zero. When natality and death rates change as a function of the population size the effect is called *density dependence*. If $b$ and $d$ do not change in response to changes in population size, the effect is called *density independence*. Density independence can really only be possible when the population density is low. At very high population densities, competition and build-up of toxic or inhibitory compounds should effect a change in the death rate.

Environmental factors that will affect populations whether there are $10^3$ or $10^8$ individuals per unit area or volume include things like temperature, pH, and other abiotic or physical characteristics. These factors can never determine the average density (as averaged over time) of a population. Density-independent factors, such as climate, are not affected by the size of the population. If the temperature drops 30°F due to a fast-moving storm front, it does not matter whether there are $10^3$ or $10^8$ organisms in a location, because the population size cannot ameliorate the effect of the storm. These factors can be categorized as *catastrophic*. In contrast, most density-dependent factors are biotic and to some extent facultative.

Density-dependent factors include disease, competition, and effects of predators and parasites. Only density-dependent factors can determine the average density of a population. This is true because only if the death-rate line has a slope (i.e., the death rate is not equal to zero) can an equilibrium population density be established. Remember that the point at which the death rate ($d$) equals the birth rate ($b$) defines the equilibrium population size. Climate and other abiotic factors cannot alter the slope of the death rate in most situations. If there are only $X$ available refuges that would protect individuals from the effects of sudden temperature change, these refuges (possible abiotic factor) would be density-dependent factors. Although density-independent factors can cause significant changes in population size, they cannot regulate populations. Variations in population size that do not move the population toward $K$ or the carrying capacity drift randomly up or down. Although the population may be high for a while, it can just as easily decrease, and with no mechanism to increase the growth rate, such a population will go extinct.

Not all ecologists agree with the designations of density dependence and density independence. For example, Andrewartha and Birch went so far as to say that there is no such thing as a density-independent factor. No factor is completely independent of the population size; all factors are density dependent. What about the effect of climate or temperature that we previously mentioned? How can population size alter the impact temperature may have on a population? Large populations have the potential to have more temperature-hardy individuals than small populations. As in the previous example, because there may be limited numbers of refuges, large populations may be forced to occupy suboptimal habitats that result in higher proportions of individuals being affected by the temperature.

Andrewartha and Birch reject the notion of independence and dependence and offer the following scheme to classify environmental variables. In their scheme, there are four separate and independent components that can effectively describe the environment: weather, food, other organisms and pathogens, and a place to live. As Krebs (1978) states, "These components of the environment are nonoverlapping . . . and the interactions between them completely describe the environment of any animal."

We have looked at two different models that seek to describe the factors that control population sizes. The first is primarily biotic and the second primarily abiotic. A third possibility exists where population size is controlled by self-regulation. Chitty (1960) proposed that "all species are capable of regulating their own population densities without destroying the renewable resources of their environment or requiring enemies or bad weather to keep them from doing so." Chitty is speaking about the species as a whole, not individual populations. Populations occupying suboptimal or poor environments may not self-regulate because densities never approach the numbers needed. However, unchecked population growth, under this hypothesis, is prevented by deterioration of the environment caused by increasing population size. Intraspecific competition is the most important factor determining population size especially in species that demonstrate spacing behaviors or mutual interference.

These three hypotheses are not mutually exclusive, and all three can and do help to explain observed patterns in nature. However, few studies have sought to determine which models best explain patterns of microbial abundance. Many studies have been made on the growth of microbial populations under laboratory conditions in both batch and continuous culture. Although these studies do provide insights, they are restricted because the conditions for growth, no matter the experimental design, are too simplistic and controlled. Estimating microbial population sizes in nature is difficult. However, these types of studies need to be undertaken.

There should not be any difference whether the organism of interest is a microbe or some higher organism when applying the three different models to explain population dynamics. Microbes both affect and are affected by biotic and abiotic components of their environments. Changes in the abiotic environment, such as a change in pH from 6.2 to 5.2, alter the population size only if that decrease approaches the lower limits of tolerance. At the lower range of tolerance, there should be fewer individuals with the ability to withstand the lower pH, and the population should decline. Regardless of whether the population is $10^6$ or $10^9$ cells mL$^{-1}$, there should be a decrease in population size. The difficulty in measuring population size changes in microbes is sampling at an appropriate interval. Although a population can decrease after some environmental change, the individuals remaining that are capable of reproduction under the new environmental conditions will do so. If the rate of increase of the remaining individuals is high, an observer could miss the population decrease and sample after the population had rebounded with, in our example, low-pH–tolerant microbes.

Biotic control of microbial populations has been observed. In one study, *Synechococcus* populations that had been maintained at equilibrium densities were decreased one order of magnitude when a predator (*Tetrahymena pyriformis*) was introduced (Figure 7.7). This change in population size was maintained for more than 150 days, indicating a new equilibrium population density had been obtained. We might expect biotic controls of specific populations to be limited in nature. Predators

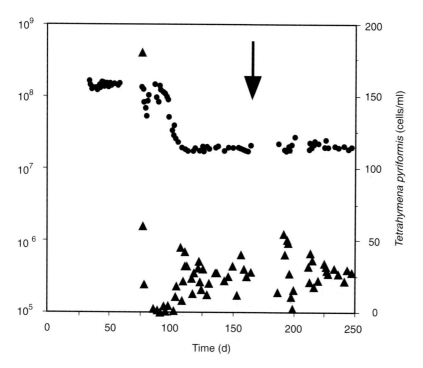

**Figure 7.7** *Synechococcus* PCC 6301 population size with and without predation. *Synechococcus* PCC 6301 (*solid circles*) was initially established at a 15-μM reservoir phosphate concentration. The equilibrium population density at day 25 is when the experimental observations begin. *T. pyriformis* (*solid triangles*) was inoculated at day 75. The *arrow* indicates an increase from 15 to 30 μM in the reservoir phosphate concentration. (From Figure 3 in Lepp PW, Schmidt TM. Changes in *Synechococcus* population size and cellular ribosomal RNA content in response to predation and nutrient limitation. *Microbial Ecology* 48:1–9, 2004.)

in natural environments are not expected to feed exclusively on a single species, but rather to feed on certain sizes of prey regardless of the species mix. However, if a population of a specific species were to increase in density relative to others in the community, this species density might be affected by predation or parasitism more than species that are less abundant. The change in population density would depend on the initial size.

Microbial populations, at least those in batch culture, do not appear to self-regulate their density to keep from over exploiting their resources. Numerous studies have shown that once inoculated, batch cultures show rapid growth and the population continues to grow until all or most of the resources have been consumed, after which the population begins to decline through death. In nature, many microbes are in a state of dormancy or quiescence. Rather than dying from lack of resources these organisms hunker down and shut down metabolic activity until suitable conditions return. How many of each species are in a dormant state at any given time is not known. Change from active to dormant may be a self-regulating mechanism to prevent over consumption of available resources.

## *r* and *K* Selection

The logistic equation is characterized by two parameters $r$, or the rate of increase, and $K$, or the carrying capacity. Some ecologists have used these two criteria to

**Table 7.1** Correlates of $r$ and $K$ Selection

| Characteristic | $r$ Selection | $K$ Selection |
| --- | --- | --- |
| Climate | Variable and/or unpredictable; uncertain | Fairly constant and/or predictable; more certain |
| Mortality | Often catastrophic, nondirected; density independent | More directed, density dependent |
| Population size | Variable in time, nonequilibrium; usually well below carrying capacity; unsaturated communities or portions thereof; recolonization annually | Fairly constant, equilibrium; at or near carrying capacity; saturated communities; no recolonization |
| Competition | Variable, often lax | Usually keen |
| Selection favors | Rapid development, high $r_{max}$; early reproduction; small body size | Slower development; greater competitive ability, lower resource thresholds; delayed reproduction; larger body size |
| Length of life | Short, usually less than 1 year | Longer, usually more than 1 year |
| Leads to | Productivity | Efficiency |

Data from Pianka ER. On r- and K-selection. *American Naturalist* 104:592–597, 1970, published by the University of Chicago.

characterize various organisms. Experience has shown that some organisms never reach their carrying capacity but always remain on the growth phase. How is that possible? Other organisms seem to always be at or near the carrying capacity. The former are generally classified as $r$ selected and the latter as $K$ selected. The designations are misleading because most organisms are somewhere in between the two extremes.

Pianka (1970) tabulated some correlates of $r$ and $K$ selection (Table 7.1). Based on these correlates Pianka categorized whole groups of organisms as either $r$-selected or $K$-selected. For example, all vertebrates were characterized as $K$ selected and all insects as $r$ selected. In many cases the difference in designation is based on whether generation time exceeds one year or not. A more general designation based on this criterion would be perennial and annual species. Perennial species equal increased tendency to $K$ selection, and annual species are more $r$ selected. How applicable are these concepts to microorganisms?

Before we can answer that question, it is important that we understand some other aspects of organisms that are correlated with reproduction. Let us consider the effect of size on reproduction or the effect of reproduction on size. Is selection acting on the life history patterns of an organism, or does selection act primarily on reproductive rate, which then selects for an appropriate body size?

It is chemistry and physics that sets the bounds on life. The smallest cells have to be large enough to hold the DNA necessary to code for everything necessary to make a new organism and the metabolic machinery required to make that happen. Protein synthesis requires some minimal room. Cells cannot be too large or the surface to volume ratio becomes too small to facilitate uptake of nutrients and release of toxic materials. Within these constraints cell size does not affect the size or complexity of an organism. Rather growth form limits both size and complexity. As macroorganisms increase in size there is a change in shape and an increasing amount of structure necessary to offset the effects of gravity (i.e., hard cell walls), arrays of supporting fibers in trees, or thicker exoskeletons or bones. Bonner (1965) discussed the relationship between size (length) of an organism and generation time. Larger size usually

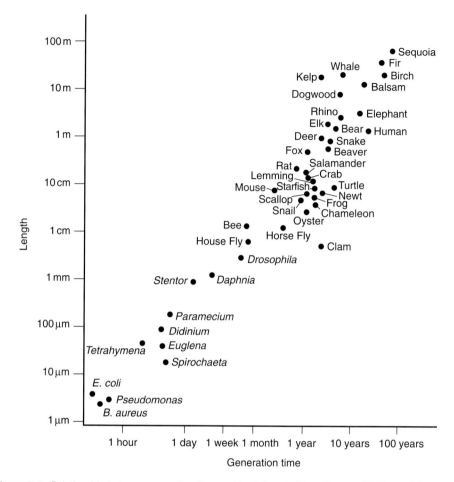

**Figure 7.8** Relationship between generation time and body length. (From Bonner JT. *Size and Cycle*. Princeton, NJ, © 1965 Princeton University Press, 1993 renewed PUP. Reprinted by permission of Princeton University Press.)

means more steps are involved in making large organisms than small ones and a longer generation time or the time to reach sexual maturity (Figure 7.8).

Microorganisms on the other hand are not affected with the problems of size and strength. Gravity is negligible even though there may be significant differences in the size of a microorganism and say its fruiting bodies. Microbes are subject to molecular phenomena such as surface tension, diffusion, viscosity, and brownian motion, all of which are related to fluids. All microorganisms, whether terrestrial or aquatic, are associated with water. Movement through a fluid is an energy demanding exercise for microorganisms and as such will affect the amount of energy available for reproduction.

Between bacteria and giant sequoia trees, generation times range from a few minutes to 60 years. With increasing size, the complexity of the organism can increase. Complexity allows division of labor among component parts. Natural selection can work on size or shape of the organism or complexity independently; however, a change in one necessarily affects the others. Large organisms are constrained into having complex development (ontogeny) with sequentially dependent stages. Each subsystem or developmental stage of a complex organism is subject to increased vulnerability from both within (e.g., translational, developmental) and from outside the organism (e.g., toxins, allelochemicals, antibiotics) with a risk of failure. Microorganisms are by definition small and for the most part not overly complex and unaffected by both the advantages and disadvantages of complexity. Colonial organisms gain some of the advantages of increased complexity (multicellularity in bacteria is discussed later).

Andrews, in observing the increase in generation times with increasing size, stated that large organisms "may be killed before reaching sexual maturity and hence leave no offspring." Large organisms colonize new habitats slower and adapt to changes much slower than do small organisms. Peters (1983) compared the length of time necessary to return populations of different organisms that had been reduced by some catastrophe back to a set biomass which he established as $100 \, kg \, km^{-1}$. He assumed a reduction in biomass to $1 \, g \, km^{-1}$ after some catastrophe and further assumed that each species population was able to grow at $r_{max}$, or the maximum intrinsic growth rate. The time ranged from days (bacteria) to 1 century for large vertebrates (Figure 7.9) to reestablish a population. One implication is that although microbes (viruses and bacteria) can respond quickly to catastrophic reduction in populations, the duration of the maximum rate of increase is orders of magnitude lower than that of the vertebrates. This has implications for the temporal scale chosen to make observations.

What is the appropriate temporal scale to observe microbial populations? Many microbial ecological studies have used time scales that are meaningful for larger organisms (e.g., weeks, months, seasons) but that have little meaning in microbial ecology because of the potential rate of change. Microbial growth rates range from minutes to days to weeks, but in general, they are much faster than those of higher organisms. Sampling at time intervals that exceed the generation time of an organism, result in comparisons across generations rather than comparisons within a population.

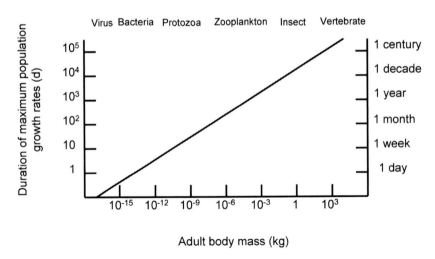

**Figure 7.9** Hypothetical relationship between body size (mass) and population growth rates and their combined effect on colonization potential. (Redrawn from Peters RH. *The Ecological Implications of Body Size.* New York, Cambridge University Press, 1983, p. 138.)

We return to our earlier discussion of $r$ and $K$ selection. Are these concepts applicable to microbial populations? Let us review the correlates of $r$ and $K$ selection as summarized in Table 7.1. Is climate variable (unpredictable or uncertain) or fairly constant for microbes? This is a very difficult question to ascertain. We do not have the analytical tools that would allow measurement at the spatial scales necessary to obtain an answer. However, climate seems less predictable at the microbial scale than at larger scales because environmental gradients can change because of physical and biological processes. Metabolism may change the pH, temperature, and relative humidity around microbes, although these changes may be small relative to spatial differences at higher scales. However, the microclimate around an individual microbe may be more important than the climate at the higher scales.

What is the mortality rate of natural populations of microbes? Based on laboratory studies, it is possible to predict a catastrophic decline after saturation and depletion of necessary nutrients and food. However, microbial densities appear to be rather constant and stable in most natural environments. Laboratory studies of microbial populations have shown that slight differences in competition can result in complete or nearly complete elimination of the weaker competitor. In nature, the levels of diversity are extremely high, suggesting that complete elimination seldom occurs. Teasing apart the dynamics of a specific population would be formidable in any but the simplest environments.

Based on body size alone, microbes are the poster children for $r$ selection. They develop rapidly and are totipotent or capable of reproduction at all ages. No cell is too old to reproduce, and daughter cells can reproduce immediately after they form. According to the Pianka scheme, $r$ selection favors short life spans that are less than 1 year in duration. Although the intuitive response is that microbes have very short life spans, this characteristic of microbes is extremely difficult to determine. How long does an individual bacterium survive? Reproduction produces two "new" cells or daughter cells that are exactly like the mother cell. Which cell is the oldest? Answer: both of them. Bacteria are therefore either the longest-lived organisms or the shortest. Bacterial cells can and do die, but that is not the same thing as maximum length of life. What is the maximum length of time an individual cell can live? Any cell that survives is essentially immortal.

For fungi, the question is as difficult if we consider the multicellular form of the fungus. Some species of fungi can live for thousands or tens of thousands of years. Returning to the example of the world's largest organism, this fungus had to be in existence for thousands of years to grow to its current size. Clearly, this particular organism is not $r$ selected.

Crowding is usually restricted or narrowly defined in macroecology as competition for various resources. We expand the definition to include all factors that change with density, including such things as predation, parasitism, and as Andrews points out, especially the production of toxic or inhibitory compounds by microorganisms. As microorganisms increase in density, the total production of toxic and inhibitory compounds similarly increases, resulting in increased negative affects on the population. However, at any moment, the production of these compounds may directly impact the growth of a few individuals that are close to the producing organism, regardless of the total density of the population.

The important terms of the logistic ($r$ and $K$) can be used to describe conditions or species. Ecologists speak of $r$-selected habitats or species. The former are habitats that select for organisms that use an $r$ strategy. Examples are old fields (e.g., abandoned farm fields) or newly exposed surfaces (e.g., sloughing from earthquakes, fires) in uncrowded conditions, in contrast to habitats that are well established and crowded with species and individuals. These $r$-selected habitats are unpredictable and transitional to more stable environmental conditions. There also are "new" habitats available for microbial growth. These include fallen logs in a forest that are susceptible to fungal colonization, surfaces of teeth after brushing, and at larger scales, the streams formed when Mount St. Helens erupted. Fallen logs in a forest are not predictable. Fungi respond to the resource by having sufficient colonizers present in the air. Brushed teeth are not a "natural" condition. After every brushing, the microflora of the mouth begins to recolonize and establish a biofilm. Before the advent of brushing, certain foods would scrape portions of this biofilm off, but much of the biofilm would remain. Completely new habitat such as that formed at Mount St. Helens is a good example of a temporally $r$-selected habitat. After eruption, the new surfaces were available for colonization; however, over time, these new habitats became the established habitats. The yearly plowing of fields is an example of a temporally stable ephemeral habitat. Each year, the abiotic and biotic relationships that had formed are disrupted, and new relationships begin to form. The new habitat formed from plowing is ephemeral because these conditions persist for a short time, but the conditions are reestablished each year.

Species that are $r$ strategists have many or most of the characteristics shown in Table 7.1 and described previously. It is important to remember that organisms cannot be selected to maximize both $r$ and $K$ strategies. The $r$ strategists have high rates of increase that favor dissemination into (i.e., discovery) and out of transitional habitats, especially as conditions become more crowded; they reproduce rapidly at the expense of materials present (i.e., no recycling or turnover of nutrients).

Microorganisms are generally similar in size, but they exhibit markedly different reproductive behaviors in response to different environments. Microorganisms have distinctly different intrinsic growth rates. If size were the variable that indirectly or directly determines which strategy ($r$ or $K$) an organism is selected to, we might expect that most microbes would fall under the $r$ strategies. Listed in Table 7.2 are a number of features or traits of $r$- or $K$-selected microorganisms. There are a number of distinct differences across several traits. The application of $r$ and $K$ is not necessarily a function of size. Comparisons across taxonomic groups can be informative but there are at least four limitations that must be kept in mind when making such comparisons. First, some concepts such as reproductive value have little or no meaning to microbiologists. Reproductive value as applied by macrobiologists is based on the likelihood of survival and reproduction for specific age classes (e.g., juveniles, adolescents, pre-reproductive and post-reproductive classes). This issue is nearly meaningless in microbiology because of the very short generation times and because most microbes can reproduce at all stages (if this has any meaning to microbes) of their life (i.e., there is no requisite maturation phase).

Second, modular organisms can increase in size indefinitely by adding new modules as opposed to growth of unitary organisms. Examples include fungal hyphae or the leaves and branches of plants. Increases in population growth for modular organisms

**Table 7.2  Life History Features of *r*- or *K*-Selected Microorganisms**

| Trait | *r* Strategist | *K* Strategist |
|---|---|---|
| Longevity of growth phase | Short | Long |
| Rate of growth in uncrowded conditions | High | Low |
| Relative food allocation during transition from uncrowded to crowded conditions | Shift from growth and maintenance to reproduction (e.g., spores) | Growth and maintenance |
| Population density dynamics under crowded conditions | High population density of resting biomass; high initial density compensates for death loss | High equilibrium population density of highly competitive, efficient growing biomass; growth replacement compensates for death loss |
| Response to enrichment | Fast growth after variable lag | Slow growth after variable lag |
| Mortality | Often catastrophic; density independent | Variable |
| Migratory tendency | High | Variable |

From Andrews JH, Harris RF. r- and K-selection and microbial ecology. *Adv Microb Ecol* 9:99–147, 1986, with kind permission from Springer Science and Business Media.

need not be delayed by postponing reproduction. The evolutionary and ecological consequences of modularity and multicellularity are discussed later.

Third, determining whether or not resources are limiting organisms requires comparisons among organisms found at the same trophic level. In ecology, it is often stated that certain nutrients or resources are limiting. For example, in many ecosystems nitrogen is the limiting nutrient. What exactly does that mean? Which form of nitrogen? Is it limiting for primary producers and their herbivores or only the primary producers? For plants the form of nitrogen is extremely important in the ability of plants to take up and use. However, there are only a few forms of inorganic nitrogen that can be taken up by plants. With microbes resources may be very diverse. Careful consideration is needed to make comparisons among organisms at the same microbial trophic level. Trophic designations of producer and consumer smear when we consider microbes. Some bacteria and fungi may break down or depolymerize very restrictive set of compounds, whereas others are able to break down many different compounds or molecules. Are these microbes all at the same trophic level? Perhaps they are not. Primary depolymerizers are not the same as organisms that take the breakdown products of depolymerization and continue the process or completely mineralize the material.

Andrews suggests that to determine whether natural selection affects reproductive output independently from size it is important to compare body sizes. Body size in microbes is not necessarily an indicator of reproductive output. Microorganisms can change their body size in response to changes in the availability of resources. Often this change in size is during times with low reproductive output (i.e., larger cells are reproducing faster because nutrients and resources are abundant).

The longevity of the growth phase (see Table 7.2) of microorganisms is a relative measure. However, *r*-strategist microbes should grow rapidly taking advantage of resources quickly before other microorganisms can get established and this rate of growth should be highest in the absence of intraspecific or interspecific competitors. *K*-strategist microbes would show long-term continuous growth at a much slower rate and little to no growth when these organisms are isolated from others. This is because of the complex interdependencies that form between $K$ strategists, with various growth requirements being met by unrelated taxa that live in proximity.

Microbial $r$ strategists must be able to shift from growth to reproduction of dispersal forms when resources begin to become limiting because of increasing densities. These strategists compensate for often catastrophic death losses by producing extremely high numbers of offspring which are able to disperse. Because new habitats are unpredictable, large numbers of dispersing forms are required to randomly find available habitat. In contrast, $K$-strategist microorganisms have slower growth rates, lower death rates and disperse at variable rates. To survive long term, all microbes must disperse at times.

Additional research needs to be conducted examining the strategies of microorganisms under various environmental conditions to determine the universality of the characteristics listed in Table 7.2. The most obvious conclusion of this section is that microorganisms cannot be lumped into a single category because of their small size and apparent reproductive rates. Understanding of the population ecology of microorganisms will require that we examine these organisms under both laboratory and field conditions to determine aspects of their life histories and survival strategies.

# Metapopulations, Multicellularity, and Modular Growth

## 8

## Metapopulations

Populations can be groups of free-living cells suspended and dispersed throughout a medium; they may be groups of groups or aggregates of few to many cells that are dispersed in a medium; or they may be colonies of cells attached to surfaces. Individual colonies may contain billions of cells. There may be many such colonies within an area. It may be as important to understand the distribution and abundance of colonies as it is to understand the distribution and abundance of individual cells. Ecologists can consider the populations of interest to be the colonies or the populations of cells that make up each aggregate or colony. The distribution of colonies within the environment is in many respects as important as the distribution of free-living cells. Studying populations of populations is a subdiscipline of population ecology known as *metapopulation ecology*.

Metapopulation ecology incorporates the distribution of habitats or patches where local populations can survive and that interact through the dispersing of individuals directly or indirectly. Patches are localized areas where a population lives or could live once colonized. Patches can be of various sizes and qualities. The structure of the metapopulation is then a function of a system of patches that have a certain distribution of patch sizes and distances between patches.

Although numerous studies have investigated the distribution of microbes, few have approached the study at an appropriate scale. Do microbes demonstrate metapopulation structures? Before attempting to answer that question, we must review more of the theoretical underpinnings of metapopulation ecology.

For many higher organisms, the impact of humans on the environment has resulted in an increased level of habitat fragmentation. Habitats that once were contiguous are more and more isolated from each other. Dispersal corridors among patches have been reduced or eliminated. The habitats themselves continue to decrease in total area. With smaller total area available, population sizes have necessarily decreased. Smaller populations, all other things being equal, have a higher probability of going extinct than do large populations. Hanski (1996) emphasizes that the relationship between population size and the probability of extinction is not linear and that this relationship is different for each species. For species for which extinction is more a function of demographic stochasticity, the risk of extinction increases only when the population size becomes very small. Demographic stochasticity is unpredictable changes in the size of the population. The population increases and decreases at random. If the population falls below the minimum needed to survive and reproduce, the population would go extinct. In contrast, species that are more affected by environmental stochasticity have a high probability of extinction, even with large population sizes. Environmental stochasticity is unpredictable changes in one or more aspects of the environment. These changes can include abiotic and biotic components. For example, if the environment makes unpredictable changes in pH, few organisms may be able to tolerate or adjust their physiology to allow them to survive in such an environment, regardless of how many individuals may be in the population. Environmental stochasticity is most pronounced in anthropogenically disturbed locations where inputs of contaminants or other modifications of the environment can occur randomly.

The effect of increasing habitat fragmentation is also related to specific species. Habitat fragmentation increases the distance between suitable patches. If corridors that are suitable for dispersal exist, some of the effects of fragmentation are reduced. However, fragmentation often results in increased interpatch distance. Species that are able to traverse the interpatch distances are not affected as much as species that cannot disperse the distance or that do not have the inclination to disperse or migrate that distance.

Metapopulation models have been used to predict how long a species will persist in a given habitat. Long-term persistence is possible only when the habitats are not too small. If the habitat becomes too small the probability of extinction becomes high due to random events, disease, or over exploitation of available resources. If the habitat becomes too small, the probability of colonization also becomes increasingly small, and the species is unable to offset localized extinctions.

Colonization and recolonization of habitat patches is directly related to the ability of the species to disperse. Dispersal according to Hanski can be examined within several complementary contexts. These include behavioral, evolutionary, and population dynamics; population genetics; and metapopulation dynamics.

The behavioral context of dispersal includes understanding proximate causes such as how organisms assess habitats, how they orient during dispersal, and which risks are associated with dispersal. For some higher organisms, the presence of conspecifics increases the colonization of patches. This phenomenon is called *conspecific attraction*, and it results in increased immigration to occupied patches with decreased colonization of empty patches. It is assumed that the presence of conspecifics is an indicator or habitat quality. Conspecific attraction results in a negatively density-dependent emigration rate (Hanski, 1996) which in turn increases the extinction rate of small populations.

In an evolutionary context, the most significant factor in determining dispersal rates is the risk of mortality during dispersal. If mortality is low, natural selection should favor increased dispersal and vice versa when mortality is high. There are two competing selection pressures acting on dispersal. The first is selection against dispersal because of the increased costs and risks. The second is selection for dispersal because of the chance of a great payoff in terms of increased numbers of offspring in an unoccupied suitable patch.

How applicable is metapopulation thinking to microbial populations? Once again we are up against the problem of scale. Few experiments have examined the dispersal, extinction, immigration, or emigration of natural populations of microbes. The exceptions are numerous studies that have reported the dispersal of pathogenic microorganisms. Epidemiology is, at one level, the study of microbial dispersal and colonization. From a microbe's perspective what is a suitable patch? Do microbes have the ability to assess patch quality, to detect conspecifics, and to increase, or decrease dispersal rates? These characteristics are often ascribed to higher organisms. With only a single cell, limited genome, and restricted morphologies, it may seem that microbes are the epitome of randomly distributed organisms.

## Dispersal

Some microbes have evolved complex behaviors and structures to facilitate dispersal. Examples include most species of fungi and bacteria that form fruiting bodies (*Myxobacteria*). In these organisms, selection has favored the production of specialized structures and cells. Spores are formed that are released into the environment in response to atmospheric currents, rain splash, or through propulsion from the fruiting body bursting apart and waft considerable distances. Settling of the spores is random, and the spores presumably are not capable of site selection.

Other microbes disperse using water and air currents through active or passive entrance into these vectors. Cells produced from crowded surfaces actively swim away or are sloughed off. Once entrained in flowing water or air currents, the microbes have little ability to determine where they will be deposited. There are some hypotheses that microbes can affect the distances they are dispersed by aiding in the formation of rain that would bring the microbes back down to the earth. However, microbes in general are not able to direct the orientation of their dispersal.

The dispersal of vertebrate pathogenic microorganisms has evolved in many cases to be a function of the host. Respiratory diseases often cause coughing and sneezing both of which force dispersers into the air. Suitable patches for these pathogens are new healthy hosts that become infected after breathing in the aerosols ejected by a sick host. Pathogen ecology is a study in metapopulations. Only pathogens that emigrate from the host will continue to proliferate. Other pathogens must be eaten or drunk before they can infect a new host. What comes in one end of a host has some probability of going out the other end. Some pathogens have evolved to withstand the environmental conditions of various digestive tracts to continue the life cycle of these types of pathogens. To be successfully transmitted to a new host, these organisms must be able to survive for extended time periods outside the host. To facilitate new infections, defecation by the host must occur frequently into or near to water or onto vegetation consumed by new hosts.

Microbes are found in the air, water, and soil; they got there through some means. Microbes are found at all elevations and depths sampled. The distribution of microbes is basically every habitat on earth. Life did not evolve in every habitat, so over the incredible long time of microbial existence, microbes have managed to colonize and establish in all of these diverse and unique habitats found on our planet.

# Modularity

Multicellularity and modular life forms of microorganisms are common among the fungi and many bacterial species. A fundamental characteristic among these types of organisms is that their growth is in modules, iterative, and usually involves branching. Table 8.1 lists some of the major characteristics of both and unitary life forms.

In general scientists, including microbiologists, have spent most of their time describing and investigating the ecology of unitary organisms. Early naturalists went out and collected organisms, and to them, single representative types were the ideal, especially for most vertebrates, invertebrates (including insects), plants, and any other living thing that could be collected. Although many plant species are modular or clonal, this fact was lost on the early naturalists. In microbiology, the emphasis before the molecular biological revolution was on the isolation of a "pure" culture, which meant that a representative single cell had been obtained. Colonies growing on plates were thought to be made up of identical cells originating from a single cell.

Growth of modules is iterated indefinitely and the number of modules is indeterminate. Modular organisms are frequently clonal but not necessarily colonial. Let us examine fungi as an example of modular organisms. The following discussion is based largely on Andrews' work (1991).

The development and growth of fungal hyphae have been examined closely under controlled laboratory conditions. Hyphae that germinate from a spore grow radially outwards with increasing branching. When two hyphae cross, they can fuse. If resources remain constant and growth is unrestricted, the ratio between the number of branches and the total hyphae length becomes constant. This consistency is species or strain specific and is known as the *hyphal growth unit*. The mycelium is the duplication (many times over) of this hyphal growth unit. Fungi and actinomycete bacteria can produce other sexual or vegetative modules.

When we consider that a rotting log on the forest floor may have been colonized by numerous species and strains of fungi, the fact that the fungi eventually form basid-

**Table 8.1** Some Major Attributes of Unitary and Modular Organisms

| Attribute | Unitary Organisms | Modular Organisms |
|---|---|---|
| Branching | Generally nonbranched | Generally branched |
| Mobility | Mobile, active | Nonmotile, passive |
| Germ plasm | Segregated from soma | Not segregated |
| Development | Typically preformistic | Typically somatic embryogenesis |
| Growth pattern | Non-iterative, determinate | Iterative, indeterminate |
| Internal age structure | Absent | Present |
| Reproductive value | Increases with age, then decreases, generalized senescence | Increases, senescence delayed or absent; directed at module |
| Role of environment in development | Relatively minor | Relatively major especially among sessile forms |
| Examples | Rabbits, birds, humans; vertebrates typically | Plants, hydroids, corals; invertebrates typically; fungi, bacteria |

Data from Andrews JH. *Comparative Ecology of Microorganisms and Macroorganisms.* New York, Springer-Verlag, 1991, with kind permission from Springer Science and Business Media.

iocarps (i.e., fruiting bodies) indicates an amazing level of communication or intense competition that results in the elimination or exclusion of genetically unrelated hyphal growth units. Although hyphae and mycelia from different species may have strong interactions, they may also be separated from each other temporally and perhaps spatially.

Fungi are not the only modular microbes. Bacteria are often observed in samples as single cells, although this growth form may be the exception and not the rule (Shapiro 1985). Microcolonies that are frequently observed under microscopy occur when cell division is not followed by dispersal. The large colonies observed when bacteria are grown on agar plates with appropriate nutrients are seldom encountered in nature. Exceptions include biofilm development and the massive growth of cyanobacteria or bacteria associated with sulfur hot springs.

Microcolonies may be both a hindrance and an aid in increasing fitness. One of the primary disadvantages is the increased intraspecific competition. Within the microcolony, every other cell has exactly the same nutritional needs. However, every cell is an exact or nearly exact copy of every other cell. In essence, competition with other cells, from the perspective of a single cell, is generally equivalent to competition with oneself. In higher organisms, with clear cell differentiation, every cell in the organism has exactly the same genetic make-up. Cooperation in feeding, removal of toxic wastes, and other processes increases survival of all of the somatic and germ cells, even though only the germ cells can send copies into the next generation. Individual cells within the organism do not compete for resources. All work together, and all survive, grow, and make new copies of the organism (i.e., reproduction), or they all die together. Growth of microbial colonies may require similar cooperation among cells.

Living in multicell colonies can provide certain advantages. For example, the cells may have an increased ability to migrate through the concerted efforts of the group. Slime molds and myxobacteria are able to move as a group. Microbes also release enzymes into their immediate environment to facilitate the breakdown of compounds and molecules necessary for growth and reproduction. Being in a microcolony would enhance the concentration of extracellular enzymes and the breakdown products of these enzymes. By remaining in multicell aggregations, pathogenic bacteria are able to resist host defenses such as antibodies and phagocytosis. Remaining as a group provides respite from various environmental abiotic threats, such as desiccation or susceptibility to ultraviolet light.

Hyphae of fungi and actinomycetes are classic examples of this multicellular life form (Figure 8.1). Branching increases the surface to volume ratio of the organism, which in turn increase the rate of transfer of materials into or out of the cells. Branching can also be seen in the colony growth forms of some species of bacteria on solid or semisolid materials (Figure 8.2). The growth patterns of modular organisms increase the capture of space which is closely tied to the amount of resources available. Colony growth is a resource-gathering unit. The architecture of the colony, the way the individual cells are packed together, in many ways determines the efficiency with which resources are obtained by the genet. Each module or colony captures resources that are close and in so doing creates a *resource depletion zone* (RDZ). When colonies or modules are packed tightly together, the RDZs in resource poor environments may overlap, further reducing the available resources to each module.

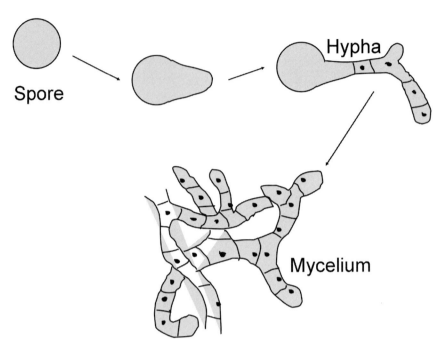

**Figure 8.1** Development of mature fungal mycelium from a spore. Notice that the spore develops into a hypha, which continues to branch forming the mycelium.

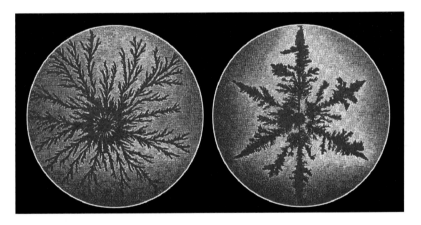

**Figure 8.2** Branching of bacterial colonies grown on thin, hard agar. (From Ben-Jacob E, Cohen I. Cooperative formation of bacterial patterns. In: Shapiro JA, Dworkin M (eds): *Bacteria as Multicellular Organisms.* Oxford, UK, Oxford University Press. Copyright 1997 by Oxford University Press, Inc. Used by permission of Oxford University Press, Inc.)

RDZs appear as circles on agar plates. In nature, they may be circular or spherical with the size being determined by the rate and direction of resource renewal and the rate the colony consumes the resources. Very active gathering modules will necessarily increase the size of the RDZ. The more diffusible a resource is, the wider the RDZ will be; if a resource is more readily consumed and weakly diffusible, the RDZ will be much narrower.

The branching growth form of the mycelium with associated hyphae can differentiate into two different roles: survival and dispersal. Both roles require the hyphae that have remained apart through some repulsive mechanism to come together to form complex structures. These structures play no role in nutrient acquisition but rather provide an exposed surface for spore dissemination. Fungi are able to exploit two major growth forms: the branched foraging mode with separate albeit connected units and the form with highly organized and constrained growth used for reproduction and other purposes.

The fact that hyphae come together to form these complex structures with amazing architecture provides the impetus for much work on what constitutes a fungal population. Much of our knowledge about hyphal growth comes from laboratory observations. In nature, numerous strains and species of fungi may interact and compete for available resources. The RDZs may vary significantly depending on who the neighbors are and how closely related they are. The complex structures used for the release of spores should all be genetically identical or nearly so, and there must be some mechanism for the recognition of self (see discussion below on swarming behavior). Otherwise, the fitness of some individuals becomes zero and they end their lives helping unrelated individuals to reproduce.

The evolutionary fitness of a modular organism is the relative number of descendants it contributes to future populations. Different parts of the same modular organism may experience very different selection regimes and pressures. As the modular organism grows, parts may encounter very different neighbors or resources from those experienced by another portion of the module. In an ultimate sense, the total contribution of a modular genet to the next generation of genets includes the integrated contribution from all of its modules (connected or separate) and the differences in their experiences (Harper, 1985). Harper went on to say that the way in which the modules of a genet are placed in context with other genets, whether fragmented or attached in loose associations or tightly packed groups, determines the selective pressures experienced. Modular organisms have many copies of their genes and gene combinations that may be an aid under certain environmental conditions, but under other conditions, all these genes and gene combinations are repeatedly at risk of being lost. If we again consider the world's largest organism, it should be readily apparent that an organism that is 5 km across is experiencing many different environmental conditions. Selection may be favoring combinations of genes in one section of the organism over those found in another part of the same organism.

# Source and Sinks

Organisms do not live in completely homogeneous environments. The exception may be the artificial environment of the chemostat, where conditions are kept constant. Outside of the chemostat, most organisms face habitat change spatially or temporally. The degree of this change is relative to the size and mobility of the organism. Because of habitat heterogeneity, individuals of the same species find themselves in habitats of varying quality or suitability for reproduction. Some habitats produce more offspring than others. Among higher organisms, larger or older individuals reproduce at different rates from those of younger or smaller individuals. Much thought and research has gone into examining age- or size-specific birth and death rates, but little thought has gone into examining how differences in habitat quality affect births and deaths (i.e., habitat-specific demography).

Ron Pulliam and others have developed a theoretical framework that allows comparisons of demography as a function of the quality of the habitat. This concept has been called the *source or sink hypothesis* (Pulliam, 1988, 1996). Much of the following discussion is based on the 1996 article.

Any population in which the births exceed the deaths and emigration exceeds immigration is considered a *source*. These populations produce excess numbers of individuals that emigrate away. In contrast, a *sink* population is one in which births are less than deaths and immigration exceeds emigration. If immigration ceases, sink populations will eventually disappear. These terms can be applied to habitats as well as populations. A source habitat is a geographic area that has a source population, and a sink habitat is a place harboring a sink population.

In "seasonal" environments or environments that vary in some predictable way over time, the quality of the habitat changes such that there are seasons more appropriate for reproduction and times that are not advantageous. Spatially within the habitat, there will be microhabitats that are better suited for reproduction than other locations. Overall, organisms will be growing at different rates temporally and spatially within the habitat.

Consider two habitats designated A and B; habitat A is the better habitat (i.e., the source). In this habitat, births exceed the deaths. We can designate the habitat-specific growth rate as $\lambda_A$, and this value will be greater than 1. In habitat B, the deaths exceed births so that $\lambda_B$ is less than 1. For sake of argument, we will maintain a constant rate of increase regardless of the population size. The population in habitat A will increase up to the maximum population size ($n_A{}^*$). Any surplus individuals above this value must emigrate or die from unsuitable conditions. These excess individuals can immigrate into habitat B, and even though the rate of increase of habitat B is less than 1, this influx can maintain the population size and prevent it from dwindling to nothing. In some cases, the population of habitat B may exceed habitat A, even though there are no births. In other words, many individuals in a population may be living in a habitat that is unsuitable for continued survival.

We can partition the variance associated with reproductive success into at least three components, which include density-related variance, or the change in reproductive success due to changes in the population density; environmental stochasticity; and demographic stochasticity. The density of other organisms (i.e., predators or interspecific competitors) other than the species of interest also affects reproductive success. Environmental stochasticity, or habitat quality, can vary from point to point or from time period to time period. Some of this variability results from large-scale meteorological events such as weather patterns, but it can also be a function of food availability within the habitat. The third component that may affect the variance in reproductive success is demographic stochasticity, or random and unpredictable events that affect birth and death.

Although habitat quality probably runs a continuum from prime to unsuitable, we consider only three cases: prime, marginal, and unsuitable. In prime habitats, the rate of increase ($\lambda$) is generally above 1 for most densities (Figure 8.3A). Lambda ($\lambda$) is only above 1 at low densities in marginal habitats and is always below 1 for unsuitable habitats. Another way of visualizing the effect of habitat quality is to consider

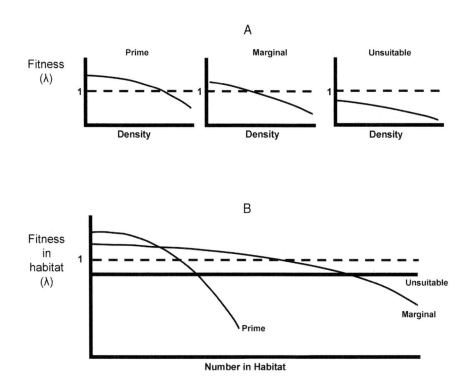

**Figure 8.3** Mean habitat quality in three hypothetical habitats (prime, marginal, and unsuitable). In the prime habitat (A), $\lambda > 1.0$ for a wide range of densities. In the marginal habitat $\lambda > 1.0$ only at very low densities. In the unsuitable habitat $\lambda$ is always less than 1.0. The number of individuals in each habitat is related to both habitat quality and habitat abundance. In average habitat (B), quality is plotted as a function of the number of individuals in each habitat. Prime habitat is assumed to be rare, unsuitable habitat is very common, and the marginal habitat intermediate. (Redrawn from Pulliam HR. Sources and sinks: empirical evidence and population consequences. In: Rhodes OE, Chesser RK, Smith MH (eds): *Population Dynamics in Ecological Space and Time.* Chicago, University of Chicago Press, 1996, pp. 43–69.)

the fitness in the habitat as a function of the number of individuals in the habitat and not the density. The total number is equal to all individuals in all local habitats. As the population grows, more and more individuals are forced into poorer quality habitats. As shown in Figure 8.3B, if we assume that prime habitats are fairly uncommon, unsuitable habitats are very common, and marginal habitats somewhere in between, we must know something about the distribution of habitat types and how many individuals are in each habitat type to predict the dynamics of a population.

Immigration and emigration are central to the source or sink concept. We have discussed dispersal as it relates to microbial populations. Species may be present or absent from particular environments because of dispersal. Finding an organism in a particular environment does not mean that that environment is suitable for that organism. This is especially true of organisms that have limited ability to control their dispersal. If dispersal is completely passive, the distribution of many organisms is related to the distance from the source population. Propagules such as seeds will settle out independent of the quality of the habitat as a function of abiotic factors such as wind speed and direction and the weight of the seed. Microbes are little affected by gravity and may be wafted or transported considerable distance before the cell attaches.

This concept is the first to relate population dynamics to environmental variability. The application to microbial populations has been minimal. This concept is a population-level concept. Before we can apply the concept to microbes, it is essential that we be able to identify species and individuals. Broad collective categories such as bacteria or α-Proteobacteria are meaningless in helping to understand the application of the concept to microbes. We must know the players and be able to sample them at meaningful temporal and spatial scales.

In marine systems, many suspended bacteria are inactive or in some state of dormancy. The same is true of bacteria being transported in many river systems. These data suggest that many bacteria are in unsuitable habitats. Because they are bacteria, they have the ability to ride out adverse conditions while they wait for prime or marginal conditions to develop. Some bacteria may be waiting significant periods for suitable conditions. However, given the appropriate conditions, these dormant, inactive bacteria are again able to replicate and reproduce.

In aquatic systems, the numbers of bacteria found in the sediments often exceeds the numbers in the water column by several orders of magnitude. Sediment bacteria can be moved into the water column by biological and abiotic events. Movement of invertebrates, fishes, or other large organisms can dislodge sediment bacteria and move them into the water. Mixing of surface waters by wind action can also move excess bacteria into the water column. Sediment bacterial populations may be sources for the microbial populations found in the water column. The waters over many types of sediment have much lower concentrations of nutrients and suitable substrates for growth. Growth rates in oligotrophic or nutrient-poor waters are necessarily very low or nonexistent. Although microbial production of suspended microbes does occur, the numbers of bacteria in the water column may reflect immigration from other habitats, including the sediments and terrestrial run-off.

Some species of normal gut flora can be found in nature outside of a host, even in the absence of known fecal contamination. Compared with the gut flora, these free-living individuals will have much lower rates of reproduction. *Regrowth* has been defined as the growth of fecal bacteria outside a host and can result in maintenance of populations outside of hosts. Are these populations of naturally occurring enterobacteria sometimes the sources of cells detected in routine monitoring for fecal contamination? It is assumed from the presence of bacteria used as indicators of fecal contamination that if the indicator is present, there must be contamination. If some of these indicator organisms are the offspring of free-living populations, the interpretation and correlation with contamination would be suspect. Indicator organisms that are truly from fecal contamination would be sink populations because they cannot grow as well in the environment. They are not growing but are the excess cells from some source population (i.e., sewage input). Indicator organisms that are present because of growth in the environment could be sink or source populations. Considering the concept of source or sink, it is very important that researchers determine whether there are natural populations of waterborne pathogens or their indicators and whether these "sources" contribute immigrants to the unsuitable sink habitats normally sampled by routine monitoring.

The source or sink concept brings habitat heterogeneity into understanding the distribution and abundance of populations of organisms. Bacterial genes are not necessarily linked to a particular species. Plasmids and other mobile genetic elements move among species of bacteria. These genes can be considered as populations, and we should be able to apply these same ecological population concepts to these gene populations. Some species may act as source populations of genes and others as sink populations. For the concept to be truly applicable, the organisms that harbor the sink population of mobile genes should not reproduce as well as the source population in the same environments, or they should be found living in marginal habitats.

In one laboratory experiment, competition along a spatial gradient of resource supply was investigated using algae and bacteria (Codeço and Grover, 2001). The spatial gradient ran from spatially homogenous to spatially heterogeneous. Spatial structure has the potential to increase local diversity because locally inferior competitors are moved from source habitats to sink habitats where they would normally be excluded. In spatially heterogeneous habitats, habitats that are divided into sources and sinks if each species has at least one source within the system local diversity may be enhanced.

# Population Ecology of Genes

In macrobiology, the distributions of genes and organisms are tightly correlated. There are similar genes found in diverse organisms, and this fact allows phylogenetic analyses to be made among groups to determine how closely or distantly the groups are related. However, there is little sharing of genes among unrelated higher taxa. For most biologists, the definition of a species (see Chapter 3) is a function of the probability of gene exchange. A population is a group of interbreeding or potentially interbreeding individuals. Interesting mechanisms have evolved in higher organisms to ensure that like organisms breed. These include premating and postmating mechanisms that are fairly effective at maintaining the genetic integrity of a species. Because species concepts are difficult to apply to microbes and many other clonal or asexually breeding organisms, the definition of a population is similarly difficult to apply.

We have discussed some of the novel evolutionary mechanisms that microbes use to share and modify the genetic make-up of individuals. These mechanisms allow the exchange of genetic information among closely and at times distantly related organisms. In this context the interspecies exchange of genetic material that is beneficial (i.e., increases the survival rate of those individuals that have the genes) requires some careful consideration.

In the previous section on the ecology of individuals, we discussed the ecology of plasmids and the impact that plasmid transfer has on the survival of individual bacteria. In the next section, we will further develop concepts that relate to the ecology of various genes, especially as those genes are exchanged among species and individuals living in biofilms. Here we apply the concepts of population ecology to genes (i.e., *ecological genetics*).

Richard Dawkins has argued the concept of the selfish gene and the extended phenotype to encourage scientists to look beyond traditional views of nature. In these concepts, focus is turned from considering how genes promote the survival of an individual organism to how individual organisms ensure the survival of various genes or gene combinations.

Evolution has been defined as any change in the genetic make-up of a population (Wilson and Bossert, 1971); however, evolution always occurs at the individual level. Changes in the genetic constitution of one individual will affect the genetic make-up of the population, albeit at a very small level depending on the size of the population. The population of genes within a group of organisms will be affected by the immigration and emigration of individuals and by the accrual of random mutations. In bacterial populations, gene frequencies are also affected by the transfer and acquisition rates of various extrachromosomal and chromosomal genes many of which can control their own dissemination.

Population genetics is the study of changes in gene frequencies. Accordingly, there are two overriding questions in population genetics. First, what are the origins of genetic variation at both the genic and chromosomal levels? Second, what are the causes of changes in gene frequency within populations?

# Sources of Phenotypic and Genotypic Variation

The phenotype of an organism is dependent on both the genotype and the environment in which the organism is living. The genotype is the sum of all DNA sequences found in the organism. In many microbes, this includes both chromosomal and extrachromosomal DNA. This genetic material is what is transmitted to new organisms by reproduction. However, in microbes, some of the genetic material may be transmitted at times other than reproduction, and the genotype of some bacteria may change significantly over time in the absence of reproduction.

The expression of some genes is affected by the environment that the organism is living in. Some phenotypic expressions may change repeatedly over time as the environment changes or is modified by the organism. In microbiology, numerous metabolic pathways have been described, the expression of which depends on the availability of various starting materials or the concentration of ending products in the immediate vicinity of the individual. Most inducible traits are a function of the environment of the organism. That is certain substances need to be present before genes are turned on or before they are repressed.

In higher organisms, another source of phenotypic variation is independent of the genes found in the offspring. For example, the amount of yolk produced by a mother may affect the size of her offspring, or one mother may provide much more maternal care after birth than another mother and alter the size, strength or ability to gain food for her offspring. These types of nongenetic effects have been called *maternal effects*. There is some evidence that fathers can also alter the phenotypes of their offspring, and to some extent, *paternal effects* can be found in some organisms. Is there anything analogous to maternal effects in microbes?

Both genotype and environment play an integral part in the phenotypic expression of microbes. Any differential division between daughter cells could be a weak analog of maternal effects. Any difference, even if slight, could provide enough of an advantage to allow one daughter cell to survive over the other. Similarly, if one bacterium sequestered more cytoplasmic resources before division than another, this increase could be considered a maternal effect because the resulting daughter cells would not have to take up or metabolize as much material as a cell with less cytoplasmic reserves. Consider the immense differences in cell size of the same species of bacteria grown in different nutrient conditions. Because the DNA content of the cells is the same, the difference in cell size must reflect greater amounts of cell products. By growing larger under plentiful conditions, bacteria may confer an advantage on their daughter cells over those cells that do not or cannot enlarge. Any cell that is able to sequester more cell products and remain slightly larger should bestow an advantage on the offspring. Having only slightly more cell products means that these organisms do not have to expend as much metabolic energy in making the material and should be at a competitive advantage. Reproducing without growth (miniaturization is discussed later) results in smaller and smaller cells.

# Sources of Genic and Chromosomal Genetic Variation

Mutations are the primary source of genetic variation and ultimately the primary source of the material that selection can act on. Any changes in proteins or enzymes are directly related to alterations in the DNA sequences that code for these products. However, most mutations have little or no affect on the expression of the genes. Many mutations are deleterious and result in a decreased capacity. Mutations are random events and occur continuously in most populations at some rate. Although an event is random, it does not mean that it is not predictable. For example, the outcome of tossing a coin is presumably random, although differences in the way the coin is tossed, the speed with which it is caught, and the amount of wind may affect the outcome, and if known, these variables would increase our chances of knowing the outcome. Mutation events also appear to be random, but are they?

In an experiment using a strain of *Escherichia coli* that was not able to use lactose, the bacteria were spread on an agar plate that contained necessary amino acids, lactose, and an indicator dye. The non–sugar-using bacteria formed colonies because lactose was provided in the medium. Mutants that were able to use lactose overgrew the parent colonies and formed papillae that turned red because of an interaction between the waste products of lactose metabolism and the indicator dye. Each papilla is thought to be the result of a random mutation in a single individual. To determine if these mutations are random, we need to ask whether each "parent" colony has an equal chance of developing one or more papillae. This can be done using a standard statistical test. The actual rate of mutation is very low; however, the total number of bacteria in each colony is an extremely large number. The number of mutations per colony would be expected to follow a distribution known as the Poisson distribution. This distribution can be expressed as

$$p(x) = \frac{m^x}{x!} e^{-m}$$

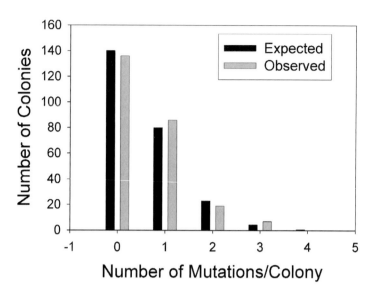

**Figure 8.4** Poisson distribution of mutations in colony morphology. (Redrawn from Wilson EO, Bossert WH. *A Primer of Population Biology*. Stanford, CT, Sinauer Associates, 1971.)

where $p(x)$ is the probability that $x$ mutations will occur per colony (where a mutation is represented by a papilla), the mean number of papillae per colony is represented by $m$, and $e$ is the constant 2.71828. For this experiment, the mean number of mutations per colony was about 0.57. We can now calculate the Poisson distribution using $m = 0.57$ and $x = 0$, 1, 2, 3, respectively, for the number of mutations per colony. Figure 8.4 shows how closely the observed distribution follows the expected. Most of the colonies have zero mutations, and only a few colonies have more than two mutations.

Evolution from mutations can only occur if the genes arising from the mutation are transmitted into the next and subsequent generations. In higher organisms, mutations can occur in somatic cells or germ cells. If the mutation is in a somatic cell, although it may cause an effect in that specific organism, it will not be passed on to the organism's offspring. Only mutations occurring in germ cells have the potential to be passed on. In microbes, any mutation has the potential to be passed on if the mutation is not lethal.

Many chemical and physical factors can cause DNA damage that, if not repaired by the cell, results is an alteration in the DNA sequence and by definition a mutation. Individual cells are fairly efficient in repairing damaged DNA through molecules such as DNA polymerase but these proofreading molecules are error prone and some mutations are not corrected. Any change in the DNA sequence is a mutation. Such changes include point mutations where a single base pair has been substituted for some other base pair (Figure 8.5). These substitutions can occur both between and within bases (i.e., purines for purines, pyrimidines for pyrimidines, purine for pyrimidine, or vice versa). Substitutions do not always result in changes in gene products because many amino acids have more than one codon that codes for it that differ in one or more bases. Any change in a codon that does not alter the transcription will not have any visible affect on the organism. Substitutions are not necessarily detrimental, because usually only one codon is affected. This is not true for other sources of mutations.

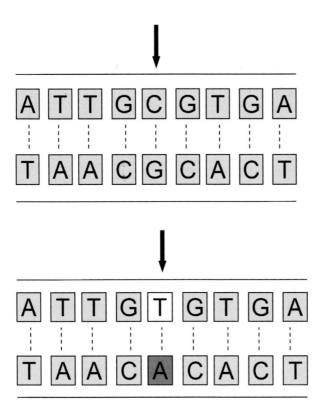

**Figure 8.5** Change in DNA caused by a substitution in one base. In this example a T was substituted for the C (*top arrow*), resulting in a change in the complementary strand of DNA from G to A. Although many substitutions do not cause changes in protein sequences, this particular substitution results in a change in an amino acid.

If one or more base pairs are inserted or deleted from a DNA molecule, the results will be much more significant because the reading frame for transcription will be off and the resulting product usually nonfunctional (Figure 8.6). The insertion or deletion completely scrambles all codons after the change. This type of mutation always has an effect, and the effect is usually deleterious. If the organism has more than one copy of the gene and at least as one copy performs its function properly, the other copy is free to change and provide the raw material for incremental evolutionary change. Some of the increments can be significant.

Another source of DNA change is transposable elements or transposons. Transposons are mobile sequences of DNA that can be inserted into other pieces of DNA like the chromosomes and plasmids of many microbes. Transposons often carry more than one type of gene and can significantly alter the phenotype of the organism. Depending on where the transposon is inserted, effects other than those coded for by the inserted genes may be manifested. In other words, the insertion may alter the reading frame of a gene by being in the middle of the frame.

Because of the large population sizes obtained in laboratory studies of bacteria, it can be shown that random mutations are sufficient to allow selection to act with a measurable response over a relatively short time frame. With large numbers of individuals, although the probability of any particular mutation is relatively constant and

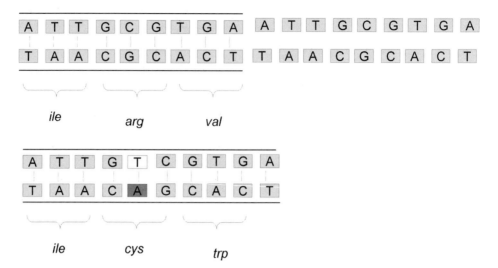

**Figure 8.6** Mutation caused by an insertion of one base pair into the DNA sequence. After translation and transcription, the old and new sequences would code for very different proteins from those in the original DNA molecule. All codons subsequent to the insertion may be different.

small for any specific gene, the chance of finding individuals with that mutation is high. If we take a pure culture of some bacteria, say a species of *Pseudomonas*, and grow that culture up over night in a liquid medium that promotes growth such that the ending population density is higher than $10^8 \, mL^{-1}$, we would expect, by chance alone, to have between 100 and 200 cells that have a spontaneous mutation for a given trait. All that is needed to identify the mutant cells is some selective medium that promotes the growth of the mutant and not the wild-type cells. For example, certain antibiotic resistance traits can arise by spontaneous mutation in the absence of the selector and then be detected after growth on a media containing that antibiotic. Before the mutation, there was no growth on the antibiotic-containing media, and after the spontaneous mutation, there is growth.

Several elegant experiments have been performed using modifications of this basic approach that have demonstrated convincingly that various adaptations and even metabolic pathways can be selected for in organisms that lacked these traits. Futuyma reports the results of two such experiments performed by others that demonstrated adaptation to temperature and acquisition of a novel biochemical pathway. In the first experiment, an *E. coli* culture derived from a single individual was grown for approximately 2,000 generations (Bennett et al., 1992). Among the progeny of this single individual were two phenotypes that made further experiments much easier. One phenotype was expressed as white colonies when grown on nutrient agar with a particular compound added. The other phenotype grew as a red colony on the same media. These were designated as *Ara*⁺ for the white colony-forming cells and *Ara*⁻ for the red colony-forming cells. This particular mutation did not affect the growth rate of the cells as determined by growing the two phenotype-genotypes together, because no change in proportions of the two phenotypes was observed over time. With this marker, the researchers chose to investigate whether *E. coli* could become adapted to grow at different temperatures.

Microbiologists have discovered that cultures of bacteria can be frozen at −80°C, at which point all metabolic activity stops, and then thawed with no apparent affect on bacterial growth. This allows researchers to take samples over time, freeze them and then thaw and compare properties of the frozen samples to the same properties of cells that have been used in various experiments or that were taken at earlier times. In the experiment we are describing, some of the ancestral $Ara^+$ and $Ara^-$ cells were frozen. Some of each genotype was then grown at different temperatures (32°C, 37°C, 42°C, and where temperatures were alternated between 32°C and 42°C on a daily basis) in replicate populations. The ancestral population had been maintained at 37°C. Each genotype was grown for 2,000 more generations, and at approximately every 200 generations, a sample was collected from each replicate and frozen. By thawing and competing the treatment samples against thawed samples of the ancestral population, they could determine whether selection had favored adaptation for change in temperature. The two genotypes allowed them to keep track of ancestral and treatment cells. By growing $Ara^+$ ancestral cells with any of the treatment $Ara^-$ cells at each of the treatment temperatures and vice versa they were able to determine fitness as the difference in density of the two types after growth over 24 hours.

In every comparison, increased fitness was observed in the treatment populations including the bacteria grown at 37°C over the ancestral population (Figure 8.7). Even those bacteria grown at 37°C for 2,000 generations had a higher fitness than the ancestral population grown at the same temperature. The variability around the means for each time period was often significant. This variability within replicate populations is further evidence of the random nature of mutation. If every population had the same number of mutations for every trait, we would have expected to see the variability greatly reduced. Variability was highest for the populations grown at 42°C, which also had the highest increase in fitness relative to the ancestral population. This indicates that bacteria adapted to higher temperatures were different from the ancestral population. Bacteria grown at lower temperatures or with the alternating temperature regime had only slightly higher fitness than the ancestral populations. In other words, the ancestral populations were better able to handle lower temperatures than high ones. Because every population used in these experiments was derived from the same individual, adaptation to the new temperatures had to be from mutations.

These experiments show that *E. coli* have the capacity to adapt to new temperatures and over ecologically realistic timeframes. Remember that natural populations of *E. coli* have evolved in the guts of homeotherms and are rarely exposed to fluctuations in temperature while in the host. These experiments show the power of selection and the potential for change in organisms that have what is thought to be an optimal temperature for growth.

In the second example, Hall (1982, 1983) performed a series of experiments that demonstrated that novel biogeochemical pathways can be selected for in incremental steps from organisms that do not have the pathway. Hall used an *E. coli* strain that was not able to metabolize lactose. Remember that the ability of *E. coli* to metabolize lactose is under control of the *lac* operon (Figure 8.8). The *lac* operon has several important genes that code for certain enzymes. The production of β-galactosidase, which hydrolyzes lactose, is coded for by the *lacZ* gene. The movement of lactose into the bacterial cell is controlled by a permease that is coded for by the *lacY* gene. These two genes are transcribed after the operator gene (*O*) has been freed from a

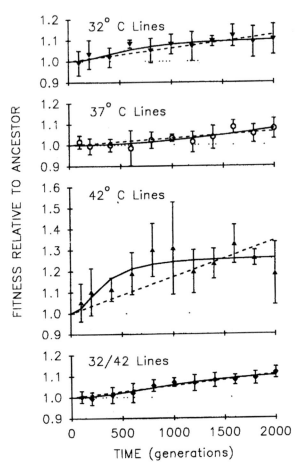

**Figure 8.7** Adaptation of bacterial populations (*E. coli*) to temperature. (From Bennett AF, Lenski RE, Mittler JE. Evolutionary adaptation to temperature. I. Fitness responses of *Escherichia coli* to changes in its thermal environment. *Evolution* 46:16–30, 1992.)

repressor protein, which permits lactose to enter cells easier and results in the breakdown into simpler sugars that are readily used by the cell.

The cells used by Hall were not able to grow on lactose because the *lacZ* gene had been mostly deleted. In this case, lactose could enter the cells, but hydrolysis could not occur and because the appropriate products were not formed *lacY* was repressed. Through a series of experiments, Hall was able to first select for the ability to use lactose under certain conditions because of a mutation on a previously unknown gene which altered an enzyme structure that allowed lactose hydrolysis. He then was able to select for alteration of a regulatory gene that permitted the synthesis of the novel enzyme in the presence of the substrate. He was able to select for an enzyme reaction that induced permease production and the movement of the substrate into the cell. This is a striking example of how selection acting on mutations has the potential to develop all the steps needed for a complex adaptation.

We have spent some time demonstrating that selection can act on random mutations that occur in bacterial populations. Each of the experiments described have been

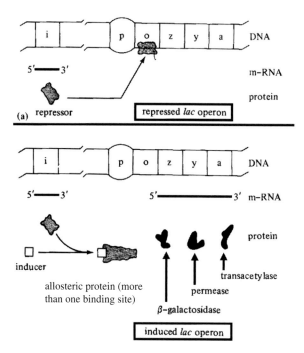

**Figure 8.8** Model of the *lac* operon. The product of the gene is a repressor protein that binds to the *o* region of the operon and prevents transcription (a). This repressor has to be removed by an inducer. (From Gottschalk G. *Bacterial Metabolism*. New York, Springer-Verlag, 1979, 2nd ed., 1985.)

laboratory studies done under nearly ideal conditions with large populations of bacteria. In nature, there are large numbers of bacteria in many environments; however, it is not clear how large the population density of specific bacteria really are (i.e., community sizes are large). Mutations of specific genes occur at fairly constant rates and are not a function of the population size. We would expect that mutations are occurring in natural populations. Most natural environments are extremely heterogeneous for physical and biological properties, and many of these properties change temporally as well as spatially. With a constantly moving target, how can organisms become adapted for any trait in these variable environments? Can the complex adaptations Hall demonstrated for *E. coli* occur under natural environmental conditions? The answer is yes, because all of the complexities of life that we can observe and measure evolved under "natural" conditions.

The answer to the preceding questions may lie in an area of population genetics that is almost totally absent from higher organisms. Bacteria have novel evolutionary mechanisms that allow a more rapid response to environmental change. These mechanisms include conjugation, transformation, and transduction. Conjugation allows exchange of genes between and among other bacteria by cell-to-cell movement of pieces of chromosomal or plasmid DNA. Transformation is the process of taking up of naked DNA from the environment by a single cell. Transduction is the movement of DNA into bacterial cells through viruses.

Microbes in nature are faced with constant change in their environments. The movement of certain genes between and among species of bacteria facilitates the organismal response to some of this environmental change. Not all extrachromosomal genes are "adaptive," and not all of this material can be transferred easily or at all. Recombination can also occur between mobile elements and chromosomal DNA, which further increases the genetic variability of the populations and increases the likelihood of advantageous combinations. However, extrachromosomal genes do confer various advantages that are of immediate value in certain adverse or stressful conditions. These elements often contain genes that provide resistance or tolerance to various agents such as heavy metals or antibiotics. Others contain genes that allow the mineralization or at least initial breakdown of complex molecules derived from human activity.

The study of population genetics in heterogeneous or fluctuating environments is still being developed for higher organisms, and few researchers have tried to apply the concepts to microbial populations. Studies done in laboratories can provide interesting data on potential interactions and processes, but because of their artificial nature may not reflect processes and interactions in nature. Microbes have evolved over most of their history in the presence of other microbes and for the last billion years with higher organisms. Few habitats in nature contain a monospecies of some microorganism. Even if there was only one species at a particular location, that species might have various strains that differed to greater or lesser extents for some traits. To complicate our understanding of microbial ecological genetics, species boundaries are questionable or nonexistent, and some genes appear to move easily among taxa. Given the fact that most of the microbial world is not accessible for direct observation, much of microbial ecology today is based on molecular biology or genetics because of the power of these molecular techniques to detect and distinguish novel microbial groups.

Casual perusal of any environmental microbiological journal is sufficient to see that much of microbial ecology is descriptive work describing the presence or absence of various microbial groups, species, or strains using various molecular techniques. Because most microbes cannot be cultured, their presence is inferred from the sequences of DNA extracted from environmental samples. Specific genes can be screened from environmental samples and give a presumptive measure of "species" diversity. Depending on the sequences used, the taxonomic resolution is increased or decreased. Often researchers may be more interested in the distribution of certain functional genes over the distribution of organisms. Molecular data on specific genes or DNA sequences is essentially gene ecology and not organismal ecology, and because certain genes are not taxon specific, gene ecology may provide interesting insights into microbial ecology.

# Gene Ecology

Let us examine the ecology of a few different genes. Genes change at different rates based on the rates of spontaneous mutations and recombination events. Dobzhansky (1970) lists the mutations per 100,000 cells for streptomycin resistance in *E. coli* as about 0.00004, for resistance to T1 phage as 0.003, and for arginine independence (does not require arginine) as 0.0004. These numbers are quite small and range over three orders of magnitude. At a population density of $10^{10} mL^{-1}$, we would expect about 40 streptomycin-resistant cells, 3,000 cells resistant to T1 phage, and 400 cells to be arginine independent.

Some genes are highly conserved because of their central importance to cell survival. Included in this category are genes that code for the synthesis of ribosomes. Because functioning ribosomes are essential for all protein synthesis, it is critical that these genes remain nearly intact. Any change that results in a nonfunctional product would be lethal. However, small changes in the sequences of these genes are thought to be good molecular clocks to determine time since divergence among taxa and can be used for phylogenetic analyses. Other functional genes, such as those used in nitrogen fixation (*nif*), methane production (*mcr*), methane oxidation, sulfate reduction, mercury resistance (*mer*), and a myriad of other genes can be used within or between taxonomic groups as indicators of gene diversity and supposed local adaptation to the environment.

Resistance to mercury is conferred by a complex set of genes collectively designated as *mer*. There are at least nine gene sequences involved; however, any one organism does not have all nine. Mercury pollution is global due to the burning of coal, and the amounts of mercury released into the environment continue to increase. Mercury-resistance genes are often carried on plasmids that move throughout the bacterial community. Because it is impossible to isolate and characterize all the species of bacteria that carry the *mer* genes, scientists generally screen a composite extraction of environmental DNA collected from specific sites. Although mercury pollution is everywhere, there are numerous locations where the concentrations are especially high. At these locations, we may expect to find higher frequencies of mercury resistance in terms of both numbers and types of *mer* genes (i.e., diversity and richness) along environmental gradients.

The population in this context is the set of *mer* genes found within a location independent of the bacterial species. Inferences were based on the number and types of *mer* genes present. Researchers in Copenhagen monitored the distribution of mercury-resistant genes along a mercury contamination gradient in a harbor and found that highest levels of resistant genes were found in the most contaminated locations. This is not particularly astounding, because we would expect the genes to be present where they are needed. In another study, researchers found that mercury-resistant bacteria were higher in monkeys that had been given mercury amalgam during dental procedures. Selection favored increased frequencies of antibiotic resistance genes with increased exposure to mercury. Each of these examples indicates that there is a relationship between concentrations of mercury and mercury resistance but they do not show whether certain forms of *mer* are more adaptive under different environmental conditions.

That each gene may have a unique ecology is not generally considered in microbial ecology. Rather, the ecology of the entire organism becomes the focus. However, understanding factors that determine the distribution and abundance of genes or alleles of those genes may be of great importance in understanding processes and functions in nature. Many genes are not restricted to a specific taxon but can move freely among different species and strains. This movement of the genes independent of the organism is an exciting area of microbial ecology that needs additional attention.

# Effects of Habitats, Genome Size, Diversity, and Bacterial Communication on Population Processes

9

## Habitats

Changes in season affect population and community structures. For individuals living in temperate zones, the continual change of seasons is expected and part of their normal life cycle. Astronomical events such as the alteration in the tilt of the earth over an annual cycle can have significant impacts on biological events. Most students are aware of the conspicuous flowering that occurs in higher plants at very definite times of the year. Some plants bloom in the spring, others in the summer months, and others late in the fall. Most deciduous trees have vibrant color changes just before they drop their leaves every autumn. These color changes are in part due to the translocation of critical nutrients from the leaves back into the stems and roots of the plant. Each of these plants has a fresh flush of leaves in the spring. Prairie grasses can also have brilliant color changes as nutrients are translocated during fall.

The density of various organisms can change significantly with changes in the seasons. Many of the density changes observed in populations are accompanied with changes in life stage (i.e., immature to adult). Organisms experience different competitors, different predators, and different parasites as seasons change and the structure of their communities change.

Temperature and light are obvious abiotic causes of seasonality in the temperate zones. In tropical regions rainfall controls seasonal patterns with wet and dry defining the major seasons. Changes from one season to another are not clearly defined and the seasons sort of blend into each other. The effect of seasonality on organisms is a function of the longevity of the organism and the life stage the organism is in during any one season. For example, most insects enter into diapause during periods when food is not available, but other consumers migrate with the seasons (e.g., birds, butterflies).

Habitats vary spatially, often over fairly short distances. Differences may include availability of water, concentration of nutrients, availability of organic resources, and electron donors and electron acceptors. How an organism experiences its environment is a function of both the spatial and temporal variability of the environment. How an organism perceives its environment may be very different from the perceptions of the observer of those organisms. For example, forest fires were thought to be something that had to be avoided to maintain wilderness. However, from the perspective of some trees, fires are essential for maintaining the diversity and health of the forest.

The generation times of the organisms in an environment help to set the bounds of various events and whether these events may or may not affect the biology of the organisms. The frequency of fires may well be within the generation time of a tree but far beyond that of a mouse. Most mice will never experience a forest fire, but many trees will.

Various terms are required to understand how organisms view their habitats. These terms include predictability or unpredictability, regularity or irregularity, favorableness, continuous, patchy, isolated, ephemeral, and constant. What may appear as unpredictable to one organism may be extremely predictable to another organism. For example, large migratory herds of mammals often feed indiscriminately on a wide variety of plants. As such their habitat, because of the large spatial scales involved is fairly predictable. Mammals can usually find something to eat because their diets are more diverse or they are able to move and find suitable resources. Any organism that relies on just a single type of food may perceive their environment as unpredictable. The chance of finding the right food source among everything else makes the environment less predictable. In other words, the organism must find the resource. Examples include insects that lay their eggs in only one fruit (e.g., fruit flies) or other insects (e.g., parasitic wasps). Any event, whether biotic or abiotic, that destroys the fruit or the host insects will have enormous consequences (i.e., death) for the larvae living in the fruit or insect. However, the destruction of a single fruit by another consumer will have little effect on the survival of the large mammals.

Although every habitat is essentially unique for the organisms occupying it, there are generalities such as those we have listed that allow characterization. Southwood (1977) developed a logical framework of templates for characterizing habitats. In this chapter, we separate time and space, with the understanding that there is usually an interaction between them.

Whether an organism is found in one habitat or another habitat is a function of the *favorableness* (*F*) of the environment. Favorableness is just another way of stating that the rate of increase (fitness) of a particular population in one habitat is compared with the rate of increase in another habitat. If there are more births than deaths in a given habitat, we can expect higher densities in the favorable habitat and greater favorableness. The length of time that a given habitat permits an organism to have more births than deaths in each generation is designated *H*. *H* depends on the generation time (*T*) of the organism of interest. The amount of time that a habitat is unsuitable for reproduction is designated *L*.

In this chapter, we consider four classifications dealing with the temporal nature of a habitat and three classifications that characterize the spatial components of the habitat (Figure 9.1). In some habitats $F$ is *constant* and $L$ periods may or may not occur and if they do, they are short enough in relation to $T$ that $H$ is very long. If $F$ fluctuates predictably over an annual cycle the habitat is considered *seasonal*. Both seasonal and constant environments are *predictable* and as such organisms can key in on needed resources. In *unpredictable* habitats, $F$ levels are variable over time which affects the period length of $L$. In unpredictable habitats, $L$ may be short enough compared with the generation times ($T$) that $H$ is still long enough for moderate levels of population increase or maintenance. *Ephemeral* habitats, by definition, last for very short periods. In ephemeral habitats, $H$ is short because the length of $L$ is usually very long and predictable.

An organism uses space to find suitable resources for growth and reproduction. This is the day-to-day range an organism travels to feed and is designated $R_t$ for trivial range. An organism can also traverse much larger spatial scales during migration or dispersal to find new favorable habitats. This movement on the landscape is termed $R_m$. Spatial classification of habitats is based on the size of unfavorable areas ($U$) between favorable patches. In *continuous* environments $F$-values are large or if they vary the frequency is small relative to $R_t$ and $R_m$. In *patchy* environments $U$-values are larger than $R_t$ but are generally smaller than $R_m$. The last spatial classification is *isolated*. For isolated environments $U$-values usually exceed $R_m$.

We can combine the temporal and spatial classifications of habitats into twelve possible types:

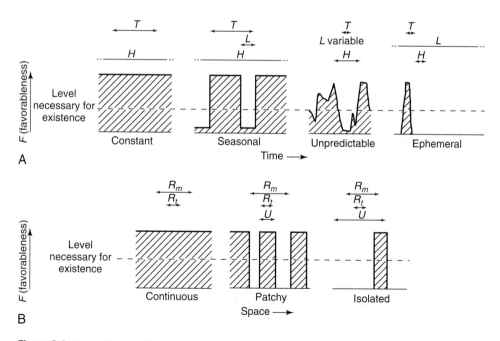

**Figure 9.1** Favorableness of habitats temporally (**A**) and spatially (**B**) as a function of different habitat classifications. Habitats may be constant, seasonal, unpredictable, or ephemeral. Patches may be continuous, patchy, or isolated. (From Southwood TRE. Habitat, the templet for ecological strategies? *J Anim Ecol* 46:337–365, 1977, Blackwell Publishing.)

1. Constant/continuous
2. Constant/patchy
3. Constant/isolated
4. Seasonal/continuous
5. Seasonal/patchy
6. Seasonal/isolated
7. Unpredictable/continuous
8. Unpredictable/patchy
9. Unpredictable/isolated
10. Ephemeral/continuous
11. Ephemeral/patchy
12. Ephemeral/isolated

Two of these habitat types (i.e., ephemeral/continuous and ephemeral/isolated) cannot support life. The ephemeral/continuous would not support life because there would be no condition where $F$ would be above that necessary for existence. An ephemeral/isolated habitat would have an extremely low probability of colonizers during the appropriate time and $H$ would be too short to allow most organisms to reproduce.

The other 10 habitat types should allow organisms to survive although the strategies associated with each habitat will be different. For example, we might well expect that dormancy would evolve in habitats where adverse seasons are found (i.e., where there are long periods of suitable conditions followed by long periods of poor conditions and these conditions cycle on some predictable basis). Dormancy can either be *predictive* or *consequential* (Begon and Mortimer, 1981). Predictive strategies occur in advance of an adverse condition and are best seen in the response of various organisms to seasonal changes. Consequential strategy, as the name suggests, occurs as a result of an adverse condition. Clearly responding to an adverse condition after the fact may be costly in terms of survival. However, if an organism can enter a dormant state after adversity it may well have an advantage by being able to respond to favorable conditions when they reappear or of only going into dormancy when adverse conditions do appear instead of at some proscribed time. If resources are still favorable an organism that does not enter dormancy on a seasonal cue may continue to garner resources and to grow and reproduce.

Most of the material presented in this section has been developed for higher plants and animals. Little attention has been focused on microorganisms. Can we apply the habitat template models to microbial systems?

The entire biosphere has microbial inhabitants. All habitats will contain some, if not very many, microorganisms, the diversity of which we have yet to determine. Most researchers using these habitat templates have way too coarse a view of the habitat. Constancy, predictability, patchy, continuous, and other classifications can be applied fairly easily to large organisms. At the microbial scale, we may well ask whether there is such a thing as a constant environment. Are there instead predictable, patchy, or continuous habitats? The answer is yes. The gut of a homeotherm may be considered as a constant/patchy environment. Any cell that passes through the gut would realize the patchy nature of the habitat. While in the patch (i.e., the host), everything is good. All cells residing in the gut would experience a fairly constant delivery of food and resources. However, once defecated, the cells run a considerable risk of not finding their way to the next patch.

Hyphomycete fungi that colonize leaves that fall into streams would find a fairly patchy environment that is seasonal. During spring and summer, this resource would not be generally available. Every autumn there would be a major input of fresh material available for colonization. Many depressions that fill with water after rain events would be ephemeral and continuous or patchy. In other words, these concepts apply to microbes and to higher organisms. An understanding of the selective pressures found in these different habitat types may give clearer insights into the distribution and abundance of all microorganisms.

Some studies have shown that microbial populations do change seasonally. However, most of these studies have taken too few samples at too few intervals or over too large of an area to be able to definitively show seasonal patterns. Studies performed by Laura Leff and students have provided insights into seasonality and compositional changes in bacteria in stream habitats. From these studies it is apparent that microbial populations do change over time. Specific groups of bacteria were not detected during some seasons but were found at high densities at other seasons. Although the overall assemblage density changed over seasons, variation was much less that experienced by specific groups, indicating that assemblage monitoring does little to provide insight into specific populations.

# Genome Size and Genetic Diversity

Beginning students in biology learn that the sum total of the genetic potential of a species is the genome of that organism. Each individual carries copies of all or most of the genes that can be or are expressed in a population. All of the evolutionary history of the organism is preserved in the genome. For most organisms, all of the future evolution of the species must begin with the present genome and this information is carried primarily on the chromosomes. Exceptions include mitochondrial DNA that is inherited maternally and is found only in the mitochondria.

Microbes because of their small size have a limited genome. Most bacteria have only a single circular chromosome that carries most of the housekeeping genes; genes that ensure that the organism feeds, grows, and reproduces. Some microbes (primarily bacteria) have additional or extrachromosomal genetic material that is carried on mobile genetic elements such as transposons, plasmids, and integrons. This additional genetic material often provides unique capabilities that help the organism through severe environmental conditions. These extrachromosomal elements are known to carry genes that provide resistance to a wide variety of toxic and otherwise lethal materials. They also can carry genes that allow the organism to consume resources that otherwise would be unavailable, especially xenobiotic compounds. The potential number and types of plasmids and other mobile elements may be quite large, but the total phenotypic diversity of microbes is restricted because of the generally small genomes. This is a hindrance to future evolution and an aid in immediate survival and fitness of the organism. However, the phenotypic diversity may be extremely large for bacteria.

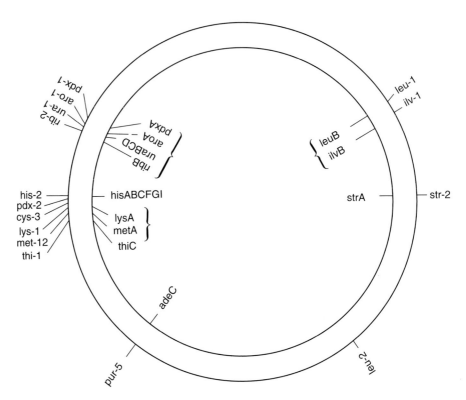

**Figure 9.2** Comparison of gene maps of *Nocardia mediterranei* and *Streptomyces coelicolor*. (From Figure 2 in Schupp T, Hutter R, Hopwood DA. Genetic recombination in *Nocardia mediterranei*. *J Bacteriol* 121:128–136, 1975.)

Microbial genomes show a remarkable degree of similarity in the general map of the genes. This is especially true for more closely related taxonomic groups. However, even some distantly related bacteria have very similar gene maps (Figure 9.2) (e.g., the Actinomycetes *Streptomyces* and *Nocardia*) (Schupp et al., 1975). Because bacteria were among the first organisms on the planet this conservation of gene order is remarkable considering the millions if not billions of years since species diverged and the total number of genome replications and possible rearrangements of the chromosomes that have occurred since divergence. We should expect little overlap and yet constancy in some characters is maintained over taxa and time.

Even though the size of the microbial genome is restricted by the size of the envelope enclosing it there is a considerable range in genome sizes. Early work on the size of bacterial chromosomes demonstrated a range that varied from 580,070 bp for nutritionally fastidious mycoplasmas to 11,593 kb for some of the cyanobacteria. Most heterotrophic bacteria had intermediate sizes with a mean of $3,604 \text{ kb} \pm 1997 \text{ kb}$ (Fogel et al., 1999).

Analysis of the data showed that the sizes were not continuous, but rather appeared to cluster at nodes, the medians of which were separated by multiples of two (Figure 9.3). The sizes appear to be the results of doublings of the DNA. Taxonomic divergence at this level may be related to periodic doublings of the genome.

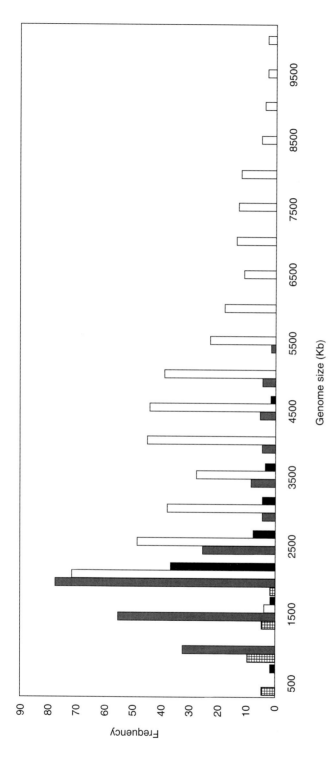

**Figure 9.3** Prokaryotic genome size distribution (*N* = 641). *Open boxes*, free-living prokaryotes; *gray boxes*, obligate parasites; *black boxes*, thermophiles; *boxes with horizontal lines*, endosymbionts. (From Islas S, Becerra A, Luisi PL, Lazcano A. Comparative genomics and the gene complement of a minimal cell. In: Origins of Life and Evolution of the Biosphere 34:243–256, 2004, with kind permission from Springer Science and Business Media.)

Why would doubling the DNA content result in species divergence? Why would we expect differences to obtain just because there are at least two copies of every gene? As long as one copy of the gene remains functional the other copy is free to change. This poses a few problems. Namely the same mechanisms that ensure that one gene remains intact are the same mechanisms that would keep the other copy of the gene intact. Mutations at one copy of the gene should, in theory, be detected and repaired as frequently as another copy of the same gene located at essentially the same place on the chromosome. The data suggest that DNA doubling is correlated with, if not the cause of, taxonomic divergence.

In many higher organisms, the size of the genome is related to the amount of highly repeated DNA (*microsatellite* DNA). This highly repeated noncoding DNA can make up to 90% of the genome size of eukaryotes and can be used for DNA fingerprinting. Microbial genomes lack microsatellite DNA sequences. Most DNA in a microbe codes for a particular function. Why would microbes not have sections of repeated DNA on their chromosomes? Microbes do not have the intracellular room to carry around noncoding DNA. Every piece of DNA must code for an essential element in the survival of the organism. Novel DNA is obtained through conjugation, transduction, and transformation. Because bacteria do not have microsatellite DNA, why do their genomes range in size over an order of magnitude?

At this point, the ecology of the organism becomes important in understanding the size of the genome. Remember that microorganisms generally do not carry around excess DNA. Why is there a relatively large size range? The mycoplasmas were among the microbes with the smallest genomes. Mycoplasmas are bacteria that do not have a cell wall and that are nutritionally dependent. Phylogenetic analysis of the mycoplasmas indicates that this group is a hodgepodge derived from many different degenerative evolutionary events in several different kinds of bacteria. The lack of cell walls and nutritional dependence are phenotypic characters that have evolved numerous times from several kinds of ancestors. The loss of cell walls made these organisms better able to adhere to the host cell membranes and opened up new niches previously inaccessible to the organism. Loss of the ability to obtain food outside the host restricted the niche breadth of the organisms but ensured survival within the host because the host provided everything necessary for survival. Each of these phenotypes is the result of loss of genetic information. Maintaining the genes for making cell walls or for being able to obtain food outside the host was evidently too costly to the organism in terms of maintenance or in terms of some of the cells expressing the traits and being considerably less fit than their progenitors. This particular example is one of genome reduction and not increase and is an example of an evolutionary *tradeoff*.

Trade-offs balance the costs and benefits of one evolutionary strategy against the costs and benefits of another strategy. Is the loss of a cell wall more beneficial than maintaining it under the selective environment of the host? Once lost, traits are often difficult or impossible to recover. Given the number of different ancestors that each evolved into a mycoplasmas-type organism it seems obvious that the tradeoff was beneficial. Trade-offs that do not provide a fitness advantage will in most cases end in extinction.

Genetic polymorphisms and the resulting genetic diversity of a microbe can be influenced by the environment the organism is grown in. For example, certain cyanobacteria, *Nostoc linckia* show a differential response in the amount of genetic polymorphisms found in the populations depending on where the organism was grown. Those cyanobacteria grown in highly exposed areas with intense solar radiation, fluctuations in temperature and desiccation demonstrated the highest levels of genetic polymorphism and gene diversity compared with populations grown in milder habitats in the same canyon (Satish et al., 2001). These high levels of environmentally induced polymorphisms are probably maintained by the combined forces of diversifying and balancing selection; suggesting again the importance of environmental stress in prokaryotic evolution and ecology.

The genetic diversity of microbial populations can be exceedingly high. To help to visualize this potential, let us consider the genetic diversity of a standard amount of organism biomass. For the sake of this example, let us set the biomass at 100 g. How many potentially different genomes are possible in a sample of 100 g of biomass taken from a population? The total genomic diversity of such a sample from an elephant, lion, human, dog, or tree would be one (Figure 9.4). That amount of biomass could be taken from a single individual and so we would not be able to estimate the true diversity with such a sample size. Such a sample of small mammal or reptilian biomass may result in $10^2$ to $10^4$. However, as can be easily seen the total potential number of microbial genomes could exceed $10^{14}$. This maximum diversity would never be obtained because of the ways microbes reproduce. Collection of any sample could have high numbers of identical individuals and lower genetic or genomic diversity. Nevertheless, the potential genomic diversity of microbes is very high. The limitations imposed by a small individual genome size are offset by the genomic diversity of an

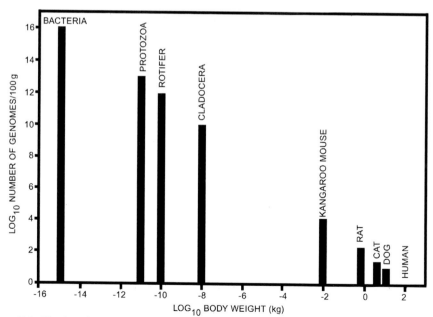

**Figure 9.4** Number of potential genotypes found in a sample of 100 g of biomass. (Redrawn from Herdman M. The evolution of bacterial genomes. In: Cavalier-Smith T (ed): *The Evolution of Genome Size.* New York, John Wiley & Sons, 1985, pp. 37–48, copyright John Wiley & Sons Ltd. Reproduced with permission.)

extremely large population size. Each clone, by definition, is slightly different genetically from other clones. Clonal diversity is one measure of a population that gives insight into the evolutionary potential of the population. Populations that have low clonal diversity may be susceptible to environmental change.

We have used the mycoplasmas as an example of evolutionary trade-offs and as one end of the continuum of genome size. From these observations, we should be able to predict the relative size of a species genome from the way the organism makes its living. For example, human pathogens might be expected to have fairly small genomes based on the reasoning used in the mycoplasmas example. Because of the constancy found in the host in terms of temperature, pH, gases, and nutrients, little extra genetic information is required for survival other than the ability to infect the host and adhere to membranes. In contrast, we might expect that the genome of a free-living freshwater pseudomonad that can process 50 to 100 different types of organic compounds under variable temperature conditions to have a much larger genome. Some of the cyanobacteria have the largest measured genome sizes. Would this be expected knowing something about the life history and ecology of these organisms? Cyanobacteria can live in extremely variable or harsh environments, such as thermal pools, prairie soils, lakes and ponds, and various fungi. Although the thermal pools may have a fairly constant thermal regime, the ability to survive requires specialized proteins and enzymes and the associated cell machinery to function under extreme conditions. Cyanobacteria may carry the genes for nitrogen fixation, genes for the ability to migrate in the water column, and genes to produce toxic or inhibitory compounds. By knowing the ecology of various cyanobacteria species, we can predict something about the size of the genome of these organisms relative to other organisms.

Is there a relationship between genetic diversity or genomic size and the expected environmental variability the organism may experience? Figure 9.5 shows two possible relationships between these variables. In one example, the level of genetic diversity increases continuously with increasing habitat or environmental variability.

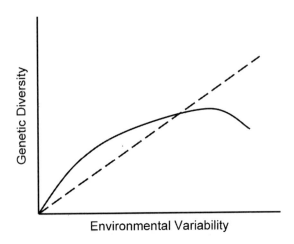

**Figure 9.5** Two hypothetical relationships between environmental variability and genetic diversity of a population. The *dashed line* shows a linear positive increase in genetic diversity with increasing environmental or habitat variability. The *solid line* shows that genetic diversity does not continue to increase with habitat variability but reaches a maximum before the maximum environmental variability.

In the other example, genetic diversity increases up to some point and then begins to decrease with increasing environmental diversity. Very few studies have investigated the level of genetic diversity with increasing environmental variability and, fewer still have investigated the relationship between genomic size and habitat variability.

The genetic diversity of one species of bacteria (*Pseudomonas cepacia*, now designated *Burkholderia cepacia*) was examined along an environmental gradient ranging from pine plantation to bottomland hardwood (McArthur et al., 1988). Environmental variability was estimated by summing the coefficients of variation of five measured variables including percent soil organic matter, dissolved organic carbon concentration, Fe, $NO_3$, and Mg concentrations. The coefficient of variation is a unitless measure and so can be summed across various variables. The larger the resulting value the more variable the habitat. Genetic diversity was estimated from multilocus enzyme electrophoresis (MLEE) for 10 metabolic enzymes. In this study, the genetic diversity of the samples was positively correlated with environmental variability (Figure 9.6). The most variable environment was the bottomland hardwood site. All sites had fairly high genetic diversity. This observation of high genetic diversity within a site, along with the observation of a relationship between habitat variability and genetic diversity, suggests that selection is acting at two levels: variation among habitats and variation within habitats. The variation differences among habitats suggest the importance of directional selection acting at the landscape level. At the within-habitat level, the bacteria must occupy an extreme position along the continuum of possible *environmental grain* (Levins, 1968). The pattern suggests that selection in variable environments leads to enhanced genetic diversity and that genetic variability is responsive to environmental differences even over fairly narrow geographic ranges (only 500 meters in this study from pine plantation to bottomland hardwood sites).

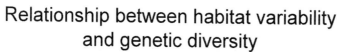

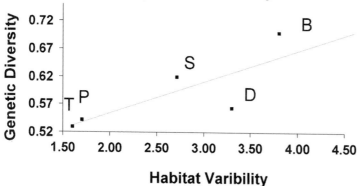

T=Turkey Oak, P=Pine Plantation, S=Swamp
D=Upland Deciduous, B=Bottom Land

**Figure 9.6** Genetic diversity of *Pseudomonas (Burkholderia) cepacia* as a function of environmental or habitat variability. (From McArthur JV, Kovacic DA, Smith MH. Genetic diversity in natural populations of a soil bacterium across a landscape gradient. *Proc Natl Acad Sci USA* 85:9621–9624, 1988.)

# Feeding Ecology and Modular Growth

In Chapter 5, we discussed the feeding ecology of individual microorganisms. Here we examine the ecology of feeding as a group or population. This concept has meaning for many microbes that are either continually or occasionally found in colonies or that exhibit a modular growth form. Modular organisms can be either single individuals or colonies of many individual organisms working together. There are four basic feeding or foraging behaviors. These four include foraging of a single (unitary) organism along a specific route (often observed in animals); feeding at a single location by a single organism that is immobile (most plants are examples of this feeding strategy); feeding at many locations by the growth of a branching system that can fragment (clonal organisms such as many plants, fungi, and some bacteria are examples of this strategy); and the foraging of a "community" of the same organism as a single unit along defined branching structure, with each feeder contributing to the benefit of the whole community. Slime molds and certain bacteria are examples from the microbes.

Finding suitable resources for growth in an unpredictable environment requires more than a haphazard or random sampling of the environment. Harper (1985) addressed this problem by considering what foraging patterns are the most efficient at capturing resources. To begin, he modeled an idealized packing of depletion zones as an array of hexagons (Figure 9.7) and asked, "What growth forms—particularly what branching habit—most effectively captures the centers of the packed hexagons?" Theoretically the most efficient and economical effort is to forage in a spiral with no branching. This is

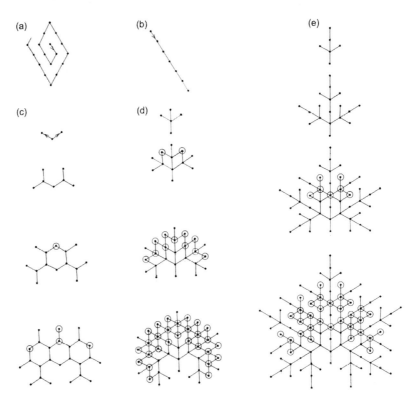

**Figure 9.7** Depletion zones as idealized by Harper (1985).

analogous to the search patterns used to find downed airplanes or sunken ships. However, there are few examples of this strategy found among living organisms. Branching systems might seem to be an effective strategy to exploit depletion zones with minimal or optimized overlap among the branches. However, any branching systems that were to continue unmodified or unchanged would eventually result in double occupancy of a zone (overlap) or zones with no occupancy. Modular organisms must modify their branching patterns and systems continually through some sort of feedback.

Only one resource can be limiting at a particular location and time, but organisms transcend both time and space, and many potentially different resources can be limiting over the life span of the organism. Resource zones may be temporally and spatially dynamic, and they will vary with respect to specific resources. Foraging strategies for one type of resource may be ill suited for another resource. Some resources, such as nutrients, have characteristics that provide clues to the foraging organism about their location and to some extent the concentration. At the other extreme, some resources can affect their capture by producing substances (e.g., antibiotics) that inhibit or kill the forager and affect the foraging.

Plant clones can exploit and explore their environments with various degrees of vegetative spreading. Those plants that demonstrate aggressive spread have been termed *guerrilla* growth forms and those at the opposite end of the continuum that demonstrate a slower expansion as *phalanx* growth forms (Figure 9.8). The guerrilla

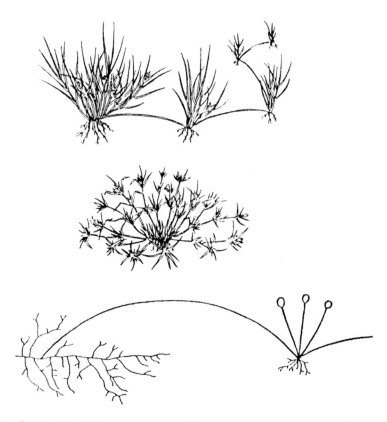

**Figure 9.8** Guerilla and phalanx growth forms exhibited by plants and other modular-like organisms. (From Andrews JH. *Comparative Ecology of Microorganisms and Macroorganisms*. New York, Springer-Verlag, 1991, with kind permission from Springer Science and Business Media.)

strategy involves long internodes, infrequent branching, and spaced modules with a minimum amount of overlap of RDZs. In contrast, the phalanx strategy is represented by short internodes, frequent branching, and considerable overlap. These terms that were coined for plants can be applied to the growth forms of other sessile organisms including microorganisms.

Microbial growth is fairly plastic and as such these two terms (guerrilla and phalanx) can be applied at two different levels to growth forms. First, these terms can be used to compare different taxa in a relative sense. Certain fungi (Zygomycetes and Oomycetes) extend in a guerilla-like mode (see Figure 9.8). *Rhizopus stolonifer* colonizes new substrate with stolon-like runners that develop rhizoids and tufts of sporangiosphores along the nodes. Notice that these extensions are like the stolons of plants with the appearance of nodes. In contrast, other fungi grow as a dense structure across standard laboratory medium or on their natural substrates. An example is apple scab.

Second, the growth descriptor terms guerilla and phalanx can be used for different phases of growth of the same organism. Basidiomycetes colonize wood on forest floors in a guerilla-like fashion sending out elongated mycelial cords or branching rhizomorph systems (Andrews, 1991; Thompson and Rayner, 1982). When a significant nutrient resource has been colonized the fungus develops a diffuse mycelial network. Slime molds are perhaps the most extreme example of different growth forms. They alternate between free-living amoebae or plasmodial networks and a migratory slug just before reproduction that is a large aggregate.

We have been using terms that were coined for plant ecology/biology and have been trying to apply them to microorganisms (specifically fungi) with some success. However, plants and microbes forage very differently from each other even though both are basically sessile and use branching systems to capture resources. The major difference is that for plants many resources move to the plant. For example, light comes to the plant, gases such as carbon dioxide and oxygen move by convection and diffusion and soil nutrients can move quickly through bulk flow caused by transpiration of the plants. In contrast fungi must move to the resource.

Fungi are classified as modular organisms based on architecture and because they grow by iteration of individual hyphal units. Bacteria, in nature, are usually found in small aggregates that can be considered modular because the individual module (cell) can be iterated forever. The bacterial modular organisms can be essentially immortal. Biologists usually refer to multicellular units as modules in higher organisms.

Several evolutionary implications are derived from modular life strategies as enumerated by Andrews (1991):

1. High phenotypic plasticity, which includes shape, size, growth occurring as a population event, and the reproductive potential. All of these characteristics can and are modified during different stages of growth.
2. Exposure of the same genet to different environments and selection pressures. Specific modules may experience increases or decreased growth or reproduction depending on the specific environment these modules have extended into or differences in resource availability. Microclimate differences, concentration gradients of nutrients or inhibitory or toxic compounds, and other factors such as increased

vulnerability to predation or competition may occur over relatively small spatial scales or temporal scales.

3. Iteration of the germ plasm and the potentially important role for somatic mutation. This particular point is very important in clonal organisms, such as plants in which somatic mutations can occur in branches. In microorganisms, each module is capable of reproduction. The development of fruiting bodies or other reproductive structures in microorganisms limits the number of cells that can actually reproduce.

4. Potential immortality of the genet. Microbial genomes can be maintained over geological, if not cosmic, time scales.

Because many microorganisms are sessile or with restricted mobility for most or all of their life the physical characteristics of the habitable space have a major influence on the size of the module and the reproductive output or potential. Because movement is limited or rare interactions with neighbors can be fairly strong. Although microbes can disperse, most do not even when environmental conditions become adverse; hence the need to be able to cope with the change. Microbes are essentially sessile, and over time, they will deplete resources and must develop search and exploitation strategies from fixed positions.

In this section, we have considered bacteria as modular organisms in loosely associated aggregates. Shapiro (1997) has argued that bacteria in general are multicellular and that colonial growth forms result in an actual differentiation of cells. We will examine this concept in more detail here as it applies to the acquisition of resources and to mechanisms for communication among and between colonies and cells. Microbial chemical communication has developed into an intense subdiscipline and will be examined in more detail below. Here we examine the role of communication in the development of colonies and the differentiation of cells and to some extent in the chemotaxis within colonies. Most of these observations are based on laboratory experiments.

# Intercellular Communication

One primary form of intercellular communication between bacteria is the exchange of DNA. Such communication has been shown repeatedly to affect both the ecology and the evolutionary response of the recipient cells. This is especially true for pathogenic strains that have acquired resistance factors. Other examples include transfer of DNA that aids in the breakdown of complex contaminants, tolerance and resistance to heavy metals, ability to reduce species of metals, and many other traits.

Communication at this level is unidirectional, with one cell donating information (DNA) to another cell that may or may not be related. Such transfer of information has interesting evolutionary and ecological consequences. By transferring DNA that increases the survival of another organism, the donor has in affect increased the number of competitors. If the selection pressure is great enough to adversely affect organisms without the information, the transfer seems to provide advantage to organisms that may be competing for the same resources, especially if those organisms have other traits that give additional advantages. Transfer of extrachromosomal DNA ensures the survival and replication of the DNA. From the perspective of the extrachromosomal DNA, it does not matter whether a donor and recipient are closely related; what matters is that additional copies of this specific DNA sequence are replicated and passed onto future generations.

For many years, it was assumed that intercellular communication or multicellularity among bacteria was the exception among a background of single-cell adaptations and ecology. The pioneering work of Shapiro and coworkers has demonstrated that the underlying assumptions about the single-cell life history of bacteria needed to be reexamined. From these studies it has been shown that intercellular communication includes molecules in addition to DNA that regulate gene expression and physiology of bacteria.

Shapiro observed two basic patterns that were visible in stained colonies of *E. coli*. The first pattern was wedge-shaped radially oriented sectors. These sectors, long observed by bacterial geneticists, represented distinct clones of bacteria that originated from a common ancestor. Colonies of bacteria are often considered descended from a single common ancestor. If this is true, all variability in colony growth and pattern is preprogrammed or the result of point mutations within the colony. The formation of sectors becomes visible when a genetic change in a cell confers a new set of properties on all descendants. However, it is still unclear how genetic changes in a progenitor translate into a very exact and complex geometry of growth (Figure 9.9).

The second pattern that was observed was sharply defined concentric rings. Bacteriologists had not observed these rings and no known explanation could be given based on vegetative growth or genetic switches. In contrast to the first element the cells within the ring were not clones but were associated by position in the growth of the colony. Cells within a ring descended from cells in the previous ring and they became the progenitors of the cells in the next ring. Each ring expressed a different phenotype. How could all cells within a specific ring express a common phenotype?

These observations suggest that microbial populations within a colony, at least those grown under the conditions imposed, are really multicellular communities of closely related but different strains with spatially organized biochemical differences. Through a series of elegant observational experiments, Shapiro and coworkers have clarified some aspects of the regulatory and physiological controls during colony formation. The large colonies found growing on nutrient agar plates have been the emphasis of numerous studies. How applicable are these observations to bacteria found in nature?

To answer that question, we must determine if large colonies of bacteria ever form under natural conditions. There are various physiologic problems that living in large groups of cells must create. Such things as transport of nutrients and gas exchange must be coordinated if there really is differentiation of cell types within the colony and the colony must deal with accumulation of wastes. How does a colony get nutrients from the leading edge of growth to the cells in the interior where all available nutrients have been depleted? One interesting observation from Shapiro is that reproduction within a colony under the close crowded conditions was as fast as reproduction of cells suspended in liquid media. Apparently, there was no nutrient limitation that affected the rate of reproduction within colonies.

**Figure 9.9** Complex geometry of growth of *E. coli* colonies. (From Shapiro J. Multicellularity: the rule, not the exception. In: Shapiro JA, Dworkin M (eds): *Bacteria as Multicellular Organisms.* Oxford, UK, Oxford University Press. Copyright 1997 by Oxford University Press, Inc. Used by permission of Oxford University Press, Inc.)

Based on these and other observations we need to reconsider what a population is in microbial ecology. Is it the collection of individual cells distributed both as free-living and colony forms? Is it only the colonies or only the free-living cells? Do we consider a single colony, regardless of size, as a single unit? Step back and consider the definition of a population as currently used by scientists studying higher animals and plants. Recall that a population is a group of interbreeding or potentially inter-breeding individuals. We have gone to some length to show that that definition falls woefully short for most of the world's living things. Is it asking too much to try and force that definition on microbes? Before we can attempt such an undertaking it is necessary that we consider how clonal microbes really are (especially the bacteria).

## Clones or Sex?

In Chapter 6, we discussed the ecology of sex as it relates to individuals. Recombination, although occurring in an individual, is a population phenomenon. Sharing of genetic information by definition involves more than a single individual. The size of the population is important in determining the rates of recombination. Bacteria with their novel evolutionary mechanisms of conjugation, transduction, and transformation require interactions and genes from other individuals, making these processes population events. In this section, we discuss recombination and sex from a population perspective.

Do bacteria have sex? MLEE has generated large data sets that allow statistical analysis of bacterial populations. MLEE examines the allelic variation in multiple primarily chromosomal genes. These data sets have shown that for many species of bacteria there is a nonrandom association of alleles (i.e., *linkage disequilibrium*). Genes are embedded in a collection of genes. No chromosomal genes exist independent of the collection of genes on the chromosome. For organisms with more than one chromosome, each gene's association with other genes is determined by how closely the genes are linked (i.e., located on the same chromosome).

For bacteria there is only one chromosome so every gene is linked with every other gene. The association of alleles of one gene with a specific allele of another gene suggests that they are closely linked genes (i.e., physically close on the chromosome). For example, consider two loci ($A$ and $B$) that each has only two alleles ($A_1$, $A_2$, $B_1$, $B_2$). If over time $A_2$ became associated solely with $B_1$ and $A_1$ solely with $B_2$, these genes would be considered to be in strong linkage disequilibrium because only the $A_2$, $B_1$ and the $A_1$, $B_2$ combinations would be found. Recombination causes linkage disequilibria to break up. The more frequently recombination occurs the greater chance that the linked genes will become disassociated. Based on the MLEE analysis of some bacterial species only a few gene combinations out of the many possible are regularly observed. These data suggest that, at least for those species, the population structure is primarily clonal (i.e., very little recombination occurring). However, Maynard Smith (1993) reexamined the MLEE data for several species bacteria and obtained some interesting results. The following discussion relies heavily on the work of Maynard Smith.

One measure of linkage disequilibrium is the Index of Association ($I_A$). This index is a generalized measure of linkage disequilibrium and has an expected value of zero (0) when there is no association between loci. The $I_A$ has been used in sexual and asexual populations. In sexually breeding populations that produce gametes, the expected value of $I_A$ is zero. In bacterial populations that produce by binary fission, the $I_A$ is affected by the rates of two different processes, first by the divergence of lineages through fixation by drift or selection of new mutations and second through the transfer of genes between lineages by recombinational events.

If recombination is not occurring the value of $I_A$ will be influenced by a number of factors. These factors include the form of the phylogenetic tree, the number of loci being analyzed, and how likely mutations that are electrophoretically identical occur independently in different lineages. Because of these factors it is best to determine which values of $I_A$ are significantly different from zero and which ones are not. However, magnitude of change in the index may be indicative of important biological change and should be considered when performing analyses.

Values of this index that differ significantly from zero indicate that recombination in the isolates analyzed has been absent or rare. Absence of recombination is likely due to either the lineages being spatially isolated either in different parts of the same host, different hosts, or living in geographically isolated hosts. Free-living bacteria may be living in the same habitat, different habitats in the same area, or geographically isolated habitats. Another reason may be that there is a mechanism that limits genetic exchange (e.g., genetically distant strains prevent homologous recombination). The index of association can be calculated for subsets of the data and for the complete data set. Two different analyses can be made.

1. Analysis of subgroups. Subgroups are formed after performing some sort of phylogenetic analysis. If the population structure is clonal at every level analyzed then $I_A > 0$ will be found for all subgroups and for the whole data set (Figure 9.10A). If recombination is occurring between closely related lineages but not between more distant ones, a pattern similar to that shown in Figure 9.10B will be observed. In this example $I_A$ will approach zero for the subgroups.
2. Analysis of all electrophoretic types, all isolates, and clusters of isolates. $I_A > 0$ can occur even when the population is sexual over the long term but where one or a few electrophoretic types (ETs) have recently become widely dispersed and abundant (see Figure 9.10C).

A continuum of possible population structures can be observed from the data presented in Table 9.1 that range from completely sexual to clonal. Within groups of bacteria there is evidence that at different scales the population structure changes. For example, *N. gonorrhoeae*, whether examined using all of the isolates or only the ETs, is effectively sexual. Recombination is taking place frequently among isolates of these bacteria. *Pseudomonas syringae* does not show any indication of recombination at either scale examined and must be considered an example of clonal population structure. *N. meningtidis* is an example of how population structure at one scale is different from that at another. When all isolates (688) are considered together the population structure appears to be clonal. However, when the isolates are grouped into ETs and the analysis performed, the results suggest that the population is sexual. This is an example of an *epidemic* population structure for which there is a high value if $I_A$ for the complete data set that is reduced when the ETs are considered as a unit. For example, of the 331 ETs observed, one was found 156 times. Maynard Smith's group predicts that this ET will, because of recombination, decrease and eventually fail to mark epidemics caused by *N. meningitidis*. Several variants of the common ET have been observed that differ at one or more enzyme loci and they are becoming increasingly more common.

Other patterns can be noted by carefully examining the data in Table 9.1 but the question arises why these patterns are important in microbial ecology? There are several reasons to consider. First, any understanding of microbial processes must include an examination of genetic controls. This is true in the cycling of nutrients, in the breakdown of anthropogenic compounds, in the use of microbes for industrial purposes, and in the treatment and identification of medical infections to name just a few. Failure to recognize that microbes (especially bacteria) do not necessarily follow the *E. coli* model has resulted in many false assumptions about how bacteria make a living in the natural world. *E. coli*, as the best-studied organism, has provided keen

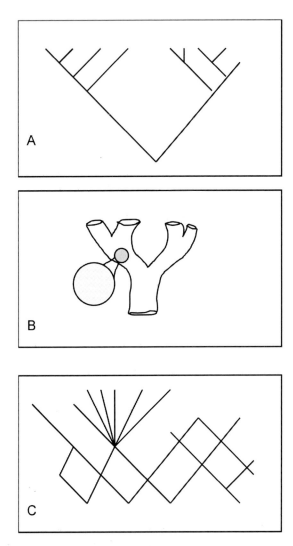

**Figure 9.10** Population structures (A and B) of isolates separated into two major branches. The population structure A is clonal at every level. No recombination occurs between any levels (i.e., isolates or branches). Recombination does not occur in population B between isolates from the two major branches, but there is frequent recombination between isolates within a branch, which results in the netlike rather than treelike pattern within the subbranches, as represented by the enlarged section of one of the major branches. Population C demonstrates an epidemic population structure in which there is frequent recombination within all members of the population so that there is a netlike structure rather than a treelike structure. Occasionally, a highly successful individual may arise and increase rapidly. (From Maynard Smith J, Smith NH, O'Rourke M, Spratt BG. How clonal are bacteria? *Proc Natl Acad Sci USA* 90:4384–4388, copyright 1993 National Academy of Sciences, USA.)

insights into many aspects of microbiology but not microbial ecology. Application of knowledge gained in the study of *E. coli* to other bacteria will further our understanding of ecosystem, community, and population processes that are under the control or influence of microbes. However, expanding our knowledge of novel microbial systems will allow greater understanding of these processes.

**Table 9.1**  Index of Association Measures between Loci in Bacteria

| Bacteria | No. of Isolates | No. of ETs | No. of Loci | I$_A$ for All Isolates | I$_A$ for ETs Only |
|---|---|---|---|---|---|
| *Neisseria gonorrhoeae* | 227 | 89 | 9 | 0.04 ± 0.09 | −0.16 ± 0.17 |
| *N. meningitides* | 668 | 331 | 15 | 1.96 ± 0.05 | 0.21 ± 0.08 |
| *Haemophilus influenzae* | 2209 | 561 | 17 | 5.40 ± 0.03 | 1.25 ± 0.18 |
| Division 1 | 2117 | 42 | 17 | 4.42 ± 0.03 | 0.90 ± 0.21 |
| Division 2 | 92 | 14 | 17 | 2.30 ± 0.14 | 1.20 ± 0.37 |
| *Salmonella* | 1495 | 106 | 24 | 3.11 ± 0.04 | 1.38 ± 0.14 |
| *S. panama* | 99 | 13 | 24 | 1.34 ± 0.16 | 0.28 ± 0.39 |
| *S. paratyphi B* | 118 | 14 | 24 | 2.68 ± 0.13 | 0.36 ± 0.37 |
| *S. typhimurium* | 340 | 17 | 24 | 1.03 ± 0.10 | 1.22 ± 0.33 |
| *S. paratyphi C* | 100 | 9 | 24 | 4.66 ± 0.15 | 4.07 ± 0.46 |
| *S. choleraesuis* | 174 | 11 | 24 | 2.57 ± 0.12 | 1.36 ± 0.42 |
| *Rhizobium melitoti* | 232 | 50 | 14 | 6.34 ± 0.09 | 4.74 ± 0.19 |
| Division A | 208 | 34 | 14 | 0.47 ± 0.10 | −0.24 ± 0.24 |
| Division B | 23 | 15 | 14 | 0.24 ± 0.29 | 0.01 ± 0.35 |
| *Legionella* | 170 | 62 | 22 | 4.63 ± 0.11 | 3.80 ± 0.18 |
| *L. pneumophila* | 143 | 50 | 22 | 1.49 ± 0.12 | 0.69 ± 0.20 |
| *Bordetella bronchiseptica* | 304 | 21 | 15 | 1.29 ± 0.08 | 0.99 ± 0.30 |
| *Pseudomonas syringae* | 23 | 10 | 26 | 18.35 ± 0.29 | 12.11 ± 0.44 |
| *P. syringae tomato* | 17 | 4 | 26 | 3.39 ± 0.33 | 1.68 ± 0.67 |
| *P. syringae syringae* | 6 | 6 | 26 | 1.42 ± 0.57 | 1.42 ± 0.57 |

ETs, electrophoretic types; I$_A$, index of association. From Maynard Smith J, Smith NH, O'Rourke M, Spratt BG. How clonal are bacteria? *Proc Natl Acad Sci USA* 90:4384–4388, copyright 1993 National Academy of Sciences, USA.

# Bacterial Sex

The examples provided above indicate that some species of bacteria are essentially sexual and others essentially clonal. Binary fission has been an extremely successful evolutionary process as bacteria and other organisms that reproduce in such a manner were among the first organisms on the planet and they are still with us today. When did bacterial sex first develop? Based on the prevalence of gene exchange in natural environments it seems obvious that the benefits of limited sexual recombination have been exploited by many different species of bacteria.

When should microbes in general and bacteria specifically engage in gene exchange with the resulting recombination and novel gene acquisition? This is a very important and difficult question to answer. In the laboratory, we can exert various selective regimes under controlled conditions and observe the exchange of genetic material between microbes. In nature, when should bacteria engage in gene exchange and how frequently? Can bacteria monitor their environment such that they "know" when to take up novel genetic elements that would confer an advantage? Is the exchange of the novel DNA under the control of the novel DNA? Do these mobile genes control their dissemination throughout microbial populations and communities?

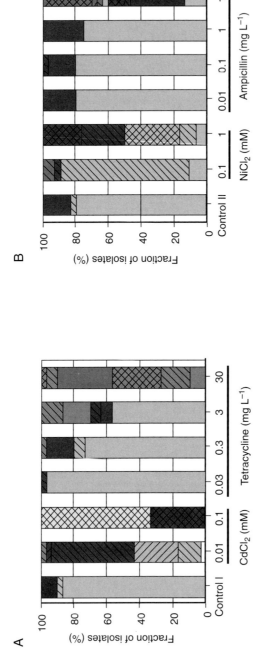

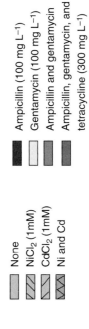

**Figure 9.11** Indirect selection of antibiotic resistance or tolerance and heavy metal resistance in bacteria exposed to antibiotics or metals. Notice that when metals or antibiotics are used as a selector, the frequency of metal and antibiotic resistance increases relative to the controls in the experimental flasks. (From R. Stepanauskas, R., T. C. Glenn, C. H. Jagoe, R. C. Tuckfield, A. H. Lindell, C. J. King, and J. V. McArthur. Co-selection for microbial resistance to metals and antibiotics in freshwater microcosms. Submitted to Environmental Microbiology, 2005.)

Maintenance of this extrachromosomal DNA has an evolutionary cost as explained in the section on individuals. Most observations on the distribution of plasmids indicate that not all of the bacteria in a sample are carrying the extra DNA. Is this by chance or by selection? Clearly some of the genes carried on plasmids and other mobile genetic elements confer a selective advantage under various conditions. For example, tolerance of toxic metals would be an advantage in various environments both natural and man-made. However, not all bacteria found in metal-rich environments have genes that confer resistance or tolerance. This may reflect how many bacteria are in contact with the metals or with the bioavailability of the metals. Not all bacteria may be under the selective pressure to harbor genes that confer resistance or tolerance, even though they may appear to be in the same environment.

When selection is conferred through artificial selection in the laboratory the frequency of various genes can be shown to increase relative to a population without the selection. For example, bacteria exposed to either Cd or Ni for twenty-four hours showed much increased proportions of resistant or tolerant bacteria compared with controls that were not exposed (Figure 9.11).

If individual bacteria have the capability to sample, take-up, and cull various genetic elements the effect of environmental change would be minimal. Any organism that had the ability to selectively sample and take-up combinations of genes that confer an advantage should have a selective advantage against other organisms without such ability. Having a trait before selection would be an advantage but maintaining that trait when there is no advantage could have a negative evolutionary cost as discussed elsewhere. Mutations being random events will occur independent of their need in a population. However, when a trait arises through mutations and that trait is favorable in the environment where the mutation occurs the trait can be said to be *preadaptive*. Preadaptive traits do arise because of the environment but rather they are selected for after the fact. Clearly much more research is needed to understand the ecology of mobile genetic elements.

# Population Spatial Stability

## 10

## Uniformity of Populations

Populations are not homogenous across their ranges because most habitats are not homogenous for most variables. Variation in higher organisms often can be observed in physical attributes such as coloration, size, number of offspring, and any other measurable trait. Depending on the size of the range of an organism, differences can be small or large. When differences occur between geographically isolated populations (e.g., continents), much information on the level of gene flow and dispersal can be obtained. Numerous studies have been performed on a wide variety of organisms, and several important conclusions can be made based on these observations, according to Futuyma (1998):

1. Species are generally not genetically uniform over their geographic range. Significant differences in allele and genotype frequencies occur among populations for many different traits and characteristics.
2. Some of these observed differences may be the results of localized selection acting on members of the species in different environments.
3. Genetic differences between populations are not different from the genetic differences found among individuals.
4. Differences among populations can be very great or very slight, depending on the trait and the selective pressure.
5. The genetic differences found within a population are the foundations for genetic differences between or among populations.
6. There is a continuum of variation among conspecific populations.

In the following discussion, we consider directly or indirectly each of these implications as it relates to microbial populations. We have demonstrated that microbial populations do vary over fairly small geographic ranges. The suggestion was that these populations might vary even more if we were successful in capturing the level of environmental difference experienced by the microbes. We continue to be limited in our ability to measure important environmental variables at scales that are meaningful to microbes.

Is there local selection and adaptation in microbial populations? What evidence would we expect if microbial populations were locally adapted? For example, if we were studying a population of bacteria along a stream course that passed through an abandoned mining operation, would we expect the bacterial populations to change along this continuum? Bacteria from upstream are constantly being imported into lower reaches. However, the bacteria from the lower reaches seldom move back upstream any meaningful distance. (However, consider the movement of bacteria attached to the surface or integument of fishes or invertebrates.) Most bacteria in transit continue through and presumably never colonize lower reaches.

In one study, genetically labeled bacteria were dripped into an artificial stream channel to determine how far they drifted and whether or not they were able to colonize wooden substrates. Sterile wooden dowels, acting as downed wood, were placed into the stream channels at various time intervals before release of the marked bacteria. Water in the mesocosms was supplied from a nearby stream and passed through the system. Freshly placed dowels were colonized by the marked bacteria at much higher rates than were the dowels that had established biofilms. Distance from the source of the drip also influenced the ability of the bacteria to colonize dowels; dowels closer to the source had higher colonization than dowels 10 m downstream (Figure 10.1) (McArthur, unpublished results). These data demonstrate that some suspended bacteria being transported in streams can colonize available substrate. The implications are that established biofilms in streams are not readily colonized by suspended bacteria, and input of new genes or gene combinations may be rare or limited under base flow (i.e., stable) conditions. Storm flows sufficient to cause scouring of stream surfaces may open new colonization locations and allow input of new genes or gene combinations. It is not known how long bacterial or microbial populations remain within specific locations.

For the stream passing an old mine, what would we predict about the bacteria found in the biofilm of rocks and cobble in this stream? To make a meaningful prediction, we need to determine what characteristics of the bacteria we feel might be under different selection regimes. Mining usually increases the levels of heavy metals that are washed into a stream. Mining often lowers the pH of adjacent areas significantly (pH < 2.0). However, for the sake of argument, we will assume that the volume of water passing through the mined section is sufficient to swamp the pH affects. We are left with metals, which may be important selectors. Heavy metals are generally toxic or inhibitory to many organisms, including bacteria. Bacteria have evolved mechanisms to increase their resistance or tolerance of heavy metals. Some of these mechanisms include efflux pumps that are capable of pumping metals out of the cells and back into the environment. If the metal concentrations below the abandoned mine are high enough, we might predict that the bacteria adjacent to and below the mine to have higher proportions of heavy metal–resistance genes. Let us say that the levels of mercury are elevated adjacent and downstream of the mine. What would we predict?

Mercury resistance is conferred through a complex and beautiful system that involves combinations of genes collectively called *mer*. We have discussed aspects of this system in other chapters. In review, the *mer* genes are often found on plasmids and are potentially mobile and transferable to other bacteria. In this example, we need to remember that we are studying populations– a single species of bacteria distributed in space and time, not the distribution of the genes in space and time (the ecological distribution of genes is discussed later). We would expect that the proportion of bacteria harboring *mer* genes to be higher in the contaminated section of the stream than

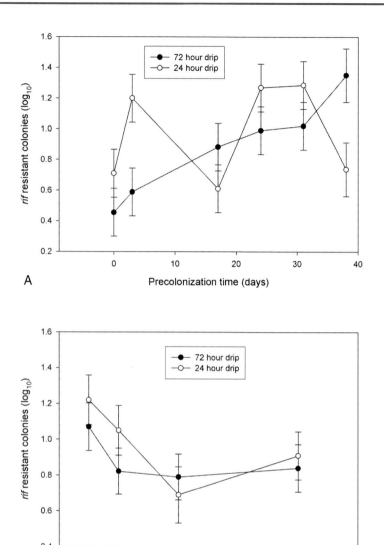

**Figure 10.1** The ability of genetically labeled (multiple antibiotic resistance) bacteria to colonize wood substrates in artificial streams as a function of amount of precolonization of the wood (**A**) and the distance from the source (**B**). Labeled bacteria were dripped into and passed once through an artificial stream. Water was pumped from an adjacent stream. All return water was sterilized by UV light before return to the stream.

above the source of the pollution. If we sample and find that the pattern meets our prediction, have we demonstrated that the genes are adaptive? The answer is no.

## Adaptation

What evidence would be sufficient to demonstrate an adaptive response? Before we can answer that question we need to define what adaptation is. Futuyma states that adaptation has three meanings in biology. Physiologists use the term to describe the phenotypic adjustments an organism makes to its environments (e.g., temperature acclimation). This usage is unfortunate because many people see this type of response and assume that that is what evolutionary biologists mean when they use the word. It is not. In evolutionary biology, the word has the two additional meanings. The first is a process. Organisms become adapted to their environments through genetic change. Adaptation can also refer to the actual features of an organism that increase its chance for reproductive success relative to alternative features. Adaptive traits increase fitness and are derived characters that have evolved in response to some environmental selective agent. Futuyma further states "that a feature is an adaptation for some function if it has become prevalent or is maintained in a population (or species or clade) because of natural selection for that function." Traits that evolve, in most cases, must have preexisting traits with phylogenetic histories that limit the potential adaptations, and potential adaptations should differ among lineages.

Bacteria can and do meet environmental challenges by taking up genes that confer selective advantage under the imposed environmental conditions. Are the traits conferred by uptake of extrachromosomal genetic elements adaptations based on the Futuyma definition? Under certain circumstances these genes can become very prevalent. Antibiotic resistance has increased significantly in most populations of pathogens and many naturally occurring bacteria. Many forms of antibiotic resistance are carried on plasmids and other mobile elements, whereas others are chromosomal. Does natural selection maintain these genes in the population?

We should examine the response to individuals within a population, not the community at large, because the genes may have distributions that are independent of any one species or group. If we examine a single population (i.e., species) distributed throughout our hypothetical stream, we can ask whether there are changes in adaptive traits that increase the fitness of the bacteria and which differ spatially? The answer is probably yes. We should expect the distribution of strains or species to represent localized selection acting on the populations that are present. Colonization of new strains may be infrequent. Predictable temporal changes are expected and include things such as changes in temperature, changes in the concentrations and types of dissolved organic matter with changes in season, and changes in inorganic nutrients in response to in-stream and allochthonous inputs.

Under the selection imposed by the heavy metals those bacteria that have genes that confer resistance should reproduce more than individuals without the trait under the same conditions. In some instances, failure to have the appropriate genes may result in death. However, if we were to sample all species within the contaminated reach of the stream, many individuals of many different species would not have the *mer* genes. Does this mean that they are not reproducing and potentially at death's door? Probably not. The strong selection imposed by the heavy metals will not be homogenous across the contaminated reach. Numerous locations will be essentially metal free. Only those microbes exposed to the selector will be under the pressure to obtain the needed genes that confer resistance. Under the strong selection of localized concentrations of heavy metals we would expect to find increased incidences of metal resistance genes.

# Populations in Time

We have presented some field data that suggests that at least bacterial populations are temporally stable for some period of time. How long that time interval is would be dependent on a variety of biological and abiotic controls. We might expect that the temporal stability of most bacterial populations to vary along a continuum of possible time frames. Some populations may remain stable for extended periods, whereas others may change much more frequently. Laboratory cultures grown under controlled conditions change from one dominant strain to another, often on repeatable intervals. This phenomenon is known as *serial or periodic replacement*, and it has been observed for different laboratory cultures (Figure 10.2).

Periodic replacement occurs because of random mutations occurring in the culture. By chance, a favorable mutant can arise that is better suited for the conditions imposed and this mutant strain increases in abundance until it replaces the previously dominant strain. After the new dominant is established, new mutants can arise that confer an edge in intraspecific competition and this strain eventually comes to dominate and the cycle continues. In nature, it is doubtful that many habitats exist where the conditions remain constant sufficiently long to see such a pattern. Instead of a single dominant strain, several competing strains coexist and vary in distribution and abundance based on the environmental constraints imposed at a specific location.

Stream ecosystems are extremely variable both spatially and temporally and the fact that certain strains can dominate for extended periods of time is intriguing and begs

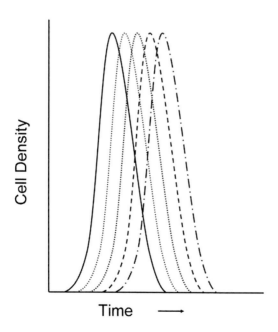

**Figure 10.2** Model of periodic selection of bacterial genotypes. At any moment, the average density of bacteria is fairly constant, but the strains present change periodically as new strains are selected for and rare strains become dominant due to changes in environmental conditions. Each curve represents one strain that grows to a maximum density and then is replaced by a more fit strain, which is subsequently replaced by another.

further examination. However, there are other habitats that are not so variable. For example, the guts of homeotherms are a much more constant environment than the environment experienced by bacteria living in the biofilms in a mountain stream. The gut environment is not completely constant because levels of nutrients and the chemical make-up of these nutrients may vary based on the feeding and wellness of the host. In this quasi-stable environment, are there changes in the microbial flora? This question is too broad because there are many species of microbes living in homeotherm guts. Let us rephrase the question. Are there temporal differences in the genetic composition of *E. coli* living in the gut of a single host?

This was exactly the question asked by Caugant et al. (1981), who examined the temporal dynamics of *E. coli* populations in a single host over one annual cycle. Isolates of *E. coli* were collected from the host and then screened for MLEE variability. Among the 550 isolates, these researchers found 53 unique electrophoretic types. Of these 53, most were transient in nature; they were isolated only once or a few times during the year. However, two electrophoretic types were found repeatedly and were considered the resident strains of this individual. Colonization was never completely successful for all of the other strains. This is an important observation and deserves to be asked of other habitats and for other microbes. Are there resident strains that are always present in a specific habitat and temporary transient strains that come and go over time but never establish? These questions have not been asked nor investigated thoroughly. As a side light these workers and others have determined that humans share a fairly high percentage of *E. coli* strains with their pets! Colonization follows the fecal to oral pattern. Mom was right; you should wash your hands.

With the advent of molecular biological techniques it is now possible to ask and answer questions about the temporal stability of microbial populations. Unfortunately, most studies do not have sufficient samples or are collected in such a way that examination of spatial and temporal dynamics is not possible. Sufficient true replicates and sampling frequencies that are appropriate for the populations under study are required to make inferences or to even detect actual population changes and not simply the natural variability among samples or locations.

## Bacterial Communication: Do Microbes Talk to Each Other?

In classic ecology, density-independent and density-dependent effects may influence populations. Some phenomena in microbial ecology seem to be a function of how many organisms are present in a particular location or in other words density dependent effects. For example, virulence appears to be a function of population size. Once a critical threshold of cells is present virulence genes are turned on and the infection takes off. It has been hypothesized that the bacteria wait to become virulent until a high number of cells are present so that the bacteria can overwhelm the host defense mechanisms. This seems plausible. However, for this to happen bacteria must monitor the density of similar cells and then on cue change and become virulent. If so, how does a single-celled organism with a very limited genome monitor the density of conspecifics? Before we address this specific example we need to shed some light on the subject. Specifically we need to discuss bioluminescence especially that found in the light organs of higher marine creatures.

Numerous higher organisms including invertebrates and vertebrates have evolved light producing structures that aid in feeding and predator avoidance. These light organs can be specialized structures that harbor high densities of certain bacteria. In the squid this bacteria is *Vibrio fischeri*. These bacteria are capable of light production using the *lux* system. However, these specific bacteria only produce light when their densities exceed some critical threshold and the light organs provide a contained volume with suitable nutrients to produce the required number of bacteria. This is a beautiful example of symbiosis and apparent cooperation between the host and the bacteria. The host provides an essentially predator and competitor free environment, plenty of nutrients and oxygen and in return uses the light produced by the bacteria to ensure its own survival through feeding or predator avoidance.

It is that critical threshold of cell density that is important. How do the bacteria "know" when the density is high enough? Possible mechanisms include monitoring levels of waste products or the concentration of food. When the wastes get too high or the food too low an individual cell could know that the density was getting high through receptors on their cell surface.

An alternative and highly studied mechanism is based on the secretion of a specific molecule by the bacteria. The cells monitor the concentration of this molecule and when the concentration is sufficiently high light is produced. Let us look at this process in more detail and then try to understand the response in terms of the evolutionary ecology of the microbes and the hosts. The study of this phenomenon provides a number of unique evolutionary and ecological scenarios that will help in the understanding of microbial population ecology as well as the application of evolutionary concepts to microbial systems.

Production of light is regulated and tightly correlated with the density of *V. fischeri* in the light organs. In these organs the density of *V. fischeri* can reach $10^{11}$ cell per mL. As the cells grow they produce and release an autoinducer hormone. Because the light organ is basically a closed system (from the perspective of the bacteria), this autoinducer begins to accumulate. The effect of the autoinducer can only be hypothesized in a closed system. In open seawater, we would not expect there to be a sufficient accumulation of the autoinducer because of diffusion and microcurrents and macrocurrents that would waft the hormone away or dilute it relative to the producing cells. After sufficient autoinducer has been produced that detection can be made by the individual cells, a cascade of effects begins that results in light production (Figure 10.3).

This effect has been termed *quorum sensing*. A quorum is a group that is of sufficient size to conduct business. In this case, a quorum is a cell density such that the signal can be initiated to begin the production of light. Since the description of the light-producing quorum-sensing model, many other bacterial systems have been studied and autoinducer molecules identified. These systems include the transfer of plasmids, production of antibiotics, production of cellulases, proteases, siderophores, surfactants, cholera toxin, and other virulence factors. Quorum sensing also has been implicated in biofilm formation, motility, sporulation, mycelium formation, and swarming.

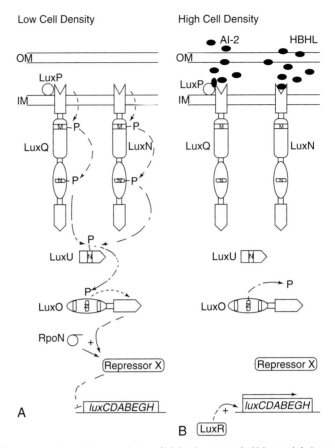

Low Cell Density          High Cell Density

**Figure 10.3** Quorum sensing and the regulation of bioluminescence in *V. harveyi*. At low cell density (**A**), a multistep phosphorelay is thought to indirectly repress transcription of the genes required for bioluminescence by activating the transcription of an unidentified negative regulator (repressor X). At high cell density (**B**), corresponding to a critical concentration of signal molecules results in inactivation, thereby preventing the up-regulation of repressor X activity. Such de-repression allows transcription of luxCDABEGH and emission of light. (From Whitehead NA, Barnard AML, Slater H, Simpson NJL, Salmond GPC. Quorum-sensing in gram-negative bacteria. *FEMS Microbiol Rev* 25:365–404, 2001, with permission from Elsevier.)

The light organ example is an interesting symbiosis. The interaction between the host and the bacteria can be very complex and exact. Consider the symbiosis occurring in the nocturnal Hawaiian bobtail squid. The adult squid is only about 2 cm long. The light organ is located ventrally next to the ink sac within the mantle cavity. The light organ usually harbors about $10^{11}$ bacterial cells/mL. Light is emitted downward at night and the squid can control the emission to match the intensity of light from the moon or stars. This is accomplished through three structures through which the squid can manipulate the amount of light being emitted. These three structures are a reflector to direct the light emission, a lens-type structure, and a shutter that can be opened or closed. Through the combination of these structures, the light response makes the squid nearly invisible to predators looking up. In addition to the mechanical structures used to direct and control the light, the squid are able to regulate the amount of light being produced by limiting the amount of oxygen that reaches the bacteria in the light organ. Oxygen is required by the bacteria to produce the light. Interestingly there is evidence that the squid can expel over 90% of the bacteria from the light organ each morning. This procedure has a number of possible effects. First, by lowering the total number of bacteria, older senescent bacteria are removed and the squid ensures that the bacteria needed for light are the most robust and healthy. Second, the squid does not need to provide as much oxygen to the light organ during the day. Third, expelling of bacteria into the water column may actually seed the water with bacteria that can infect new hatchlings of the squid. Newly hatched squid do not have the bacteria, nor do they have well-developed light organs.

Light production by an individual bacterium has little biological function, because the amount produced would be undetectable by other organisms. *V. fischeri* can be found in small colonies and as free-living cells outside of their vertebrate and invertebrate hosts. These non-symbiotic cells are capable of light production (i.e., they have the genes to make light, but presumably they are unable to detect the appropriate quorum size to begin the production of light). Light production from this vantage point appears to be adaptive only in the context of a symbiosis. However, recently researchers have shown that light production can benefit individual cells by recycling reducing equivalents and providing appropriate wavelengths of light for DNA repair.

Quorum sensing is not limited to Gram-negative or Gram-positive bacteria but can be found in all types. However, the specific types of molecules are different between Gram-negative and Gram-positive cells. Within a group, the basic structural chemistry of the communicating molecule is similar among the different species of bacteria yet each molecule elicits a species-specific response. A wide array of specific responses is possible. Certain molecules have been found that can elicit a response across taxonomic lines. These molecules have been suggested to be a sort of common language.

Based on the number of papers being written on quorum sensing and the detection of numerous autoinducer molecules it would appear that this phenomenon is widespread if not universal among the prokaryotes. Such a widespread process should be susceptible to cheaters and to other organisms coming up with ways to disrupt or circumvent the desired affect. Cheaters are organisms that obtain benefits without the associated costs by taking advantage of other organisms. We discuss cheating in depth in the next section. However, quorum sensing is vulnerable to cheaters because an organism that does not expend the energy to produce the autoinducer molecules could obtain the benefit by being in close proximity to those that do.

Implicit in the arguments for quorum sensing is an ability of cells to detect or monitor the density of like organisms. In a pure culture such as that experienced in the light organs of marine organisms, it seems plausible that detection of density can be accomplished. However, outside a closed system, such as the light organs, the ability to monitor the density of conspecifics would be a daunting task.

# Quorum Sensing and Infections

In infectious diseases, it has long been known that the invading cells need to reach a critical density in order to swamp the host defenses. Those invading cells that can control the expression of virulence as a function of cell density would have an advantage during infection particularly if the host's defenses were keyed to the virulence factors. Many pathogens have been shown to regulate various physiologic processes in a density-dependent manner. Cell-to-cell communication in infected tissues will be affected by the density of the infecting cells and the ability of the quorum sensing molecules to diffuse through the specific tissues. However, studies have shown that these molecules can be detected in vivo and indicate that bacterial communication may be possible. The role of quorum sensing molecules in infections is not limited to animals or humans but has also been studied with plant pathogens. *Erwinia carotovora* is a plant pathogen that infects potatoes. This bacterium has two sets of genes that are expressed based on the apparent density of the bacteria. One set of these genes codes for the production of an antibiotic carbapenem. The other set of genes produces by a number of pathways exoenzymes, pectinases, and cellulase that break down the potato tissue and release the nutrients that are then used by *E. carotovora*. The coordinated release of the exoenzymes swamps the plant defenses, and the antibiotic eliminates potential competitors. The combined affect of the two distinct gene sets is the availability of a resource that is shared only by other *E. carotovora*. In this instance, cell–cell communication seems like a benefit to the group that cannot be obtained by an individual or for numbers of individuals that are below the critical threshold for the release of the exoenzymes and antibiotic.

## Evolutionary Implication of Quorum Sensing

Communication between individuals has received considerable evolutionary thought. When should evolution generate mechanisms that allow communication? For example, why do birds flock? Why do certain rodents give warning signals? If the whole purpose of evolution were to increase the probability of having genes or progeny in subsequent generations, would there not be some advantage in not warning your neighbor about a possible predator? If your neighbor is eaten and you are not, then you have won, right? If the fitness cost of giving a signal were great or the receiver of the signal only experiences the fitness benefit, we would expect communication to be only between closely related individuals. The level of communication between genetically different individuals is a function of the coefficient of relatedness, $r$, between the communicating individuals. In sexual populations, the value of $r$ is $\frac{1}{2}$ between parents and offspring. Cousins are only related by one eighth. The number decreases geometrically with decreasing levels of relatedness.

Within a multicellular individual each cell is genetically identical to all other cells. We would expect and do find significant levels of cell–cell communication in a variety of forms and functions such as hormones and other steroids. It is the combined effort of all the cells that ensures survival of their genes in subsequent generations. The cells of one organism seldom if ever communicate with the cells of another organism.

## Cell–Cell Communication in Bacteria

How closely related are bacteria within a population? Because many bacteria are clonal and reproduction is by binary fission, bacteria in close proximity to each other are probably very closely related. They may be identical. This observation has made some researchers consider microbial populations like a multicellular organism.

However, as Brookfield (1998) pointed out there is a fundamental difference between the cells in a true multicellular organism and the clonal cells in a bacterial population. A multicellular organism has genetically identical cells but the individual is genetically different from all other organisms in the populations with which it is competing. The clonality of bacteria imposes the genetic identity between communicating cells and the corresponding identity on the rest of the population. If this is so no evolutionary change is possible. For clonal organisms what matters most in determining the genotypic correlation between communicating cells is the distribution of times to common ancestry (i.e., how long since divergence) of the communicating cells relative to the corresponding distribution of the whole population. In other words, imagine a population of bacteria made up of a number of different genotypes where the proportions of those genotypes are a function of both colonization events and time since last mutations. We would expect that if the number of colonizing and presumably more distantly related cells is higher than the frequency of identical or nearly identical cells that communication would be reduced between and among cells.

In nature, when a cell successfully colonizes a new habitat, say *E. carotovora* on potato plants, most of the cells in the immediate proximity to that cell will probably be related to the founding cell. In this case, the benefits of quorum sensing are basically to the founding cell. The success of quorum sensing will be a function of the level of successful colonization by different cells. If the mean number of colonizing bacterial cells is low and most cells on a surface post-colonization are genetically closely related, we can expect quorum sensing to be favored. If there are many colonizers, each of which is genetically different, the genetic relatedness among individuals on the surface would be low and quorum sensing would not be favored.

Colonization success is just one parameter that should affect whether quorum sensing is an adaptive strategy. Other factors include genetic relatedness among cells as defined above, the individual investment in signaling (i.e., autoinducer molecule production—some cells that produce exceptional quantities of the inducer may be favored because they are more likely to get cooperation from nearby cells), the total concentration of the molecule, and the individual investment in cooperation. There also are costs for cooperation.

Brown and Johnstone (2001) developed models based on these parameters and have generated what appears to be mathematical support for quorum sensing under certain conditions. Figure 10.4A predicts that as density increases, the potential for a group benefit also increases, until a threshold is reached above which the group benefits exceed the individual fitness benefits. Similarly, as relatedness increases, the level of cooperation increases, but at levels of cooperation that exist even between largely unrelated groups when the population size is sufficiently large.

In Figure 10.4B, signaling effort has been modeled as a function of mean colony size and relatedness. Both parameters have a predicted significant impact on signaling effort. The level of relatedness is important in regulating how much conflict is occurring within a colony. This model predicts that as the relatedness approaches $r = 1$ then the strength of the signaling effort should also decrease. There is some level of relatedness above which signaling should decrease to the lowest level that still allows effective functioning as an indicator of cell density. The model predicts that as relat-

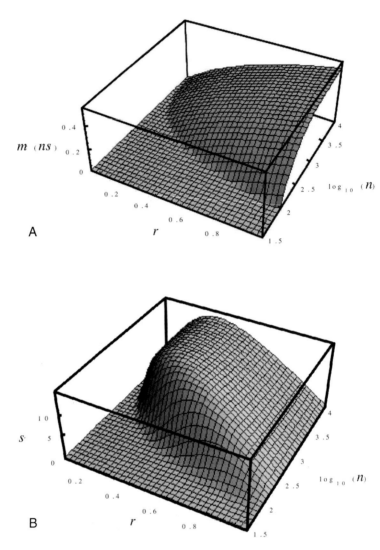

**Figure 10.4** Equilibrium signaling effort as a function of the pairwise relatedness among colony members (r) and $\log_{10}$ mean colony size (n). Colony sizes are assumed to be drawn from a normal distribution, with coefficients of variation equal to 0.2. As density increases (**A**), the potential for a group benefit also increases, until a threshold is reached above which the group benefits exceed the individual fitness benefits. The signaling effort (**B**) has been modeled as a function of mean colony size and relatedness. (From Brown SP, Johnstone RA. Cooperation in the dark: signaling and collective action in quorum-sensing bacteria. *Proc R Soc London B* 268:961–965, 2001.)

edness decreases and conflicts increase, signaling strength should increase to a maximum. This maximum suggests that in poorly related colonies, each individual is signaling intensely to manipulate other cells (competitors) into higher levels of cooperation and at earlier times. With continued decreases in relatedness, there is a subsequent decrease in signaling due to decreased benefits of investment in the group. At these levels the costs associated with competitive signaling are greater than the benefits of the competitive reward. The relationship with colony size supports the notion that there is little to be gained from signaling unless individuals have some chance of experiencing high cell densities.

If evolutionary theory were robust, we would expect it to be applicable to all organisms. The degree of relatedness among groups or individuals has important evolutionary meaning and should affect the level of cooperation. Much of the observational work on quorum sensing has been done on laboratory cultures under controlled conditions. Nature is seldom as cooperative as the laboratory, and most of the interactions in nature, although similar to those found in the laboratory, are much more complex. For example, biofilms in industry and in the laboratory are much simpler than biofilms found on the surfaces of rocks or wood in aquatic and marine ecosystems. These complex environments do not have single species biofilms, but rather extremely complex and dynamic assemblages that vary spatially and temporally. Biofilms in nature are also subject to predation and disruption from other biological and abiotic events or activities and increasing the variability. Epiphytic communities are seldom single-species assemblages. In the *E. carotovora* example, the fact that antibiotics are produced is indicative that unrelated species are expected. Quorum sensing has been readily accepted by many researchers but not by all. Some caution is suggested based on simple evolutionary principles.

## Quorum Sensing and Evolution

The annual output of papers dealing with bacterial quorum sensing continues to increase. The first paper on the subject was published in 1994, and by 2001, more than 125 papers were published in a single year. It has been said, "If everyone thinks your idea is good, it probably isn't; if everyone thinks your idea is bad, it probably is; and if there are some for and some who are against your idea, then you probably have something important." It is the concept of controversy that spurs science forward. Quorum sensing has a devoted following, but there are those whose see things differently. These opposing ideas are germane to our discussion because they bring to bear evolutionary concepts that need to be addressed, especially because they affect the ecology of the organism.

Quorum sensing confers an advantage on the group, not the individual. An individual's fitness is secondary to the group's fitness. In quorum sensing, there is a shared mechanism that involves the regulation of the release of an autoinducer into the immediate environment around the cell. Each cell produces some of these molecules, releases the substance into the environment, and simultaneously measures or senses the concentration of the extracellular substance at the cell surface. After a threshold level of the substance is sensed, the cell begins the production of other extracellular compounds. Because the concentrations of the autoinducer substance in liquid cultures exceed a threshold only at certain densities of cells, researchers have hypothesized that the system evolved to allow cells to monitor cell density. In so doing, the cells are able to find the optimal expression of function that is most beneficial at large population densities. Redfield (2002) takes exception to this generally accepted model and much of what follows is based on her arguments.

In pure cultures of clonal organisms, selection could favor a cooperative response that benefited the entire group over clones that act selfishly. However, in nature where mixed populations (mixed in the sense of different strains and different populations of species) are anticipated to behave through quorum sensing, selection would favor selfish cells. Selfish cells could potentially obtain all of the benefits of quorum sensing for free (i.e., no metabolic costs). Cooperation by signaling within species, especially between closely related cells, may be susceptible to cheaters but not necessarily. However, hypothesized interspecies communication is much more difficult to envision because cooperating members of the communities would be under constant competition from non-cooperators that reap the benefits of the extended groups. Non-cooperators have yet to be reported. This suggests that the regulation of autoinducers confers an advantage or benefit to each individual cell and not to the entire group.

If the benefit is to the individual cell and not to the group as a whole what functions other than quorum sensing can be postulated. Redfield suggests that one candidate function is the detection of the extent of diffusion and mixing that occurs within the immediate environment of the cell. Why would detecting the extent of diffusion be important to a bacterial cell? The answer lies in the mechanisms bacteria must use in order to obtain food.

Bacteria must attack all their food items outside of the cell. There is no phagocytosis in bacteria. Bacteria must secrete exoenzymes into their immediate environment. Through the action of exoenzymes, food is broken down and then taken up by the cell. Substances that are released into the environment by a single cell will be effective to the cell that released them only if the material stays close by and is not removed by diffusion or mixing. Often, the effective distance of an exoenzyme to the producing cell is overlooked (Figure 10.5). Although it is recognized that exoenzymes are released, the effect of diffusion is not considered.

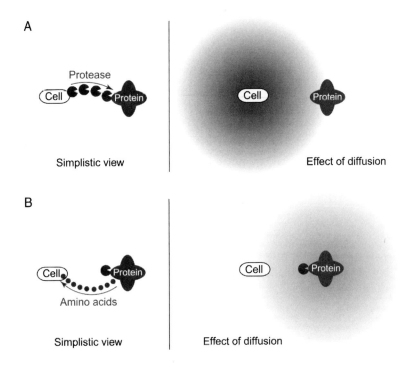

**Figure 10.5** Effective distance of an exoenzyme. The diffusion distance of an enzyme affects the response of the bacteria producing the enzyme. If many bacteria were near, an individual would be able to detect increased levels of the enzyme. (From Redfield RJ. Is quorum sensing a side effect of diffusion sensing? *Trends Microbiol* 10:365–370, 2002, with permission from Elsevier.)

Diffusion limits the amount of the enzyme reaching a target and the amount of product that reaches the cell. In the diffusion model, the autoinducer serves as a molecular sensor. The local concentration of the inducer is constantly being monitored by the individual cell and indicating the molecular processes that are occurring near the cell. Under natural conditions, the effect of diffusion and other mechanisms that affect concentration of materials may be very unpredictable so that any cell that can produce and detect an inducer would be able to monitor these changes directly. If high levels of the autoinducer are detected, diffusion is restricted, and the molecules excreted into the environment should stay near the excreting cell. Substances that may help an individual cell include antibiotics, surfactants, and other secondary metabolites that increase the potential resource capture by the secreting cell. The breakdown products also remain close to the cell, making it easier to be taken up into the cell. Any molecule that is used to detect a physical process such as diffusion should be relatively easy and metabolically cheap to produce, and the molecule should not be found under ambient conditions.

This model provides a viable alternative to some of the observations and predictions of quorum sensing. However, it does not claim to address all aspects of quorum sensing. For example, competence in *Bacillus* and *Streptococcus* is regulated by autoinducers and other unrelated processes are also regulated by autoinducers (biofilm formation and acid resistance in *Streptococcus*, antibiotic formation and surfactant production in *Bacillus*). Regulation of motility by autoinducers may reflect other benefits of being able to sense the environment and not just for diffusion. In this case, being able to detect proximity of a solid surface can cause the autoinducer to accumulate, which then may induce a shift to a different mode of motility.

## Disruption or Manipulation of Quorum Sensing Response

Organisms such as bacteria, algae, and higher plants may benefit from disrupting quorum sensing and may be selected to develop ways to disrupt the effects of autoinducing molecules. Biofouling, or the formation of unwanted biofilms on any structure, is a major problem in as disparate areas as medicine and the shipping and fishing industry. The first known quorum sensing mimic was discovered by researchers trying to prevent biofouling of ships in marine waters. It was discovered that the red alga, *Delisea pulchara*, produced substances that inhibited biofilm formation. It appears that these substances produced by the alga are successful in preventing bacterial biofilm development on algal surfaces. The algal compounds are effective in shifting the bacterial communities from Gram-negative bacteria that normally colonize to Gram-positive bacteria that are poor colonizers of marine surfaces.

In the previous example, the algae produced compounds that inhibited the autoinducer molecules of the Gram-negative bacteria. However, some higher plants produce compounds that may actually stimulate quorum sensing. Organisms that have been shown to produce compounds that manipulate the quorum sensing response include peas, rice, tomato, soybeans, and many isolates of bacteria. Higher plants can both negatively and positively affect quorum sensing in bacteria and by so doing affect the diversity and function of bacterial epiphytes. Although being able to control colonization of potential pathogens seems straightforward, it is not so clear why plants would wish to stimulate a quorum sensing response. One possible benefit may be detection of virulence factors before the bacteria have sufficient numbers to cause an infection and allow the plants to increase defenses.

## Eavesdropping by Bacteria

If some organisms are capable of disrupting or manipulating the autoinducer signal, it seems plausible that some bacteria may be selected to read the signals of other bacteria and use that information to their advantage. In essence, some bacteria may be listening in or eavesdropping on other species and using that information to illicit responses that block or take advantage of the induced response. The ability to respond to signals from both intra- and interspecies would allow a bacterium to alter its phenotype in response to both an increase in same-species density as well as changing its species proportion in mixed communities. One example of eavesdropping has been described for the plant pathogen *Pseudomonas aureofaciens*. This bacterium appears to be able to monitor the signal production of competitive bacteria and when necessary produce phenazine, an antibiotic, which effectively eliminates all competitors. Bacteria could be selected to monitor the autoinducer signals of competitive species or the density of species whose metabolic products are the "food" of the monitoring species. Rather than respond to the accumulation of the metabolic products the bacteria would induce production of enzymes necessary to metabolize these products based on the density of the producing cells.

Some bacteria could produce signals that cause a miscue to another species and cause them to begin a quorum sensing response before it is advantageous to do so. Such premature responses, for instance in a plant pathogen, may not swamp the plant's natural defenses and would result in the elimination of the miscued species. Subsequently, the eavesdropping species would have reduced competition. Although these scenarios are for the most part hypothetical, if selection can favor the response seen for *P. aureofaciens*, other possibilities exist and need the attention of researchers to tease out the interactions and effects.

## Quorum Sensing: Final Thoughts

Quorum sensing is an appealing hypothesis because humans can relate to sociality as a means of increasing feeding and reproduction. The careful research of a number of highly qualified individuals and teams has resulted in the identification of many autoinductive molecules. Each of these molecules can illicit very different responses. Certain classes of molecules are found in different groups of bacteria and there seems to be some molecules that can be sensed by distantly related bacteria (i.e., Gram-positive and Gram-negative bacteria).

Quorum sensing is an excellent example of how a concept can take rapid hold on the scientific community. Few alternative hypotheses have been put forward to offer a different explanation of the role of these autoinducer molecules, even though the "group" level response flies directly in the face of current evolutionary thinking.

Do bacteria talk to each other? Is there a language of molecules that provides bacteria with meaningful, up-to-date information on the density of their conspecifics or mixed communities? The identification of molecules that can be sensed by bacteria and that initiate various metabolic responses is intriguing and worth additional thought and experimentation. However, science progresses by considering new observations in the light of previous theory and study. Altruistic communication among bacteria may be occurring, but it needs carefully executed experiments to prove that this repackaging of group selection is more beneficial to the group than to an individual bacterium.

# Cannibalism, Miniaturization, and Other Ways to Beat Tough Times

Microbes have successfully navigated most of the earth's history. During this immense time frame the availability of resources has gone through periods of boom or bust. For most microbes, the availability of resources probably changes over ecological and geological time frames, and for some organisms, availability may change on a diel basis. Regardless of the spatial and temporal scales, many microbes face conditions that are favorable for growth and reproduction and conditions that are adverse. What mechanisms have evolved to allow microbes to survive when times get really challenging? Here we will discuss some novel ways that microbes meet these challenges and ensure survival over time during harsh times.

Microbial population densities can range over several orders of magnitude within the same environment depending on time of year or time of day. The survival of microbes, specifically bacteria during times of nutrient limitation is accomplished in a couple of ways. Certain species of Gram-positive bacilli have evolved the ability to survive starvation conditions through a process called *sporogenesis* or the formation of an endospore. *Endospores* are a unique cell type that is formed within vegetative cells by a complex sequence of morphological changes that are under the control of a cascade of genetic events. Endospores are unique in that they have no metabolic activity and are able to survive long periods. After conditions are suitable, the endospore can be transformed into a vegetative cell through the process of *germination*. Endospores are usually resistant to extreme temperatures, drying, various chemicals, and radiation all of which can cause cell death of vegetative cells.

The formation of endospores is energy expensive and once started cannot be stopped. The change from vegetative to endospore is accompanied by a series of profound changes in intracellular enzymes and cell chemistry. Some enzymes found in vegetative cells are completely absent in the endospore and vice versa. In other words, many genes are expressed during the process of sporulation that are not expressed during vegetative cell growth.

Spores are usually formed as nutrients become depleted and in response to lack of energy after competition for available nutrients. Sporulation is a very effective means of survival under harsh conditions. The cells that are capable of forming spores will have some advantage over other cells if the environment significantly changes. However, if the environment is extremely unpredictable, such that conditions vary quickly between suitable and harsh, the cells that form spores may be at a disadvantage. Because sporulation is not reversible, the cells that sporulate are hedging their bets that the environment is going to be adverse for some lengthy period. Cells that delay sporulation sometimes may have a competitive edge over the spore-forming cells if the environment changes back to good conditions after sporulation has begun.

Programmed cell death (i.e., *apoptosis*) is a phenomenon that has been observed in both eukaryotic and prokaryotic cells and involves the death of the cell through an internal genetic program. It has been suggested that programmed cell death can benefit bacterial populations under certain circumstances. These circumstances include starvation conditions. Under these hypotheses, some fraction of the cells die, and their death provides nutrients that ensure the survival of the remaining cells.

These hypotheses have been extended to spore forming bacteria. González-Pastor and coworkers have proposed a model in which programmed cell death is an integral part of sporulation. This model is an extension of quorum sensing discussed earlier.

The model by González-Pastor and colleagues proposes that some cells resist sporulation by killing and consuming the released nutrients of sister cells. The delay in sporulation would allow these cells a longer sensing of the environment to determine whether conditions required sporulation or not. Under nutrient limitation a key regulatory protein is activated but it seems only in a subpopulation; those cells that are active for the regulatory protein. All other cells remain inactive. If starvation conditions seem imminent for the entire population then a cascade of processes are initiated in the active cells. First these cells produce a killing factor and a pump that exports the killing factor out of the cell and protects them from programmed cell death (Figure 10.6). The inactive cells do not produce the killing factor or the pump and are susceptible to being lysed by the killing factor. The active cells also produce a signaling protein. The delay in sporulation gives the active cells a window of opportunity if the environment were to change back to being more favorable. Sporulation is irreversible soon after initiation. Those cells that delay sporulation have a clear advantage if resources were once again available. However, if the environment continued to worsen then those cells that delayed sporulation may not be able to finish the process.

This is a very interesting study, but it leaves many questions unanswered. Engelberg-Kulka and Hazan (2003), in reviewing the model, asked: how such a system could evolve to benefit bacterial populations. Based on our understanding of evolution and population biology it should seem apparent that evolution does not benefit populations but rather individuals. The question becomes what evolutionary benefit is there to individuals to be active or inactive for the regulatory protein? Is the frequency of active and inactive cells random? Is the killing factor specific for this particular species of bacteria or is it a more general antibiotic? This is a very important question. How closely related are the cells that are killed to those that are releasing the killing factor? Are the active cells derived from a different strain or the same strain as the inactive cells? Consider the former.

If the active cells are from a different strain what we are seeing is intense competition/predation. Under favorable conditions when resources are abundant all cells are able to obtain whatever is needed for growth and reproduction. When starvation sets in then competition is at its most intense levels and selection should favor those cells that are best able to garner the remaining available nutrients even if it requires killing of other cells to capture those nutrients. Those strains affected by the release of the killing factors may be closely related to the releasing cells and as such are the strongest competitors. Selection should be strongest against those cells that have nearly the same

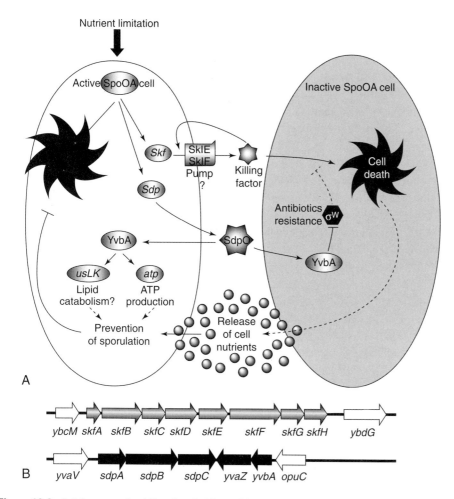

**Figure 10.6** Defying starvation. When faced with conditions of decreased nutrient availability, a subpopulation delays sporulation by activating a regulatory protein, which then switches on a series of operons. (Reprinted with permission from Engelberg-Kulka H, Hazan R. Cannibals defy starvation and avoid sporulation. *Science* 301:467–469, copyright 2003 AAAS.)

nutritional requirements to reduce competition. In this instance individuals are benefited and not the population as a whole.

Let us consider now the situation where the active and inactive cells are derived from the same cell line. Again the question needs to be addressed as to whether active cells are a random subset of the cells or whether active cells are produced by certain cell lineages. In the study by González-Pastor and coworkers, the initial strain was a homogenous mixture of wild-type cells. These populations of wild-type cells developed the active and nonactive phenotype. The antibiotic produced was targeted at siblings. This type of antibiosis (i.e., the killing of genetically identical cells) is unique among microbes. Normally, antibiotics are used against other species. If the colony is considered as a multicellular unit rather than a collection of individual cells, the destruction of some cells to ensure the survival of the multicellular organism is analogous to cell differentiation in higher organisms. There is some evidence that wild strains of this bacterium form multicellular structures in which spore formation occurs primarily at the apical tips similar to other species such as *Myxococcus xanthus*. In these multicellular structures, the release of killing factors and signal molecules may influence the location and timing of spore formation.

The life cycles of many prokaryotes have been shown to have developmental programs that allow them to respond to environmental stress. These programs include formation and differentiation of swarmer cells in *Caulobacter cereus*, spore formation in *Bacillus* and *Streptomyces*, development of bacteroids in *Rhizobium*, heterocysts in *Anabaena*, fruiting bodies in *Myxobacteria*, and the production of viable but non-culturable cells for many Gram-negative bacteria (Hochman, 1997). One interesting observation is the release of "gene transfer agent" from the photosynthetic bacteria *Rhodobacter capsulatus*. These gene transfer agents contain DNA of $3.6 \times 10^6$ molecular weight that is representative of all parts of the genome. This DNA can then be taken up by other strains of *R. capsulatus*.

Not all microorganisms produce spores and yet most microbes have successfully survived physiologically stressful times. The *starvation state* has been proposed as the physiological state that is analogous to the spore in non-spore forming bacteria. The starvation state is discussed in detail by Morita (1997). Here we examine the concept as it affects the population biology of microorganisms—especially bacteria. Our discussion will rely heavily on the concepts and ideas presented in Morita (1997).

## Oligotrophic State of Nature

Our image of nature in many parts of the world is an environment with abundant resources and nutrients. For example, grasslands, forests, tundra, tropics, the littoral zones of some lakes, estuaries, and other coastal habitats all have what appears to be an excess of organic matter. We see large trees, green plants, or an abundance of organic litter produced by these plants. There are exceptions, such as many desert environments and the open oceans. However, most people live nearer to the former list of environments than they do to the latter, and our view of the world is biased to some extent. This bias is probably larger than most people consider.

The world is green in many places. Green is an indicator of plant production. In the temperate regions, this green remains throughout the summer months and is a good indicator that herbivores are limited. In autumn, plant production in the form of leaves accumulates as litter or debris. The forest floor has a variable layer of this litter, but invertebrates and microbes rapidly consume much of this litter.

Plant production is a measure of the amount of organic matter in an environment that is potentially available to support higher trophic levels. However, this notion of the availability of organic matter and other nutrients is misleading. Most of the plant biomass is locked up in large long-term storage units (i.e., the plants). Plant litter is usually composed of fairly recalcitrant, refractory, or hard to break down compounds such as lignin and cellulose. Nitrogen and phosphorous concentrations in plant litter are initially fairly low and increase only because of microbial activity, growth, and biomass accumulation. Leachable compounds include sugars, amino acids, and many other structurally complex molecules that can be inhibitory or toxic to microbes. The readily usable substances such as sugars and amino acids are consumed quickly, and the remaining material is eventually metabolized, although at fairly slow rates.

There is a major difference between a substance being bioavailable and being biodegradable. Given time most organic compounds can be broken down by one or more microorganisms working in concert. Although most organic substrates can be broken down eventually (biodegradation) not all organic substrates even labile ones are necessarily bioavailable. Substrates such as sugars and amino acids that are readily taken up by bacteria in laboratory cultures may not be available because of adsorption to humic substances or clay particles in nature. The total amount of organic matter or nutrients that can be measured in a habitat is probably a very poor estimate of the available material. What may appear to humans as replete with available substrates may be quite the contrary to microorganisms. At the scale of an individual microbe, the amount of available energy and nutrients may be very restricted.

Given the time that higher plants and algae have been producing biomass, if heterotrophic organisms were not somewhat successful in breaking down this material, we would be covered in plant debris. Every time a microbe consumes some material, the environment becomes more *oligotrophic*. The more efficient microbes are at capturing the energy or carbon from plant production, the less material is subsequently available.

Under ideal conditions some bacteria can divide every 10–20 minutes. If even a few species of bacteria grew unlimited at those rates the earth would be covered in bacterial cells. At unchecked growth in 24 hours starting with a single cell there would have been 72 generations and there would be $2.36 \times 10^{21}$ cells and an ending biomass of $1.18 \times 10^8$ g, or more than 8 metric tons! After 100 generations (or only 28 more divisions) of unchecked growth, the total biomass would be $3.17 \times 10^{16}$ g, or nearly 32 trillion metric tons of bacterial biomass.

Nothing grows at those rates unchecked. Energy, carbon, and nutrients become limiting and slow or halt growth. Production of toxic wastes and inhibitory substances would further limit growth. Most organisms do not live where suitable conditions exist to promote even a few generations of accelerated growth. Exceptions include some pathogens and commensal species. Most adult people experience these exceptional growth rates every day. Bacterial populations growing in the gut produce high densities and biomass as part of feces. This excess biomass is continually removed from the gut and replaced by new growth. Few environments have a continuous delivery of food like the gut and so most bacteria either are not active or are only in short spurts with significant times of no growth or reproduction.

Bacteria in nature are much smaller than the same species in laboratory cultures. Miniaturization occurs in some species of bacteria during starvation. Other physiologic changes have been observed in bacteria that are being starved. Given the extreme heterogeneity of most environments even at the scale that researchers have measured, it seems likely that most bacteria live under energy poor conditions often over extended periods of time. Natural selection has probably favored the ability to weather nutrient deprivation because much of the earth's early history was in the absence of higher plants. This means that the diversity of plant produced organic substrates and their input as resources for microbes has been a fairly recent event. Considerable evolutionary events could have transpired before the origin of higher plants and their products so we should probably expect microbes to have mechanisms for surviving tough times especially during periods of nutrient or organic carbon deprivation.

Microbes evolved before the advent of plants and in much more oligotrophic conditions, and they are probably fairly well adapted to chronic starvation conditions. However, based on their incredibly high surface-to-volume ratios, they can react quickly to increases in food availability.

From microscopic observations of samples taken from nature and from laboratory cultures it is evident that most bacterial cells from nature are much, much smaller than the cells from laboratory cultures. Most organisms collected from oligotrophic environments are ultramicrocells, which are cells less than 0.3 μm in diameter. However, some starved bacteria reduce their size to between 0.4 and 0.8 μm when food is not available.

## Starvation-Survival

Morita (1982) defined starvation-survival as a physiologic state that is the result of insufficient amount of nutrients (especially energy) to allow cell growth and reproduction. In some species of bacteria, it has been observed that there may be an actual increase in cell density as starvation begins but the mean size of the cells decreases with each division. Initially, there is reproduction but not growth. This has the combined affect of increasing the number of genetic copies of the species and increasing the surface to volume ratio such that if and when conditions improve the species is perfectly ready to take advantage. Long-term survival under starvation conditions involves a shutdown of metabolic processes.

Laboratory studies have clearly shown a starvation-survival response due to the ability to completely eliminate energy sources. In nature there is usually some energy, although in short supply, available that may be used by some component of the microbial assemblage or community. However, many microbes may be sitting around waiting for the appropriate energy source or nutrients to be made available through the metabolism of other organisms. For some it is a very long wait.

## Aging, Senescence, and Death

Bacteria and other microbes die for the same reasons higher organisms die; for some reason the proper metabolic cycle has been altered either through a cessation or disruption in enzyme or protein synthesis. Microbes also die when eaten by other organisms or when attacked and lysed by bacteriophage. When some process such as oxidation destroys cell functions, death by senescence occurs. However, some authors (e.g., Postgate, 1989) do not feel that there is anything akin to natural senescence in bacteria, because most bacteria do not grow old and die, but rather they seemingly vanish.

At cell division, the two cells that are formed are the same age. There is no older or younger, no parent or child. Both cells have replaced the parent, the parent has vanished, and only the progeny remains. Microbial death by natural mortality has been defined as loss of functional and morphological integrity. Natural mortality includes lysis and breakdown of genetic material.

Bacteria are not immortal, although they can live for a very long time. Within any given microbial population, there are individuals under different stresses that are better able to survive than others. The absence of suitable food or energy (i.e., starvation) is probably a fairly common stress experienced by most microbes. Natural selection will favor those individuals in the population that are best able to withstand starvation conditions. Senescence perhaps should be defined as a cell that is nonviable. However, nonviable cells can retain all of the membranes, enzymes, and genetic material of live cells and if given the right conditions these nonviable organisms can be resuscitated and are capable of subsequent reproduction.

Aging, like senescence, is a difficult concept to grasp in the microbial world. Microbiologists talk of aging in bacterial cultures but the reference is to time of inoculation. When the culture is first inoculated all the cells are of similar age and physiologic condition. Aging is not restricted to cells grown in a batch culture where the media can become exhausted but also includes lag phase cells grown in continuous culture. Physiologic aging may be a more accurate designation. In some cases physiologically old bacteria are better able to withstand stresses than are physiologically new ones.

Are we able to define death in bacteria? Some microbiologists have defined death to be the inability to reproduce. For example, if one were to examine a sample through a microscope and could count some number of cells and then tried to isolate those cells or otherwise get them to grow and divide and none of them grew would you consider all the observable cells to be dead? The immediate response would be no! The researcher just had not found the appropriate media to allow the cells to reproduce. Consider that pure cultures often can enter into a dormant state in which reproduction seems to have ceased. Researchers sometimes have been able to resuscitate some of these cells, which have successfully reproduced. Were the cells dead and revived or in some state other than death? Postgate (1976) has stated, "The basic difference between fissile microbes and organisms with more complex mechanisms of multiplication is that, in the former, death only results from some environmental stress." Environmental stresses are those conditions that lead to a reduction, often sharp, in fitness (Hoffmann and Hercur, 2000).

Can bacterial cells be killed? This is more than a rhetorical question if we consider that the killed state is relative and as stated above indicates an inability to reproduce and not to the complete termination of metabolic activity or potential activity. Some "killed" cells have been resuscitated. Were they dead or just nearly so? Morita (1997) lists several types of killed cells, all of which have been brought back from the dead to living with appropriate conditions. These cells were supposedly killed by means of irradiation, heat, chlorine, hydrogen peroxide, and ethyl alcohol. All were revived by incubation with the proper metabolic components of the tricarboxylic acid cycle. The question remains–what does *dead* mean to bacteria?

Perhaps the best answer lies in considering survival rather than life or death. Survival of an individual bacterium has a clearer meaning than survival of a clonal population. For a species to survive only a single cell somewhere in the environment has to remain or become viable under favorable conditions.

## Dormancy or Resting State and Miniaturization

Dormancy, or the resting state, needs to be considered in any discussion of survival. Dormancy can be under the control of various constitutive genes and includes the formation of spores and cysts. However, some *exogenous* conditions can also cause dormancy. Exogenous control involves changes in the chemical or physical environment in an unfavorable way. Starvation-survival is an example of an exogenous response to adverse environmental conditions. Accordingly, resting cells are those cells where metabolic respiration does not occur or is greatly reduced and where division does not take place. The resting state must be a part of the natural life cycle. If starvation-survival is, as suggested, an ancient condition that has repeatedly been experienced by microbial populations then the cells resulting from this physiologic condition would be considered part of the life cycle of the microorganism.

After nutrient challenge and deprivation, a number of responses have been observed for different microorganisms. These responses can differ among species or genera and within the same species, and the difference seems to be related to how starvation is induced. For some bacteria, several generations may be needed for complete adaptation to nutrient poor conditions. Morita (1997) lists several different responses of bacteria that have been observed to take place when the bacteria are faced with oligotrophic conditions. These responses include

1. Chemotactic or phototactic response by flagellar movement or formation of gas vacuoles
2. Attachment to a surface through adhesion
3. Colonization of favorable ecological niches
4. Dispersal by specialized phases
5. The formation of prosthecae
6. Slower growth rates
7. The ability to take-up nutrients quickly when locally available
8. Formation of ultramicrocells
9. Induction of low endogenous respiration
10. Use of endogenous reserves
11. Regulation of protein turnover

This list is probably not exhaustive but clearly shows a diverse array of responses that different species can take to increase their chances of survival. Of these, the formation of ultramicrocells is difficult to understand on a bioenergetic basis. However, the increases in cell number without a concomitant increase in cell biomass have been observed numerous times. Miniaturization does increase the surface to volume ratio and the potential to respond to fleeting increases in nutrients. These changes in cell volume can be quite significant and approach a 96% reduction in volume (Moyer and Morita, 1989). In every comparison, the starved cells were smaller than the unstarved cells. Within the unstarved or starved class, cells raised under batch conditions were larger than those raised in the chemostats. The size of cells grown in chemostats depended on the dilution rate, with fastest rates having larger cells. At dilution rates as low as 0.015, the cells were still viable and retained the ability to grow.

We have discussed the killing of genetically identical cells in Gram-positive bacteria as a possible mechanism to delay or avoid sporulation. Sporulation is a very affective mechanism to survive harsh conditions. Spores do not require inputs of energy for their maintenance. On the other hand Gram-negative bacteria do not form spores and yet they are faced with harsh conditions. The starvation-survival response may be a way for Gram-negative bacteria to counter these conditions. Starved cells have reduced metabolism but some level of energy input is needed as the physiology of the cells enters the starvation condition and at other times during the apparent quiescent periods.

Cannibalism has been proposed as a mechanism for Gram-negative bacteria to survive under starvation conditions. This type of growth has been called different names over the years. It has been called *cryptic growth*, *regrowth*, and *cannibalism*. However, many studies have been unable to detect DNA or cell lysis products suggesting that perhaps this source of nutrients is somewhat limited in scope. Some have argued that cryptic growth takes place long after growth has ceased and accounts for the survival of populations. Based on the lack of consistent or definitive evidence it is unlikely that cryptic growth contributes much toward the survival of natural populations. Most studies have involved pure cultures of some species grown under laboratory conditions. It is difficult to accept that certain cells "die" and release their intracellular contents so that other like cells can continue to survive. Based on some estimates it would take nearly 50 lysed cells to support the regrowth of one cell. In the mixed complex communities of microbes found in nature it is unlikely that lysed cells of one species would be preferentially used as a nutrient source by other members of that species and not by others.

Another source of energy is the endogenous reserves in the intercellular components of the cell. It is intuitive that in the absences of an exogenous source of energy that a cell would use intercellular reserves. In other words the cell cannot help using this material. Use of endogenous energy is limited and can only provide minimal energy for very short periods of time. Most estimates of endogenous metabolism have been done using laboratory cultures under optimal conditions where energy was plentiful. These conditions probably do not exist in nature. Under laboratory conditions, various intercellular reserves have been observed for several different molecules, including lipids, carbohydrate polymers, and poly-β-hydroxybutyrate. Analyses of cells taken from nature seldom have these intercellular components.

Other intercellular substances could be used by bacteria as an internal source of energy during starvation. Basically all cellular components have been examined including RNA, ribosomes, proteins, and to some extent DNA. However, DNA synthesis is usually favored over the synthesis of other macromolecules. This too seems intuitively obvious as conservation of the genome is essential for survival after starvation. There is some evidence for genome reduction during starvation of some laboratory cultures, which may indicate the ability of the cell to eliminate redundant DNA.

Why do all bacteria not have resting stages as spores or as ultramicrocells? The answer may be that they do, and we have not been able to detect them yet. Alternatively, many species may have sufficient resources to survive under the variability imposed spatially and temporally in their environments. In other words, there are always enough resources for at least one individual to survive without having to produce a resting stage.

# Taxis: Light, Chemicals, Water, and Temperature

The movement of microbes toward or away from a stimulus is called taxis. Movement away from a stimulus is denoted as negative taxis, and movement toward a stimulus is positive taxis. The use of these designations can be confusing because movement toward a stimulus may result in cell death and that result is clearly not positive relative to the cell. Taxis can be in response to many different cues, including light (i.e., specific wavelengths or light in general), various organic and inorganic compounds or molecules (i.e., chemotaxis), pH, redox, and crowding.

Taxis requires motility and the ability to detect the stimulus. Some microbes are able to detect and perhaps create cues that cause behavioral responses in other microbes. Chemotaxis is used primarily as a means of finding food or energy. However, being able to detect harmful or toxic materials and then to move away from that material is clearly an important phenotypic characteristic.

Development of flagella as a means of locomotion can occur in response to low nutrient concentrations. Movement of bacteria toward a source of food ensures growth and increases the chances for genetic exchange and recombination among bacteria, and it increases the dispersion of bacteria. Taxis also moves bacteria away from toxic substances, or it brings assemblages of like cells together to form resistant aggregations or biofilms.

Chemotaxis toward readily metabolizable substances like sugars and amino acids requires a chemoreceptor. The chemoreceptor is always found in the cells even at low concentrations of the substance it can detect. The organism is always responsive to changes in the substrate concentration and does not have to wait for synthesis of the chemoreceptor. Motility of cells growing in habitats with high concentrations of nutrients do not need chemotaxis, and chemotaxis under these circumstances is superfluous. Movement under these conditions wastes energy that could be used for growth and reproduction.

Brownian movement is the random movement of particles in liquids. The net movement of particles by Brownian movement is toward randomness. Particles will, over time, disperse to a nearly random pattern given enough time. Chemotaxis is the movement of the whole organism (as opposed to diffusion of molecules) against the random movement of brownian action. Cells may concentrate or disperse.

Bacterial movement has been described in detail and involves a few different moves that result in different outcomes. Typically, the cell moves rapidly in slightly curved lines. This type of movement is termed a run and it generally lasts approximately one second. After a series of run events, the process is interrupted by a series of short (less than one tenth of a second) tumbles. After each tumble, the bacterium usually starts off in a new direction on its next run. The overall direction of movement is controlled by the frequency of tumbling events. Tumbling and runs are controlled by the direction of movement of the flagella. Counterclockwise movement results in flagellar synchrony and directed movement. When the flagella move in a clockwise direction, the flagella rotate independently and the cell tumbles. This type of movement allows sampling of the environment for both food and detection of toxic or inhibitory substances. When moving toward an attractant, the periods of runs are longer, allowing the cells to make progress toward the material. In contrast, movement away from an attractant or toward increasing concentration of a repellant increases the number of tumbling events, slowing the progress and allowing the cells to reorient toward more favorable conditions.

Bacteria may also move through the process of gliding along some hard surface. Although all of the exact mechanisms involved in this type of movement have not been worked out, much is known that sheds insight on the evolution of movement in microbes. In contrast to the *swimming* behavior of bacteria, which must occur in a liquid medium, *swarming* is organized movement over a solid surface and it involves specialized cells that appear to be adaptive to this phenotypic characteristic. Comparisons between swimming and swarming cells are presented in Figure 10.7. Swarming has been identified in a number of different bacterial taxa including both Gram-negative and Gram-positive organisms. Robert Belas (1997) provided an overview of swarming using *Proteus mirabilis* as a model organism. Most of what is summarized here is provided in much more detail in the review article by Belas (1997).

*Proteus mirabilis* is a member of the Enterobacteriaceae and related to *E. coli* and *Salmonella typhimurium*. When grown in liquid culture, the cells are 1.5- to 2.0-μm motile cells with 6 to 10 *peritrichous* flagella (i.e., flagella distributed over the surface of the bacterium). They are capable of directed taxis using swimming behavior either toward or away from substances. Shortly after being placed on a solid media like agar the cells begin to elongate and can reach lengths of 60 to 80 μm, with little change in cell width. This change in morphology is accomplished by growth without cell division. The result is an elongated polyploidy cell with between $10^3$ and $10^4$ flagella per cell and with the ability to move over a solid media. Interestingly individual swarmer

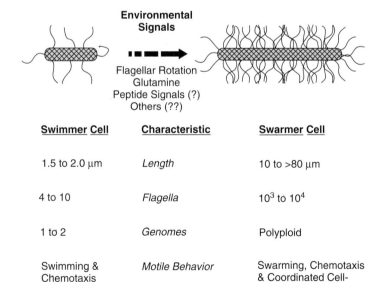

| Swimmer Cell | Characteristic | Swarmer Cell |
|---|---|---|
| 1.5 to 2.0 μm | *Length* | 10 to >80 μm |
| 4 to 10 | *Flagella* | $10^3$ to $10^4$ |
| 1 to 2 | *Genomes* | Polyploid |
| Swimming & Chemotaxis | *Motile Behavior* | Swarming, Chemotaxis & Coordinated Cell-to-Cell Communication |

**Figure 10.7** Swimming and swarming. Swarmer cell differentiation is controlled by the combined sensing of the environment that reduces wild-type flagellar rotation and, through sensing, the amino acid glutamine. The differentiated swarmer cell is elongated and polyploid. (From Belas R. *Proteus mirabilis* and other swarming bacteria. In: Shapiro JA, Dworkin M (eds): *Bacteria as Multicellular Organisms*. Oxford, UK, Oxford University Press, pp. 183–219. Copyright 1997 by Oxford University Press, Inc. Used by permission of Oxford University Press, Inc.)

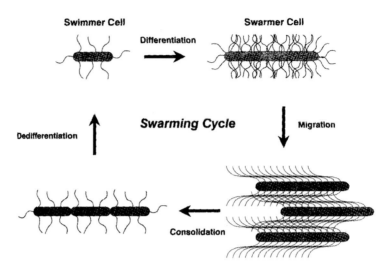

**Figure 10.8** Cyclic nature of swarming or swimming cells. Swarming cell differentiation and motility are cyclical in nature. Once differentiated, the swarmer cells move together. Movement is punctuated with periods of dedifferentiation (i.e., consolidation). (From Belas R. *Proteus mirabilis* and other swarming bacteria. In: Shapiro JA, Dworkin M (eds): *Bacteria as Multicellular Organisms.* Oxford, UK, Oxford University Press, pp. 183–219. Copyright 1997 by Oxford University Press, Inc. Used by permission of Oxford University Press, Inc.)

cells do not have the ability to swarm. The morphology appears to be adaptive only for the population. Movement begins in mass and continues until the swarming mass decreases because of loss of cells that remain behind on the surface. Other losses occur when the swarm changes directions and not all cells are able to stay up with the swarm.

The process of changing from swimming cells to swarming cells is cyclic, that is the swarming cells can return back into swimming cells (Figure 10.8). The process of changing from swimming to swarming is called *differentiation* and as described above involves growth without division or without the forming of septa. Once formed the swarming cells begin to move together in unison. On an agar plate the movement is outward for several hours and then the movement ceases and the swarmer cells *dedifferentiate* back into swimming cells. This process is called *consolidation*. As long as the cells are in contact with the solid surface the process begins again with continuous cycles of alternating swimming and swarmer morphologies until the agar surface is completely covered. If the cells are removed from the solid surface and placed in liquid media they change to swimming cells and remain that way indefinitely.

*Proteus mirabilis* is not the only species that have swarming cells. Other species that have been extensively studied include *Serratia marcescens*, *Vibrio alginolyticus*, and *Vibrio parahaemolyticus*. Swarming has also been observed in *E. coli*, *S. typhimurium*, *Yersinia*, *Aeromonas*, *Rhodospirillium*, *Bacillus*, and *Clostridium*. The cyclical pattern of alternating swimming and swarming cells has not been observed in *V. parahaemolyticus* although cells in the center of the colonies appear to have dedifferentiated.

Through a number of elegant experiments using transposon mutagenesis to obtain mutant strains of swarmer cell genes, some of the mechanisms involved in the process have been elucidated. These studies have shown that physical contact is the primary environmental stimulus for the initiation of the swarmer morphology.

This phenomenon of moving together begs the question: to what are the cells moving to or from? Is the response a chemotactic behavior? Some evidence suggests the importance of chemotaxis in both swarming and differentiation. The data indicate that swarmer cells are chemotactic but not as much as swimming cells. In *V. parahaemolyticus* iron limitation is important in swarmer cell differentiation.

Although these and numerous other genes involved in cell differentiation have been identified, very little is known about the mechanisms involved in controlling swarmer cell movement or motility. Little is known about the factors or genes that control the cycle of differentiation, migration, dedifferentiation, and consolidation.

Swarming has been observed in several taxa of unrelated organisms. The phenomenon may yet be observed in many other taxa of eubacteria. When a phenomenon is observed across many different taxa under similar environmental conditions one can begin to accept the evolutionary importance of the process. Is swarming a mechanism for surface translocation and the colonization of natural environments by these microorganisms?

If we consider that individual swarmer cells are not capable of movement but rather that movement is only possible as part of a group of similar cells, we are left with an important evolutionary question: what is the benefit to the individual cell to be part of a migrating group? Consider the negative aspects first. Being in a large group would necessarily increase the level of intraspecific competition to its maximum. Swimming and swarmer cells demonstrate chemotactic behaviors, but it seems that the number of attractants is higher for the swimming cells. Although chemotaxis by the group may move the colony toward a better food supply, individual cells can also move toward attractants and more of them. As the colony moves or swarms, some cells are left behind. Because individual swarmer cells cannot move, they are at a disadvantage relative to the migrating group because the group may have already consumed much of the available resources.

One further interesting aspect of swarming, found so far only in *P. mirabilis* cultures, is what has been called the *Dienes* phenomenon (named after the discoverer). If two strains of *P. mirabilis* are inoculated onto an agar plate, when they eventually meet, they sometimes form a thin line of demarcation between them. If the line of demarcation forms the strains are said to be incompatible or to display mutual inhibition. If the line does not form but rather the strains merge imperceptibly the strains are designated compatible and are considered identical. The recognition of self or of organisms that are essentially identical is important in an evolutionary context. Regardless of the mechanisms involved in the recognition, the fact that they can know self from non-self suggests a few important points relative to the make-up of the colony. Each colony must be started from a single cell or from a few cells of identical origin. In which case, the colony is essentially many copies of the same individual (excluding random mutations). Movement of the colony by swarming may be a more efficient mechanism to cover relatively large areas quickly and identify and colonize new suitable habitat. The purpose of any behavior or biological mechanism is to produce more copies of the individual into subsequent generations.

Being part of the group helps ensure that some of the cells find suitable habitat for continued growth and reproduction. This phenomenon has been studied in great detail on organisms growing on essentially a uniform surface with a homogenous mixture of nutrients and energy sources. In nature there are probably few surfaces as large as a Petri dish that are uniform in all directions. This behavior may be adaptive over fairly short distances. Since each swarming cell can dedifferentiate into many swimming cells this would be an effective mechanism for moving many copies of a genome toward suitable resources as the colony forages in all directions at once. Although some cells may not find suitable resources, if any are successful, the genome is preserved.

Another bacteria that has a gliding behavior are the *Myxobacteria*. In addition these bacteria form a fruiting body. These bacteria also seem to have the ability to recognize self from nonself. When swarms of *Myxobacteria xanthus* and *Stigmatella aurantiaca* encounter each other they merge but later sort themselves out and eventually form separate fruiting bodies. This level of recognition should be expected. In nature, the mix of various taxa may be very large and being able to sort out who is related and who is not would be important to organisms that form fruiting bodies that produce spores which become the start of the next generation. Failure to recognize self may result in helping totally unrelated organisms reproduce. What happens when the taxonomic resolution is not so clearly defined? What happens when closely related species are allowed to mix?

When *M. xanthus* and *M. virescens* were mixed together, they formed fruiting bodies only with members of their species. They somehow sorted themselves out and formed successful fruiting bodies that produced species-specific spores. It appears that each species produces a bacteriocin that kills the closely related species but not individuals of the same species. In this case, the species demonstrated some level of territoriality by eliminating competing species in the area where reproduction (i.e., fruiting body formation) was to take place.

# Living Together in Communities

**IV**

# Characteristics of Communities and Diversity

11

## Community Structure and Energetics

It is difficult to ascertain the true nature of microbial communities. This problem is more acute for communities than for populations or for the study of individuals. Through modern molecular techniques, we are able to perform in situ hybridizations that can provide visual estimates of the numbers of various species and perhaps some understanding of their spatial distribution. However, in all cases, sampling is destructive, and knowing the actual level of association between organisms identified in a particular sample is difficult at best and perhaps impossible. The act of sampling destroys habitat integrity and increases the homogeneity within a sample. Fine-scale interactions between individual cells or micro-colonies is lost, and although we can with some confidence identify members of different species or groups, we cannot know exactly their spatial distributions or which species were interacting before sampling. Undaunted, we press forward into the study of these important aspects of the ecology of an organism: the interspecific interactions and what factors create and maintain community structure.

Are associations of microbes communities, assemblages, or random haphazard groups? Before we can answer that question, we must define community and assemblage. We also must determine what important ecological and evolutionary interactions take place in these mixed-species groupings. There are levels of organization above ecological communities. It is important to know whether microbes interact with the environment at the ecosystem, biome, or biosphere scales and the extent of any measurable effects. These important considerations require separate treatment.

The study of biology ranges from molecules to astrobiology. In previous chapters, we examined molecular, individual, and population effects brought about by the interaction of molecules and organisms with other molecules and organisms and with the abiotic environment. Extremophiles are a subset of population, community, and ecosystem interactions that possibly demonstrate unique traits and challenges. However, the basic ecological and evolutionary principles discussed in this chapter are relevant to and can be demonstrated in extreme environments.

What is a biological community, and how does it differ from an assemblage? Everything is a matter of scale, and nature can be examined through various nested sieves. What an investigator examines, measures, or counts and the level of inference that can be made from the study depend on the methods used and the coarseness of the sampling regime. For example, the methods used to study large primates are necessarily different from those used to study mice or microbes. Most researchers focus exclusively on a particular group of organisms and to some extent on a few organisms outside the group that may interact with or affect the study group positively or negatively. For example, studies on small mammals such as mice must include attention to their predators (i.e., birds and other mammals) and to their food (i.e., plants) to understand some of the factors that control the distribution and abundance of the mice.

Most researchers sift the environment through a mesh size that captures the organisms of interest. Few small mammal researchers examine the herptofauna (i.e., reptiles and amphibians) or the annelids (i.e., worms), insects (except those used as food), nematodes, or fungi, or bacteria. Plants may or may not be included. For many ecologists, a community is specific to groups of organisms: plant communities, small mammal communities, or insect communities.

A human community is the distribution of humans within a specific geographic area. Within a human community, several sieves can be used to sort the people into groups. Religion, political persuasion, ethnic group, working class, sex, and age are examples of such filters. Within a group, numerous subgroups can form populations, and because of the nature of the filters used, there may be significant overlap between and among groups or populations. If we subdivided the community into males and females and then subdivided those sets into ethnic groups, every individual would be cross-classified into at least two groups and perhaps more, depending on the ethnicity of their parents.

*Community ecology* is the science of groups. Inherent in the study of communities is some estimate of the diversity and abundance of species found within the community. These measures (i.e., species diversity and abundance) are affected by the scale chosen by the researcher. The actual species diversity of a specific location is determined by the size of the area and the taxonomic resolution desired by the researcher. For example, the species diversity of bacterial plankton in the entire Atlantic Ocean is probably very different from the bacterial plankton diversity of a cove along the Maine coast, but the species found in the cove would be a subset of the total species found in the ocean.

We define *taxonomic resolution* in this instance as the breadth of the taxa examined. For example, a narrow taxonomic resolution refers to a study in which only methanotroph bacterial diversity is of interest. A study investigating the diversity of all heterotrophs has broader taxonomic resolution than the methanotroph study, but it still has narrower resolution than a study examining all heterotrophs and their predators. The resolution can continue upward until all prokaryote and eukaryote species are identified. The complete description of species diversity would be an immense undertaking and would require the integration of scientists across all taxonomic groups.

Principles of community ecology should be applicable to any characteristic of a living organism that comes in groups that potentially differ among populations (e.g., genes, proteins). Because of the difficulties in sampling microbes, various molecular techniques have been employed that rely on the detection of genes or DNA sequences. When do we have a community, and when is it just an assemblage? The concept of a community implies interaction among or between groups. Populations can be considered communities if individuals or colonies interact with each other. However, there are environments in which we can sample numerous diverse microbes but in which interactions are probably nonexistent. One such environment is the air. Bacteria and other microbes are distributed throughout the air column and are being transported together, but these organisms never have opportunity to interact. Another habitat in which interactions between the microbes happen infrequently is exemplified by bacteria transported in rivers and streams. They are not growing, reproducing, taking up nutrients, antagonizing other organisms, or participating in myriad other conditions that would indicate interactions. They are just being transported. These groups of species-diverse organisms are assemblages, not communities.

# Species Diversity

Species diversity of a specific location is also a matter of scale. The ability to sample the environment is restricted only by the creative abilities of the researcher. However, there are a few constraints on environmental sampling:

1. Seldom can the entire area be sampled, which means we must subsample or take representative samples across the environmental diversity present in the area.
2. The size of the sampling apparatus affects the outcome (discussed later).
3. The number of samples collected per unit time limits the scope of the study. If the area of interest is large, taking many samples may be logistically impossible. Samples collected from widely spaced locations may differ between locations and because of diurnal patterns. A sample collected at 7:00 AM may be very different from a sample collected at 1:00 PM, even at the same location.
4. A certain amount of time is necessary to process the samples. Although our interest may be in obtaining the best estimate of species diversity, the estimate will be influenced by the degree of difficulty in processing the samples. If we are counting and cataloging flowers, the ability to visually recognize differences among taxa is important and can be applied relatively quickly to samples by trained personnel. If we must rely on various molecular biological techniques to identify representatives of different species or taxonomic groups, the time required for processing can become overwhelming. Often, only a few microbial samples are processed, and our extrapolations to species diversity remain extremely conservative.

The simplest estimate of species diversity is the number of species per unit area or the *species richness*. Species richness can be used to compare and contrast habitats. However, all species, whether they are abundant or rare, are given the same weighting. Species lists from a single location are of minimal value in ecology. It is only through spatial or temporal comparison that these lists provide any meaningful information. Consider a habitat that has 38 species of bacteria. Is this a good habitat or a bad habitat? It would be impossible to tell from that single estimate of species richness. If we were to sample the same habitat multiple times over several years and found 200, 124, 78, and 38 species of bacteria for years 1, 2, 3, and 4, respectively, the estimate of 38 has a context. In this context, it appears that the quality, as determined by species richness, of this habitat is degrading over time. We could as easily compare different habitats sampled within the same time frame. Estimates of bacterial species richness obtained from pools or runs along a stream continuum or comparisons between grassland rhizosphere bacterial communities in short- and tall-grass prairies or between planktonic bacteria in the Sargasso Sea and adjacent open-ocean habitats could be used to make statements about the relative quality of the habitats of comparison. Comparisons between two very different habitats, such as deep-sea hydrothermal vents and glacial ice, can be made but are ecologically meaningless even if both habitats have high or low diversity. Because different evolutionary processes and selective forces have been acting on these communities, we would expect very different species lists.

Species richness fails to convey information on the abundance of each species in an area. Consider the data presented in Table 11.1. Each if these habitats has the same number of species (i.e., five). However, the species diversity of habitat 1 is very different from that of habitats 2 and 3, which are similar to each other.

In this example, species diversity was calculated using the Shannon and Weiner information theory expression

$$H' = \sum_{i=1}^{s} p_i \log p_i$$

where $p_i$ is the relative abundance of species $i$ in the area, and $s$ is the total number of species. The Shannon-Weiner expression is only one of several equations that estimate species diversity. The Shannon-Weiner expression takes into account the abundance of each species and assumes that all species are known from a large community. $H'$ reaches its maximum ($H_{max}$) when each species has the same number of individuals or when the probability of sampling an individual of any species is the same. $H_{max}$ depends on the number of species present and the number of species known; care therefore must be taken in comparing different habitats, especially if the total species are different between the habitats.

In our example, all three habitats have the same number of species, but the total number of individuals per species varies, and each habitat therefore has a different diversity estimate. In habitat 1, species A dominates the samples, and this drives $H'$

**Table 11.1  Density of Different Hypothetical Species in Three Habitats**

| Species | Habitat 1 | Habitat 2 | Habitat 3 |
|---|---|---|---|
| A | 960 | 200 | 100 |
| B | 10 | 200 | 200 |
| C | 10 | 200 | 400 |
| D | 10 | 200 | 200 |
| E | 10 | 200 | 100 |
| Total | 1,000 | 1,000 | 1,000 |
| Species richness | 5 | 5 | 5 |
| Species diversity H' | 0.0178 | 0.698 | 0.640 |

Units for density are numbers of individuals/unit area.

toward zero. The probability of encountering an individual of species A is almost 100%. In habitats such as 1, in which one or a few species dominate, there is a high probability of missing individuals of the less common species. The important question is whether missing those species results in the loss of critical information about the habitat and a loss of ecological understanding. It may seem intuitively obvious that rare organisms do not impact a habitat much because there are not many of them to effect any measurable change. Unfortunately, that assumption is incorrect. Failure to sample rare species may greatly affect our understanding of the community and greatly underestimate our ability to predict community functions. Among higher organisms, large predators are always rare, but their impact can be seen in population fluctuations of the prey species measured over long periods.

In habitats 2 and 3, the diversity estimates are fairly similar. In habitat 2, all the species are evenly distributed and $H' = H_{max}$. The slight differences in species abundance found in habitat 3 results in $H'$ being slightly lower than the maximum. $H_{max}$ varies depending on the number of species. Consider three habitats that have 50, 200, and 400 species, respectively. $H_{max}$ for these three habitats would be 1.69, 2.3, and 2.6, respectively. $H_{max}$ is based on equal abundance of all species. If in the habitat with 400 species there were 20 species that dominated, the diversity estimate would drop accordingly and might be lower than the $H_{max}$ for either of the other habitats. Comparisons made between these three habitats may falsely conclude that the habitat with 400 species is less species diverse than the habitats with 50 or 200 species.

In habitats with clearly dominant organisms, many of the processes performed in this habitat are driven by these organisms. It is only when the dominants change due to environmental alternations or ecological interactions that new processes can emerge. The new processes sometimes are beneficial, and at other times, the products of the new processes may exert a negative influence on the other species. For example, inputs of untreated sewage into streams or rivers can result in a species shift from "native" bacteria to a system dominated by facultative and obligate anaerobes.

Organisms can be distributed randomly throughout an area, or they may be aggregated. Organisms may live in exclusive groups, or they can be interspersed with few or many other species. Figure 11.1 shows a hypothetical community of smiley faces. Each color represents a different species of smiley face. Some of the faces are clumped together (i.e., green, yellow, and dark blue). Some are found only in association with others (i.e., yellow and dark blue). In this example, there are only seven species of smiley faces, and our challenge is to sample this environment and get the best estimate of species diversity. For argument's sake, the area is too large to entirely sample. In this example, we use two different sampling devices (i.e., blue circles) that differ from each other by the area sampled. The smaller device is five times smaller in area than the larger device. If we randomly place a single sampling device within the study plot (panels A and C) and count the number of species present, we obtain two and three species for the small and large samples, respectively. From our omniscient position, it is obvious that a single sample taken by either device is insufficient to capture the species diversity of this habitat. How many samples would we have to take to get a good estimate of the true species diversity of this study area?

Figure 11.2 shows a species area plot. Along the x-axis are increasing numbers of samples. The y-axis is the number of species found in the samples. Each curve is cumulative. The larger sampling device captures the true species diversity much faster than

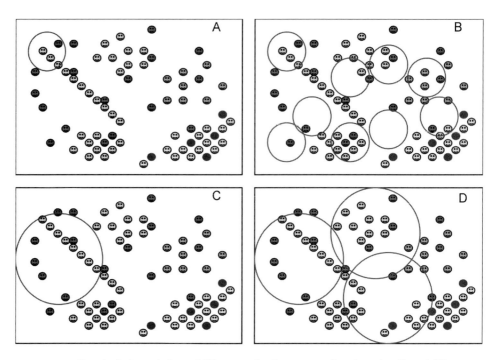

**Figure 11.1** Hypothetical populations of different smiley faces are used to show the effect of different-sized sampling devices on capturing the true level of microbial diversity. (See color insert.)

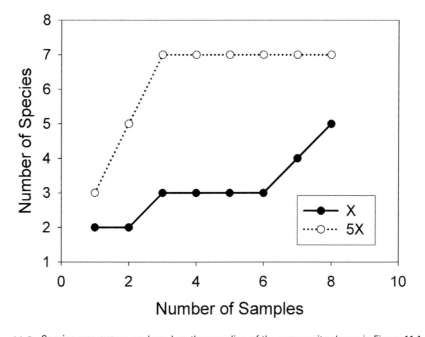

**Figure 11.2** Species area curves are based on the sampling of the community shown in Figure 11.1. *Open circles* are for the large sampling device, and *closed circles* are for the smaller sampling device.

the smaller device. With only three samples using the larger device, all of the diversity of this habitat is known, whereas even after eight samples, the smaller device has captured only five of the species found in the hypothetical habitat. Why not always sample with the large device? The answer is time and money and lack of understanding about the distribution and abundance of the organisms of interest.

In microbial ecology, we have the inherent problem of being unable to see the organisms we are trying to capture and count. Although large samples may detect numerous species, it is impossible to know from environmental samples whether the individuals captured are interacting to any degree with each other because the subtle delicate relationships are disturbed through the sampling process. We are generally unaware of the complexity of the environment we are sampling at the microbial level. Samples may be too coarse or not fine-tuned enough to sample all of the niches occupied by species of microbes. For example, if we are trying to determine the species diversity of a benthic microbial community, how large of a sampler can we effectively manage in the field and process in the laboratory? Is this sampler adequate to sample all of the environmental heterogeneity found at our study site? If our study site is composed of sandy sediments, we may choose one type of sampling device, but if the sediment is cobble or boulders, we would need a very different sampling device. Various types of organic matter may be found on the surface and buried in sediments that may be colonized with different microbes. How do we sample this material? How deep should we sample? Should we restrict our samples to aerobic areas, or do we go deeper and sample anaerobic habitats? Sometimes, the transition from aerobic to anaerobic occurs over millimeters. These and many other questions need to be considered in sampling microbial communities.

Understanding the limitations of the sampling device and the samples themselves is very important in understanding the nature of the communities we seek to describe. Knowing how many samples are needed and where to sample are critical in allowing meaningful inferences to be made about a particular community or population.

Let us say we are interested in describing the bacteria being transported in a stream. If our stream of interest is a small headwater stream that is 0.5 to 1.0 m across and only 10 to 30 cm deep, with a flow rate of $2 \, m^3 \, s^{-1}$, our sampling protocol would be very different than if we were trying to sample the Amazon River at its mouth. At its mouth, the Amazon River is 250 km wide and discharges $178,500 \, m^3 \, s^{-1}$. In the small stream, ten 500-mL samples would be 0.25% of the water passing through the study site at a particular moment, but this amount may still be a sufficient integration of the stream at the time of sampling. However, ten 500-mL samples from the Amazon would be only 0.00028% of the water volume. The question is not whether we believe that such samples are reflective of the Amazon, but rather whether we know that they are not. Differences in water mixing, inputs from other rivers, depth of water sample, season, time of day, and the level of interactions with bank or benthic sediments all would increase the environmental heterogeneity and reduce our ability to provide a true estimate of bacterial abundance and diversity.

Diversity estimates made on higher-level taxa are usually meaningless or nearly so. Genus or family diversity usually gives little insight into a habitat. By definition, these higher taxa integrate across often fairly diverse species. Within a genus, we can have organisms that differ significantly in their physiology and metabolic capabilities.

Consider the Pseudomonads. The *Pseudomonas* genus contains species that can process more than 90 different organic compounds as their sole carbon source and other species that are restricted to a single carbon compound. By restricting our resolution to the genus, we lose significant information about the metabolic potential of the community. The problem is exacerbated as we move up the taxonomic ladder.

Microbial ecologists often screen their samples using molecular techniques. The level of resolution, although theoretical at the species level, is often restricted to available molecular probes. Species-specific probes can be designed, but the effort is time consuming and assumes that the species mix is known. Many studies report data for major groups of bacteria such as the alpha, beta, and gamma Proteobacteria. Other groups have been screened for and are under the same constraints for making meaningful comparisons among or between studies. The data are too coarse. Researchers can circumvent some of these problems by constructing gene libraries and then sequencing the genes in these libraries.

## Maintenance of Species Diversity

Because habitats vary temporally and spatially, it is important to understand how species diversity is maintained. Several hypotheses examine the effect temporal variability of habitats has on species diversity. The "storage effect" model may relate directly to microbial communities. In the basic form of this model, developed by Chesson (1994) and discussed by Hairston et al. (1996), the storage effect is the result of three components: competition between species, existence of a long-lived life history stage, and variation in the temporal recruitment of individuals into the long-lived stage. Species diversity is maintained in this model when temporal fluctuations in the recruitment of the competing species is negatively correlated—one species is doing well while the others are not. However, the long-lived stage permits each species to persist through the periods when its recruitment is poor.

The initial version of the model has been modified such that storage could occur in any life-history stage capable of introducing new recruits into the population. Dormant propagules such as seeds, insect eggs, spores, or cysts all can act as the storage unit. With long-term storage of potential recruits, there can be overlapping generations as recruits are released from dormancy. For microbes that do not form spores, other mechanisms for long-term survival are possible, including miniaturization, as discussed in the section on Living Together in Populations.

The presence of a long-term resting stage allows multiple species to coexist in a habitat, because at any one moment, only a few species are dominant, and the rest are in storage. Any change in the environment that would promote the growth of one species over another would have an available set of new recruits waiting to come out of dormancy. Sampling of these environments and detection of species by any method would show numerous species living together. The species diversity of interacting microbial species and total potential species diversity of the habitat would become almost impossible to tease apart. High levels of diversity could be detected even when the actual number of competing or otherwise interacting species may be much lower.

# Origin and Maintenance of Communities

The noted microbiologist Beijerinck is credited by Baas-Becking, a Dutch microbiologist, with saying "Alles is overall, maar het milieu selecteert" ("Everything is everywhere, but the environment selects"), which is one of the commonly accepted postulates in microbiology. Bacteria-like organisms first came into existence almost 3.8 billion years ago. That is a long period for organisms to disperse and for the environment to select.

The distribution of any organism is a function of biological interactions with other organisms, its ability to disperse, and its likelihood to colonize new habitats. Other nonbiological factors can affect an organism's distribution, such as the movement of continents, creation of or subsiding of islands, volcanic activity, earthquakes, and the formation or elimination of lakes, rivers, or oceans. These large-scale events have clearly affected the distribution of higher organisms, but what about microbes?

Microbes have been "getting there" for 3.8 billion years. During this incredibly long time frame, most of the movement of continents, island formation, and aquatic and marine change has happened. Microbes or at least the progenitors of all extant microbes have been exposed to all of these events and successfully contributed copies of themselves to subsequent generations. Are the assemblages of microbes found throughout the diverse habitats of earth haphazard associations, or are they highly organized, complex systems with finely tuned interactions? Is everything really everywhere?

The spatial heterogeneity of microbial populations and communities affects the results of sampling and therefore the inferences made from the data and their interpretation. Figure 11.3 shows the two-dimensional and three-dimensional distribution of a hypothetical community of microbes. The two-dimensional presentation is the same as that in Figure 11.1 of the smiley faces. From a two-dimensional perspective, there appear to be groupings of organisms with considerable overlap among some of the groups. Sampling in two dimensions may be a valid design for communities such as biofilms, for which the depth of the biofilm is minimal. However, when sampling habitats in which there can be a significant depth effect, we must be aware of the limitations of our sampling regime and tools. Figure 11.3B demonstrates that what appeared to be groupings of species does not really occur. Many of the species are found at unique depths. If we were to sample only one depth, we would miss much of the diversity found in this habitat.

How do we know how many samples to take and at what depths? This is a very important question. Any distributions that are dispersed throughout a three-dimensional medium such as water, sediments, or air require consideration of heterogeneity within the volume being sampled, not just the cross-sectional area. Although inferences can be made from multiple samples taken at a single depth, these inferences are restricted to phenomena occurring at that depth and to no others.

Few microbial studies have investigated the fine-scale distribution of the organisms. Organisms and other parameters may be distributed in space in at least three patterns: homogeneous, random, or aggregated. After an investigator has determined the distribution of an organism, it is possible to infer various ecological and evolutionary scenarios to describe these distributions, their maintenance, and their origins. Danovaro et al. (2001) sought to answer a fundamental question in microbial sampling: Are samples taken at random sufficient to capture the variability of certain

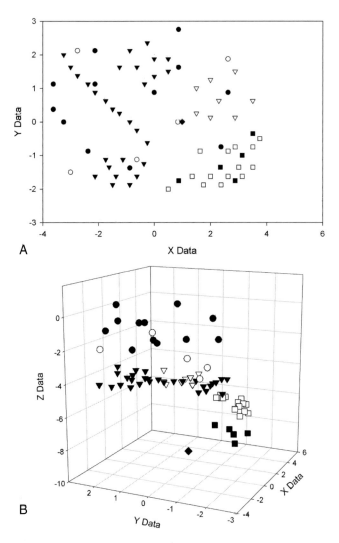

**Figure 11.3** Two-dimensional and three-dimensional distributions of individuals for the same community as shown in Figure 11.1. Notice the differences in species composition with depth.

abiotic and biotic variables? In this study, the authors set out a grid that was 42 × 42 cm and subdivided into 49 6 × 6 cm subplots. Each subplot was sampled by removing the top 1 cm of sediment and analyzed to determine concentrations of sedimentary pigments (i.e., chlorophyll *a* and phaeopigments), proteins, and carbohydrates. Total bacterial counts and the frequency of dividing cells were measured, as was the activity of various extracellular enzymes.

The distribution of every variable measured showed unique and distinct patterns across the study grid. An important but unanswered question is whether these patterns are stable over time, or more importantly, how stable this pattern is spatially. Although the sampling grid was only 42 × 42 cm and the authors sought to describe *fine-tuned* or *small-scale* distributions, we must keep in mind the matter of scale. To a single average microbe, such as one 0.2 μm long, the amount of habitat sampled by taking the top 1 cm would be equivalent to taking a $6.2 \times 6.2$ km sample for a mouse 12.5 cm long. In other words, the sample is not very "small scale."

Forty-nine samples were taken from the grid in this study. What would the patterns have been if we were to take 49 samples from a single sector of the grid ($6 \times 6$ cm)? Would we find unique spatial patterns over that sampling scale? Could we go back 10 minutes, 10 hours, or 10 days later and find similar patterns? Unfortunately, we cannot sample the same exact location twice, which brings us back to the statistical problem of getting enough representative samples to allow us to make inferences about similar habitats. We can never know whether our samples provide a true estimate. However, we can have a certain level of confidence that our samples are representative. Most microbial studies are not able to determine the confidence of their results because too few samples are taken or because no replicates are taken. This study examined 49 samples, but no replicate samples were taken.

To get a good estimate of whether there are spatial patterns, we would need to set out at least three replicate grids and then look at the average response. In reality, we would hope that the placement of the three grids would be in as similar as possible locations so that differences and patterns observed can be attributed to the gradients inherent within the grids, not to major differences among the grids. It is not the number of samples taken that allow us to determine patterns, but the number of replicates of each sample. Assemblages or communities exist spatially and temporally, and it is essential that we get meaningful estimates of both.

Many studies seek to compare and contrast different communities within, between, or among study sites. Several molecular biological techniques have been developed that allow comparisons to be made without culturing the organisms. The methods rely on patterns in RNA or DNA to be used to discriminate samples. These methods are particularly important because some of them allow meaningful evolutionary relationships to be detailed.

Bacterioplankton community structure may at first glance appear to be among the easiest to determine what factors are controlling the observed distributions. However, many factors may contribute to differences in community structure, and they are often difficult to measure. Sometimes, researchers do not find relationships between the variables they measure and the components of community structure. Let us consider freshwater bacterioplankton communities. In one study (Lindström, 2000), community structure of bacterioplankton was measured using denaturing gradient gel electrophoresis (DGGE) profiles. These patterns were then analyzed in relation to physical, chemical, and biological data that had been concurrently collected. Three variables were found to strongly correlate with the DGGE patterns: biomass of microzooplankton (i.e., predators), cryptophytes, and chrysophytes. Two of these factors were significantly correlated with the trophic status of the lakes, suggesting that the nutrient content of the lakes may indirectly influence the bacterioplankton community.

# Effect of Diversity on Ecosystem Services

Biodiversity, or "the variety of life at all levels, from the level of genetic variation within and among species to the level of variation within and among ecosystems and biomes" (Tilman, 1997), should be considered as an asset. Conservation of biological diversity is an investment and the disregard of conservation a path to reduction of valuable services and increased economic costs (Fromm, 2000). The ecological relevance of biodiversity as applied to long-term stewardship and land management must be considered from the services of biodiversity in the context of specific complementary relationships (Fromm, 2000):

- A complementary relationship of species in their habitat
- A complementary relationship of biotic and abiotic components of ecosystem structures
- The complementary relationship of the ecological functions of ecosystems and the contribution of the services of ecosystems to human welfare

Long-term stewardship or land management decisions or practices that affect these complementary relationships may result in decreased ecosystem services, including desirable remediation services. However, only the third relationship is generally considered in management decisions. Because of the complex, highly interconnectedness of ecosystems, many interactions are inherently nonlinear. Because of these nonlinearities, minor modifications in dynamics can become magnified through interactions and lead to uncertainties in the rates of various processes and the direction of change of the system (Chee, 2004).

Resilience of an ecosystem is the ability of the system to maintain certain characteristic patterns, functions, structures, or rates of certain processes despite disturbance or perturbation. Certain partially redundant control processes act at various scales to mitigate the effects of perturbation (Carpenter and Cottingham, 1997). Thus the relative contribution of a particular ecosystem service might be variable over a range of time and spatial scales (Carpenter and Cottingham, 1997). Alternative states can exist for some ecosystems and the response of those ecosystems to the impacts of anthropogenic activities can vary from smooth to discontinuous (Scheffer et al., 2001; Scheffer and Carpenter, 2003). If normal resilience mechanisms fail, new mechanisms may supersede them and produce qualitative changes in the ecosystem. These changes may be stable but undesirable to humans because the system no longer performs the desired services.

Certain ecosystem functions appear to be maintained through more than one species. The concepts of functional redundancy and functional diversity are critical components of certain remediation strategies (e.g., monitored natural attenuation). However, the stability (temporal or spatial) of these apparent redundancies is not known. Specific functional types may be used to predict ecosystem responses to perturbations or where management strategies seek to manipulate certain species to increase specific functions (Loreau, 2001). A major unanswered challenge is to determine the interactions among biodiversity, ecosystem processes, and abiotic factors as a means of predicting effects of disturbance, climate change, or other perturbations.

Obtaining estimates of bacterial biodiversity is a prerequisite for the study of bacterial community assembly and maintenance and for estimating the limits of functional redundancy. These estimates are central to understanding the ecology of surface waters, oceans, soils, and global elemental cycles, including cycling and remediation of important contaminants (Curtis et al., 2002). However, there are few reliable data sets on the relative abundance of even the most abundant members of microbial communities at any level of scale (Curtis et al., 2002). The relative abundance of sequences in clone libraries is the best information available to estimate relative abundance of bacterial species. Unfortunately, almost all reports on relative abundance in clone libraries are bascd on a single sample (Curtis et al., 2002), and there are no estimates of between- or within-sample variation. Any meaningful interpretation of the diversity estimates is not possible or does not have suitable context to determine whether the estimate is high, low, changing, or remaining constant. Because the context of the measurement is not available, it is not possible to determine whether there are any effects of perturbation on potential services.

# Molecular Techniques and Microbial Community Ecology

The field of microbial ecology has made significant strides forward because of the development of techniques that rely on fundamental evolutionary and biological processes. It is no longer necessary to culture a microorganism to detect its presence in a sample. This is a blessing and a hindrance. Early microbiologists spent much time and energy in developing appropriate media that would allow a specific organism to be cultured. To isolate a pure culture of some microorganism, some knowledge about its physical environment was required. Microbiologists also had to determine what substances were used for energy and carbon. These tasks were daunting, and it is a credit to the early microbiologists that many organisms were isolated. The need for pure cultures was especially important in medicine. If a pathogen could be isolated and identified, the probability of successful treatment was increased. However, many other microbes were isolated because of the intense interest and perseverance of various scientists. Included among these nonmedical isolates were bacteria involved in nitrogen fixation or sulfur reduction, the colorless and red sulfur bacteria, and many other aerobic bacteria involved in mineralization processes.

Despite the best efforts of researchers, the numbers of bacteria determined by plating and the numbers observed through direct microscopy were always very different. On average, less than 10% of the bacteria in a directly counted sample could be cultured, and that number is often less that 1%. For example, if a researcher collected a water sample from a pond and prepared and stained the sample with some nucleic acid stain (e.g., DAPI) and then counted the number of fluorescing bacteria by epi-fluorescent microscopy, he or she might expect between $10^5$ and $10^6$ bacteria per milliliter. The variations in actual counts would depend on the physical and biological conditions of the sampled environment. If the researcher plated an aliquot of the sample on some medium, the expected number of colonies forming would be between $10^3$ and $10^4$. What happened to all of the other bacteria seen under the microscope? Are the bacteria grown on the medium representative of the total bacterial community or a subset? If they are only a subset, are the forms that grow important in processes, or are they dormant forms that have been revived through culturing?

When a new habitat opens up in a terrestrial system, certain "weedy" species are the first to colonize. Are the bacteria that grow on the various selective growth media "weedy" forms of bacteria? This concept is germane to this discussion because it relates to the problem of determining the nature of the microbial community. If the colonies that form are representative of the type and diversity of the entire community, we need only use some scalar to multiply the counts by to obtain a fairly reliable estimate of the community structure. If the types that grow on the media are not representative of the entire community, we would not be able to make inferences about the structure and functioning of the community.

Under a microscope, there are some observable differences in morphology of microbes. However, these differences are minor and do not reflect the true diversity of the samples. It has been said that if the shape of bacteria is unknown, choose rod-shaped bacteria, and you will frequently be right. The size and shape of an organism are but two aspects of the organism's phenotype that may be used to detect similarities and differences. It has been known for many years that there are fundamental differences in the cell morphology of bacteria based on a simple staining procedure: the Gram stain. The designation of Gram-positive or Gram-negative types gives a coarse but consistent separation of bacterial taxa.

As researchers sought to determine the "true" diversity of microbial systems, their studies led them to examinations of various molecules. All living things have certain molecules, and it has been known since the 1950s that these molecules maintain their integrity (i.e., structure and conformation) within a taxon. The information molecules such as DNA, RNA, and proteins can be used to determine levels of relatedness.

The field of molecular ecology is expanding at an ever increasing rate. New techniques are continually being developed that expand the capacity, reliability, and speed of these methods in microbial ecology. In the following sections, I discuss many theoretical and practical applications of these techniques because it is important to understand what kinds of questions can be answered using molecular techniques and what the constraints are. It is important to understand the limitations and capabilities of a method before using it. Failure to understand limitations often hinders scientific progress. Students in microbial ecology must learn how to perform the different techniques used in research and become familiar with terms so that they can understand most of the papers appearing in the literature.

Many techniques are based on the same basic molecular biological principle: Differences in the structure of a specific gene or DNA sequence indicate taxonomic differences. The resolution of the various techniques is method specific, as is the ease of application. Most techniques are elegant but time consuming and to some degree labor intensive, and aspects of the procedures can be expensive. These factors combine to limit the sample size used for many determinations in microbial ecology. Numerous studies have been done on single environmental samples. The paucity of samples severely limits our ability to truly describe the extent of microbial diversity, and in many cases, it questions the inferences about community structure.

Most of these new techniques are technically and visually impressive. Unfortunately, much research is technique driven and not question driven. An old adage states that to a man who only has a hammer, everything looks like a nail. Students must first learn to ask interesting questions and then seek out techniques that will allow them to answer those questions. Sometimes, the "best" approach may be very low tech, and at other times, extremely sophisticated methods are needed to answer the questions. Applying a sophisticated approach simply because the researcher knows how to do it is not good science and does little to further our understanding of nature and the world. I will now step off the soap box and describe some of the new wonderful methods that have allowed insights into the microbial world. However, because I cannot spend too much time on this topic, readers are encouraged to seek out the many reviews and methods papers available.

## Methods Based on DNA or RNA

In the early 1990s, scientists began to use molecular techniques to elucidate relationships in microbial ecology. Based on pure evolutionary and genetic principles, DNA was an ideal molecule to use to detect microbial diversity, to assess population changes in space and time, and to compare and contrast specific microbial communities. The base-pairs that make up a DNA strand, especially those found in specific gene *sequences*, are maintained through repair mechanisms that remain fairly constant over time. This constancy is referred to as *genetic conservation*. Most deviations have a negative impact on the organism. However, over time, differences in gene sequences and in the location of the gene on the chromosome have occurred. By sequencing the genes or strands of DNA and comparing between and among different sources of DNA, we can determine the amount of homology between the samples.

The cloning and sequencing of *16S ribosomal RNA* (rRNA) is among the most powerful techniques to determine microbial diversity (Figure 11.4). However, cloning and sequencing are labor intensive, and even with the availability of high-throughput instrumentation, they are expensive and do not give information on populations or community dynamics in space or time. These techniques can be performed on extracted DNA or RNA, or they can be performed in situ. One problem associated with in situ or extracted nucleic acid hybridization is the lack of sensitivity. If the sequences of interest are not present in high enough copy number, they probably will not be detected. The use of the polymerase chain reaction (PCR) eliminated this problem. PCR-based techniques can use DNA extracted directly from the

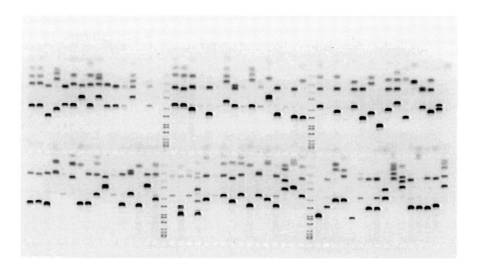

**Figure 11.4** Polymerase chain reaction gel of 16S rRNA for eubacterial DNA cut with restriction enzymes. The patterns are the migrations of different-sized pieces of DNA cut with restriction enzymes.

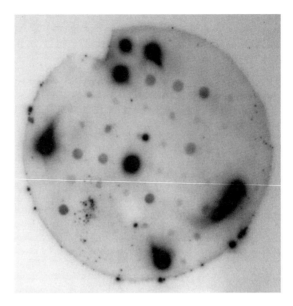

**Figure 11.5** Dot blots resulting from DNA:DNA hybridizations. The intensity of dot indicates the amount of similar DNA found between the sample and a species specific probe for *Burkholderia cepacia*.

environment as the template, or mRNA can be reverse transcribed into cDNA and then amplified. The amplified product can then be screened by hybridization to a variety of oligonucleotide probes that are specific to various levels of information captured in the environmental DNA.

Hybridization uses specific *oligonucleotide* probes to screen environmental DNA samples to determine the presence of various groups of organisms. Oligonucleotides (*oligo* means few) are short strands of DNA with known sequences. The level of resolution offered by this approach spans the range of too specific to too general. Each probe is based on sequence data (i.e., you must know something about the DNA). The choice of probes limits the information obtained (Figure 11.5). If the probes are too specific, the researcher may not detect any sequences, and he or she would know that the specific species is probably not in that specific sample. However, if the probes are too general, ecologically important differences among strains or species would not be detected. For example, if researchers had an oligonucleotide probe for primates and they had DNA from chimpanzees, apes, and humans, the resolution would be that primates are in the sample. Differences in the ecological impacts of the various primate species would be lost. However, if the researchers had a probe only for lemurs, they could conclude only that lemurs were or were not in the sample. They could not tell whether chimps, apes, or humans were there or not.

To offset these limitations, techniques have been developed that take advantage of other properties of the DNA molecule. One such method is genetic fingerprinting. Genetic fingerprinting is a set of methods that rely on the *melting* property of DNA. DNA strands separate at different rates based on the proportions of the various nitrogenous bases that make up the DNA strand. Melting can be accomplished by

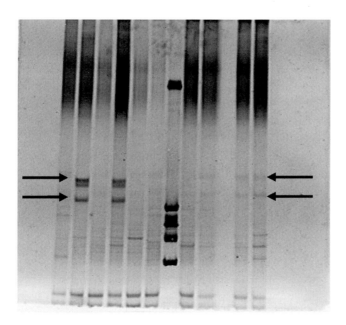

**Figure 11.6** Denaturing gradient gel electrophoresis gel of environmental DNA. Bands indicate different-sized pieces of DNA. The *dark-stained bands* on the left can be seen as faint lines on the right side of the gel.

the application of heat or by various denaturants. DGGE and temperature gradient gel electrophoresis (TGGE) are two similar methods based on the electrophoretic separation of PCR-amplified 16S rRNA gene obtained from environmental samples (Figure 11.6). The environmental DNA is extracted and then amplified using PCR with universal primers for the 16S rRNA sequences. In theory, these techniques should be able to resolve single base-pair differences in DNA. The techniques are relatively inexpensive, reliable, usually reproducible, and relatively rapid. These characteristics make it possible to investigate spatial and temporal changes in microbial populations or communities.

Restriction fragment length polymorphism (RFLP) relies on polymorphisms in DNA. PCR-amplified DNA is digested or cut with specific restriction enzymes, and the resulting fragment lengths are detected using agarose or non-denaturing polyacrylamide gel electrophoresis. RFLP patterns can be used to screen clones or estimate microbial community structure. The method can be used to detect community structure but not diversity or specific groups of organisms. Terminal restriction fragment length polymorphism (T-RFLP) has been developed to address some of the limitations of RFLP analyses.

T-RFLP uses the same basic methods as RFLP, except that on PCR, the primer is labeled with a fluorescent dye. Only the labeled restriction fragment can be detected, which simplifies the banding patterns. Each visible band represents a single operational taxonomic unit, or ribotype. The method has been used to measure spatial and temporal changes in microbial communities, to study complex communities, to detect and monitor populations in population ecology, and to estimate diversity.

## Methods Based on Fatty Acids or Lipids

Every bacterial cell (and every other cell) has a membrane that contains various lipids. The cell wall of microorganisms also contains assorted lipids. Many of these lipids are indicative of specific groups of microorganisms and can be used as signature bio-markers or as indicators of specific groups of microorganisms found in complex mixtures such as particulate organic matter in streams (Table 11.2). The detection of one of these biomarkers is a culture-independent method of characterizing microbial communities. The method is based on the extraction of the lipid biomarkers from the cell walls and membranes of microorganisms. The technique can be applied to most samples collected from nature. The method is quantifiable and provides estimates of viable biomass and community structure, including prokaryote and eukaryote organisms, and it provides an indication of the physiological state of the microbes through the analysis of various stress indicators.

Table 11.2  Concentration of Major Fatty Acids in Stream Suspended Particulate Matter at the Upper Three Runs Creek Site at Each Sampling Date

| Fatty Acid[a] | Mean Concentration ($\mu L^{-1}$) | | | | |
|---|---|---|---|---|---|
| | January | March | April | July | November |
| 12:0 | 0.14 | 0.20 | 0.18 | 0.17 | 0.26 |
| 14:0 | 0.36 | 0.72 | 0.50 | 0.45 | 0.51 |
| i-15:0 | 0.29 | 0.48 | 0.47 | 0.56 | 0.70 |
| a-15:0 | 0.21 | 0.35 | 0.35 | 0.36 | 0.52 |
| 15:0 | 0.09 | 0.12 | 0.15 | <0.05 | <0.05 |
| β-OH 14:0 | 0.11 | 0.19 | 0.19 | 0.17 | 0.16 |
| 16:1Δ9 | 1.09 | 2.07 | 1.09 | 1.00 | 2.76 |
| 16:0 | 2.04 | 3.19 | 3.09 | 2.73 | 4.16 |
| 17:0 | 0.08 | 0.25 | 0.10 | 0.10 | 0.14 |
| 18:2Δ9,12 | 0.45 | 2.70 | 0.62 | 0.42 | 1.24 |
| 18:1Δ9 | 1.95 | 1.06 | 2.57 | 1.58 | 4.94 |
| 18:0 | 0.37 | 0.61 | 0.70 | 0.52 | 0.55 |
| 20:0 | 0.24 | 0.74 | 0.47 | 0.41 | 0.45 |
| 21:0 | 0.07 | 0.09 | 0.43 | 0.11 | 0.15 |
| 22:0 | 0.35 | 1.07 | 0.77 | <0.05 | <0.05 |
| 23:0 | 0.13 | 0.21 | 0.62 | 0.38 | 0.41 |
| 24:0 | 0.53 | 1.21 | 1.42 | 1.29 | 1.04 |
| 25:0 | 0.09 | 0.20 | 0.22 | 0.22 | 0.22 |
| 26:0 | 0.34 | 1.20 | 1.23 | 1.11 | 0.77 |
| 27:0 | 0.06 | 0.20 | 0.20 | 0.17 | 0.17 |
| 28:0 | 0.21 | 1.24 | 0.27 | 0.81 | 0.35 |
| 29:0 | 0.04 | 0.29 | 0.23 | 0.16 | 0.22 |
| 30:0 | 0.08 | 1.00 | 0.42 | 0.34 | 0.70 |
| Total identified[b] | 9.6 | 19.4 | 17.3 | 13.6 | 21.2 |
| Total FA[c] | 13 ± 1.6 | 24 ± 0.4 | 21.6 ± 2.0 | 22.0 ± 1.5 | 25.8 ± 4.4 |
| Total FA (μg) SPM (mg) | 0.35 | 0.31 | 0.25 | 0.16 | 0.53 |

FA, fatty acid; SPM, suspended particulate matter.
[a]Fatty acid designations are given as the carbon number:number of double bonds; Δ indicates the position of double bonds; i indicates isomethyl branched forms; and a indicates ante-isomethyl branched forms.
[b]Total identified includes major components listed in the table and minor components not listed.
[c]Mean ± standard error.

Detailed procedures are readily available to aid researchers attempting to use these methods. Several laboratories can perform the extractions and analyses at a price.

## Methods Based on Function or Physiology

Microbial communities can be described independent of the specific organisms. Communities are collections of organisms that each perform various metabolic functions. Some functions appear to be shared among many groups. By characterizing the functional breadth of a community, differences become apparent. Not all communities of microbes are capable of the same metabolism. This is especially true of specific communities, such as methanotrophs, the community of ammonifiers, or the community of thermophiles, but what of the whole assemblage of all microbes within a location or habitat? Would we expect the bacteria in a water sample from a northern lake to have the same metabolic potential as bacteria collected from a tropical lake in India? Probably not. These habitats are so different that without testing we might expect differences. At what scale would our intuition begin to fail—latitude, elevation, vegetational gradients, littoral versus pelagic, Arctic versus Antarctic? What about differences by depth? Or season? Or stress level?

To fully characterize any community using molecular techniques would be prohibitive in cost and time. Because genetic differences should be captured, with certain variance, by the phenotypic profile, we should be able to describe similarity and differences among microbial communities by describing differences in metabolic profiles. This is the theoretical basis of procedures that use various metabolic tests to discriminate among communities.

Most of these technologies were developed to describe and characterize individual isolates of bacteria taken from clinical or environmental samples. Based on the response of the isolate to a series of metabolic tests, a numerical response profile was generated. After many isolates of many strains and species of bacteria had been screened, certain repeatable patterns were found in the data. Metabolic profiles of strains were more similar than metabolic profiles of related species and very different from totally unrelated species.

Under clinical conditions, it is important to correctly identify an isolate. However, there are many microbes in nature that are not found in infectious disease units. Microbial ecologists and environmental microbiologists were interested in characterizing the microbial organisms important in their studies. It was not long before these scientists began to characterize environmental isolates and found repeatable patterns. Could this technology be used to characterize an entire community?

Before answering the question, a description of the technique is needed. Although different in detail, the basic idea of these techniques is the use of a series of wells that each contain a specific organic compound. If an introduced suspension of bacteria is able to mineralize this compound, a tetrazolium dye is released that is visible to the naked eye and to various sensors that can capture minute differences in various wavelengths of light. The unique pattern is then given a numerical score. Figure 11.7 shows a hypothetical response of a bacterium.

From this example, the organism is able to use the compounds found in wells A1, A2, A5, B3, B4, B5, and B6. However, there appear to be differences in the intensity of the response, with well A2 being much more intense and well B4 much less. Is the

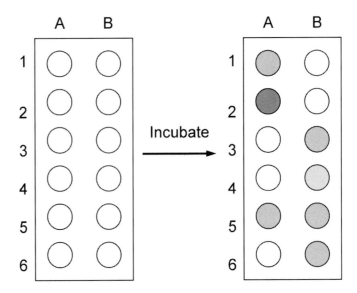

**Figure 11.7** Physiological test strip. Each *circle* represents a well that contains a unique carbon compound. *Shaded circles* indicate growth of bacteria on the substrate in that particular well. The intensity of the shading represents a graded response to the substrate.

level of color intensity diagnostic of the sample, the isolate, or the community of interest? If we repeated this test for several unique isolates of our target organism taken from similar samples and found that the response did not vary, we could use the numerical score obtained as an identifier for that organism. If we took an unknown isolate and screened it with the test strip and obtained the same numerical score, we might place a putative identification on the unknown because of the similar metabolism to the known.

To compare communities, representative subsamples are required. This is an extremely important assumption. With a single isolate, the number of cells going into the well can be known. Each well on the plate has the same or nearly identical inoculum. However, homogeneity is assumed between inocula with community screening. In other words, every sample should have the same distribution and abundance of microbial species. We can assume with some level of certainty that the number and types of dominant organisms in the community may be similar. However, rare individuals may differ between subsamples. If a rare bacterium is capable of using the substrate in the well under the specific incubation conditions, some differences among "replicate" plates may be attributed to these rare bacteria. In statistics, replicates are identical experimental units. The fact that the initial inoculation mix may vary between plates and among wells puts the equality of replicates in doubt.

The use of functional physiologic response plates to discriminate communities has provided some useful insights into various aspects of community ecology. For example, Merkely et al. (2004) compared the microbial communities found in high desert wetlands in the Great Basin of the United States. In this study, samples were collected from wetlands that had been fenced to prevent cattle grazing and from other wetlands where cattle were allowed complete access. Comparisons were made using metabolic test strips and using DGGE. Unique metabolic patterns were observed for the general categorizations of grazed and ungrazed, even though these wetlands differed in location, size, and to some extent, vegetation. The strips indicated that there was higher metabolic diversity in the grazed wetlands than in the ungrazed area. This probably resulted from a number of factors, including the input of semiprocessed organic matter that had passed through the cow's digestive systems; stirring up of the bottom sediments, with the subsequent release of nutrients; and changes in the submerged and emergent aquatic vegetation. The DGGE analyses, which were much more time consuming, basically demonstrated the same patterns. This is a good example of how a simple, straightforward technique can provide meaningful information about microbial communities as long as the limitations of the method are understood and appreciated in the interpretation of the results.

# Successional Theory

Successional theory suggests that site availability, differential species availability, and species performance all affect the outcome of succession. Site availability is a relative notion. What is an available site? Succession can occur over large geographic scales. Mt. Saint Helens is one example. On the other extreme, small sites such as tree gaps can open on the side of a mountain. At the microbial scale, the number of available sites for colonization may be infinite. All activities, whether biological or abiotic, that remove microbial species from an area would result in microscale succession occurring at that site. For example, certain invertebrates have specialized mouth parts or modified forearms that allow them to feed by scraping biofilms. These animals can effectively clean small areas of all biofilm or aufwuchs. These open areas would then be available for colonization by transported bacteria or by movement from bacteria in adjacent undisturbed areas. In either case, the diversity of bacteria first entering the open area would be a subset of all available bacteria being transported or in adjacent areas. Biofilm formation is discussed later and so mention here only that biofilm development often follows certain patterns. In other words, bacteria are not all equal in their abilities to colonize new habitat.

It may be difficult or impossible to tell whether a particular habitat or location is subject to further colonization. Site modification by resident organisms may increase or decrease the suitability of a particular location to continued colonization. However, given enough time, all locations are susceptible to microbial colonization. In the short term, many habitats are not available for colonization because of biotic or abiotic factors that do not promote further colonization.

Differential species availability is the second aspect of successional theory that affects the outcomes of succession. Even though a species may be detected in a habitat, that species may not be available to colonize new habitats. Our ability to detect microbial species requires that our environmental sample loses its structure, and native heterogeneity therefore decreases or is completely lost, depending on the sample and the sample processing. Although we may detect a species, that species may in nature be unavailable for colonization because of spatial distances or biological interactions or because the species is incapable of directed or undirected random movement toward the new habitat. Microbes have differential abilities to disperse. Let us consider some of the mechanisms that microbes use to disperse.

Although bacteria and other microbes are widely dispersed in nature, it appears that not all bacteria are found everywhere. Whether transported and imported bacteria are capable of survival under new or novel environmental conditions is unknown. In freshwater lotic ecosystems, many bacteria in transport are *allochthonous*, having originated from neighboring terrestrial systems and washed into the aquatic system. Many of these bacteria are not actively growing and presumably contribute little to any ecosystem process.

Bacteria and other microbes that successfully replicate within a system can take advantage of dispersal mechanisms to move both longitudinally within a habitat and to escape a habitat. Bacteria can also actively or passively use dispersal vectors such as formation of aerosols, invection, naturally occurring organic foams, arthropods, and vertebrates.

The third aspect of succession is species performance. By chance, a species may find an open habitat. Can that species effectively compete with the microbes already established? Do they have the ability to take up nutrients in the forms and concentrations found at the site? Perhaps they can use the available resources, but their growth rate is too slow to capture space. New habitats present numerous challenges that must first be met by the genes carried by the microbe. Movement of genes from related or non-related taxa to an individual is one way to cope with new or changing environments. However, not all microbes can take advantage of lateral gene exchange, and not all species are capable of transferring genes. Species performance is therefore a measure of the biological and ecological potential of a species.

Species must have the ability to compete for space and food, to ward off predation, to move toward better resources, to take up or cull out genes, to take up naked DNA, and to withstand infections. Along with this list of their requirements, the temperature maxima and minima of species and their salinity requirements make it clear that surviving in a new habitat is not an easy undertaking. Species may disperse into a habitat, but the likelihood of successful colonization may be remote. An exception to these constraints would be when a new habitat or sufficiently large habitat opens up such that species that randomly reach the new area have little to no competition for space or resources. One interesting question is whether the final species assemblage or community can be reached from a variety of starting points. Increases in species diversity follow patterns similar to the species area curves presented earlier. Instead of the area sampled, the x-axis is time. At increasing time scales, the number of new species reaching the habitat decreases and eventually becomes asymptotic. The question is whether the same species mix can be achieved through multiple assemblage trajectories. Given large amounts of time, we might expect that all microbial communities in similar habitats to approach similarity, but this has yet to be determined.

## Abiotic Mechanisms of Dispersal

Long-distance dispersal depends on the movement of bacteria within a habitat and on whether they can exit and survive outside of the habitat. Bacteria can effectively escape aquatic and terrestrial environments in several ways.

The *formation of aerosols* is a function of the geology of a watercourse. Any turbulence caused by rocks, boulders, and woody structures that make water splash or wave action results in aerosols being formed. Depending on the size of the droplets, the aerosols are transported to various degrees into the atmosphere. The types of bacterial species found in aerosols should be proportional to those normally found and those being transported in the water. This has been shown by sampling the air over known sources of enteric bacteria. Aerosol formation below a sewage treatment plant outfall would be expected to have higher proportions of enteric bacteria than aerosols created upstream or far downstream of an outfall. Rosas et al. (1997) sampled the air over sewage treatment plants and at locations various distances from the plant, and they found that highest numbers of pathogenic microorganisms were found closest to the plant.

*Organic foams* can be found in pristine and contaminated streams and beaches. These foams can contain up to three orders of magnitude higher concentrations of bacteria than the underlying water (Hamilton and Lenton, 1998). Bacteria aid in the formation of these foams, and selection may have favored this process as an aid in their dispersal. Air sampled immediately over naturally occurring foams had much higher densities of bacteria than air sampled over open water in two streams in South Carolina (McArthur, unpublished data). The numbers of bacteria were 1000 times higher in the foam than in the stream water.

*Arthropods and vertebrates* may assist in the transport and dispersal of bacteria in aquatic systems. The movement of juvenile or adult aquatic insects exiting the water through hatching may be one mechanism of moving waterborne bacteria out of the water and into the air. Insect activity may also increase the release of bacteria from biofilms (Leff et al., 1994), whereas fish have been shown to have many species of bacteria associated with their surfaces (Pettibone et al., 1996; Son et al., 1997). For example, fish that feed in or disturb sediments have higher proportions of antibiotic-resistant *Aeromonas* bacteria on their surfaces than fish that feed primarily in the water column (McArthur, unpublished data).

Open water and shore birds can transport bacteria long distances to different drainage systems or lake systems. If we consider the large numbers of wading and water birds and the millennia over which they have been migrating, it may not seem unlikely that waters visited by these birds should have similar microbial communities. Inoculation and cross-inoculation have been occurring for centuries. However, this subject needs to be examined in much more depth and with statistically meaningful sampling.

# Community Development

There are two types of community development. Temporal change in communities is called *succession*. Succession can occur in habitats that have previously had biological activity (e.g., an old farm field), or it can occur on newly formed habitats, such as new lava formations or the freshly exposed rocks on Mount St. Helens when much of that mountain was blown away during the volcanic blast.

Buckley and Schmidt (2001) examined the structure of microbial communities found in cultivated and uncultivated fields. The fields were close to each other but differed in the amount of cultivation, crops, and the levels of chemical inputs of nitrogen fertilizers and herbicides. All of the cultivated fields had been under cultivation for more than 50 years. However, one of the fields had been abandoned 7 years before sampling. The uncultivated field had never been cultivated. Vegetation on the abandoned field and on the field that had never been cultivated was similar. This study looked at community composition from two different scales. First, the investigators compared the amount of RNA found in each soil sample for the occurrence and abundance of alpha, beta, and gamma Proteobacteria, the Actinobacteria, the Bacteria, and the Eukarya. Second, subsets of the soils, including the never-cultivated field, the historically cultivated but abandoned field, and a conventionally managed field, were compared based on patterns generated from 16S rDNA T-RFLPs. T-RFLP provides a much broader estimate of diversity. This study is restricted to a single snapshot of time but gives insight into the formation and maintenance of microbial communities as superimposed on the cultivation history of the fields.

The major groups of microbes screened in this study by 16S rDNA can be found in almost every habitat. Clone libraries generated from soil samples collected on three continents found that the average abundance of clones of the alpha, beta, and gamma Proteobacteria were roughly 16%, 4%, and 3%, respectively. The similarity of microbial communities among widely separated sites suggests that there are, at least at this level of community discrimination, certain characteristics of soils that promote common associations. Buckley and Schmidt (2001) found that the percent and abundance of 16S rRNA of the major bacterial groups differed between the cultivated and uncultivated fields (Figure 11.8). The uncultivated field had much higher abundances of rRNA, and each group was a higher percentage of the total rRNA extracted from the soils. Lower amounts of rRNA were also found in the field that had been fallow for 7 years. Comparisons between the subset of fields using the T-RFLP data showed that there was little overlap in diversity between the uncultivated field and the other

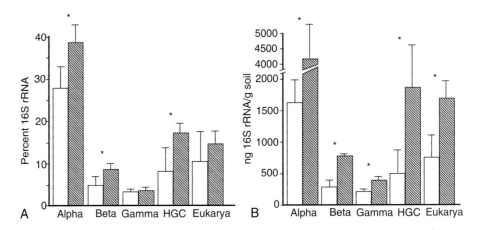

**Figure 11.8** Abundance of 16S rRNA from agricultural fields that were cultivated or never cultivated. (From Figure 2 in Buckley DH, Schmidt TM. The structure of microbial communities in soil and the lasting impact of cultivation. *Microb Ecol* 42:11–21, 2001.)

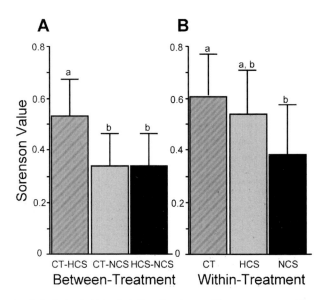

**Figure 11.9** Levels of similarity in biological diversity observed between communities of bacteria isolated from different agricultural fields. (From Figure 4 in Buckley DH, Schmidt TM. The structure of microbial communities in soil and the lasting impact of cultivation. *Microb Ecol* 42:11–21, 2001.)

two fields (Figure 11.9). The T-RFLP patterns from the cultivated fields were significantly less variable than those observed in the uncultivated field (i.e., there was higher diversity in the uncultivated field). Cultivation lowers the diversity and abundance of microbial groups, and the effect of cultivation continued for at least 7 years in this study. Even though the fallow field had vegetation similar to the uncultivated field, the microbial assemblages had not returned to what may be presumed to be the natural condition.

In contrast, Girvan et al. (2003) compared the bacterial communities found in soils collected from eight fields on three different farms in England. These farms were separated from each other by more than 65 km. Each of the fields had had different management practices for extended periods and had different crops planted on them. Bacterial communities were compared using DGGE and T-RFLP analyses. The results showed that fields with the same soil type, regardless of management practice or crop, had essentially identical bacterial communities. Despite variations in chemical and biological soil characteristics and variations in organic and inorganic inputs, these fields yielded almost identical rDNA-derived DGGE profiles. Significant differences in bacterial communities were observed only between very different soil types.

If "everything was everywhere," we might expect the fallow field microbial community to rebound to predisturbance conditions. In this situation, 7 years was not sufficient to return the environmental conditions to those before disturbance. However, 7 years is insignificant compared with the billions of years microbes have been dispersing and colonizing. The data from the English farms is intriguing because it appears that despite very different farming practices, the soil microbial communities were nearly identical, even though they were more than 65 km apart. How can we explain these apparent differences where two adjacent fields have observable differences in diversity and abundance but fields separated by kilometers do not have any such differences?

The previous examples were of secondary succession or the lack thereof. The fallow field had had previous biological activity that had conditioned the soil. Although the community did not return to an undisturbed condition, we might expect that the trajectory of community change to eventually lead to that seen in the uncultivated state.

Community development is controlled by the physical nature of the habitat, the amount and duration of prior disturbances, the potential and realized species pool, the types of species interactions, and other large- and small-scale processes and events. In the next chapter, we will examine different communities.

# Seasonality

In the section on populations, we discussed the impact of seasons on the structure, distribution, and abundance of microbial populations. The effect of seasons may be much more extreme on the structure of ecological communities. Because each species has a range of tolerance for temperature, it would not be surprising that species groups drop out from samples over annual cycles. There should not necessarily be any patterns of association based on temperature preferences. Even within the same genus, species may have very distinct temperature profiles. Differences in the time of day that various samples are collected may result in measurable differences in community composition, even though those differences are a function of diurnal sampling and not annual biological processes.

The fact that repeatable differences can be found across annual cycles is extremely interesting and deserving of increased sampling efforts and analyses. The fairly repeatable patterns of blue-green bacterial blooms in various lakes are evidence of community-level temporal patterns. These blooms occur generally in the same seasons and subside at approximately the same time each year. At an extremely coarse level, bacterial densities in many temperate lakes and streams follow an annual repeatable pattern. Highest densities are usually found in the summer months, with much lower densities in the winter. However, this is not always true, and site-specific differences have been observed in association with increased inputs of terrestrially derived organic matter in the fall and winter and some temperate sites. These coarse measures indicate that the numbers of bacterial cells is changing over time, but it does nothing to let us know whether the changes in density reflect proportional changes in all microbial species or a few microbial species or whether there is a complete community change with each season. The species mix of the community may affect the movement, availability, or cycling of various nutrients. Various species interactions also can affect the proportions of individuals and the actual species found. Community ecology is orders of magnitude more complex than population ecology.

Seasonality in the tropics is usually based on rainfall, not on changes in temperature. Rainfall does more than increase the amount of moisture in soils or the amount of water in streams and rivers. Rainfall changes the species of plants and the input of terrestrially derived organic matter through leaching or leaf litter. Rainfall alters the amount and movement of other nutrients through soils and therefore the amount of nutrients reaching other habitats.

# Concepts in Community Ecology

## 12

## Open-Water Communities

Isolated lakes provide a natural experiment to observe patterns in microbial communities. This is especially true for lakes in the same geographic regions that have no surface water connections. In some regions of the world during glaciation, large chunks of ice were forced into the ground, and as the glaciers retreated, these buried chunks of ice melted and formed what are known as kettle lakes. Kettle lakes are usually isolated, but depending on their proximity to one another, they may have surface connections across bars. Konopka et al. (1999) studied a series of kettle lakes in Indiana to determine community diversity of bacterioplankton. Ten different lakes were sampled at various depths and on different dates over a 2-year period. Bacterial communities were compared using denaturing gradient gel electrophoresis (DGGE) analysis of polymerase chair reaction (PCR) products of a portion of the 16S rDNA gene. Comparisons between communities were done using Sorenson's index:

$$C_s = 2j/(a+b)$$

where $j$ is the number of bands in common between two samples and $a$ and $b$ are the number of bands in samples A and B, respectively. The theoretical maximum of this index is 1.0 when all bands are in common, and the minimum is 0 when no bands are in common. For this study, the index ranged from 0.15 to 0.85. The greatest dissimilarity between epilimnetic communities was found between two lakes that were separated from each other by 24 km. However, the next lowest dissimilarity was found between two lakes that were only 4 km apart. Lake-to-lake differences, if a function of the lakes, should be maintained through time. Samples collected on one date should be proportional to samples collected on a subsequent date. Although the actual counts and species may differ between sampling dates, we would expect the proportion of taxonomic groups to be similar within a lake and among lakes. In this study, the similarity among lakes did not remain constant through time. Similarity in July decreased with increasing geographic distance, but in August, the reverse trend was observed (Figure 12.1). There was an inverse relationship between similarity of samples collected in July and those collected in August. How can we interpret these results?

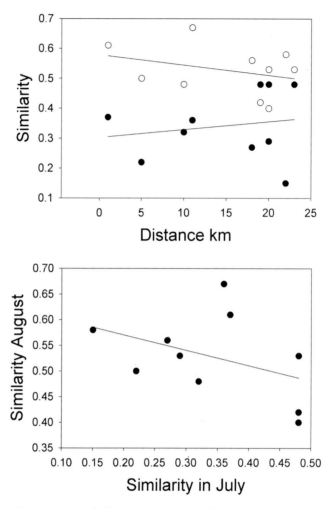

**Figure 12.1** Spatial and temporal similarity among lakes in Indiana separated by distances of 1 to 24 km. (Data from Tables 3 and 4 in Konopka A, Bercot T, Nakatsu C. Bacterioplankton community diversity in a series of thermally stratified lakes. *Microb Ecol* 38:126–135, 1999.)

These lakes were selected based on various abiotic factors that suggested that the lakes were similar. These results are an excellent object lesson in the inherent variability associated with environmental samples. No two natural systems are exactly alike! Inferences made from observational data cannot be made. If each lake was exactly similar in chemistry, aspect, depth, geology, hydrology, and riparian vegetation, we might expect them to be biologically very similar. Given that they are different in probably every characteristic, it is not surprising that they are not too similar biologically.

DGGE captures some of the taxonomic variability found in bacterial communities. At this level of resolution, there were several bands (i.e., presumptive unique taxa) that were found only in certain lakes. Given the time since the last glaciation, we may expect that colonization of these lakes may have reached some equilibrium. The fact that numerous DGGE bands were found in every lake supports the idea that colonization of some types is ubiquitous. However, the lake-specific taxa suggest that lake to lake variability in abiotic and biotic characteristics selects for certain taxa over others.

Marine and estuarine habitats have been the subject of numerous studies on microbial communities. Some of the pioneering work in culture-independent methods was performed on marine samples (Giovannoni et al., 1990). A review of marine microbial diversity (Suzuki and DeLong, 2002) gives a history of these methods. In short, the bridge between culture-dependent and culture-independent methods has been made in just the last decade. Suzuki and Delong (2002) summarize the diversity of more than 1100 bacterial determinations made using 16S rDNA sequences (Figure 12.2) since the early 1990s. In every sample collected from sea water, novel sequences continue to be identified and added to the extant database. Although species diversity has not been even closely mined, only a relatively few major bacterial divisions (Hugenholtz et al., 1998) seem to be found across all samples (Table 12.1). Comparisons of samples identified using culture-dependent or culture-independent methods indicate the same major groups of bacteria being identified. However, there is little overlap in the specific species identified using the two methods. This is not surprising if we consider the sampling device used to capture the diversity, and it supports the concept that cultured organisms are not representative of the bacterial community.

These patterns are not surprising and are actually expected. Microbial divisions are artificial taxonomic designations based on criteria established by researchers. As discussed in Chapter 3 on species and speciation, any taxonomic designation above species is a man-made construct. Patterns of relatedness exist in these higher taxa, but geographic patterns of higher-level taxonomic designations may give little insight into the species and functional diversity of a habitat. We are overwhelmed by the level of microbial diversity as expressed through modern molecular techniques and rightfully so.

The universal tree of life based on small subunit rRNA sequences graphically depicts this incredible diversity. Most of the branches on the tree of life are microbial. A single branch represents each of the higher organisms (i.e., a single branch for animals and a single branch for all higher plants). However, within the animals, for example, there are many taxonomic levels and groups.

Let us consider only one group within the higher animals for example the class Mammalia. The Mammalia include only about 4500 species, but among the higher animals, they are perhaps the most biologically differentiated group in regard to size, shape, form, and function. We need only consider the great blue whale (about 100 tons) and hognosed bat (1.5 g) to get some appreciation of the diversity of form. Contemplate a researcher who sent out teams of technicians with the task of obtaining DNA samples of every vertebrate in a haphazard sampling of every continent. Would we expect the researcher to be amazed that after screening these samples they discovered there were mammals represented? Hardly an eye would be raised.

The length of the branch in the universal tree is proportional to the genetic diversity of the group. The animal branch is fairly short relative to some of the microbial branches, but we expect unique differences even within a small subset of the animals; specifically the mammals in our example. The fact that major groups of bacteria have been found in every ocean is not that unique and highly expected. Mammals are on

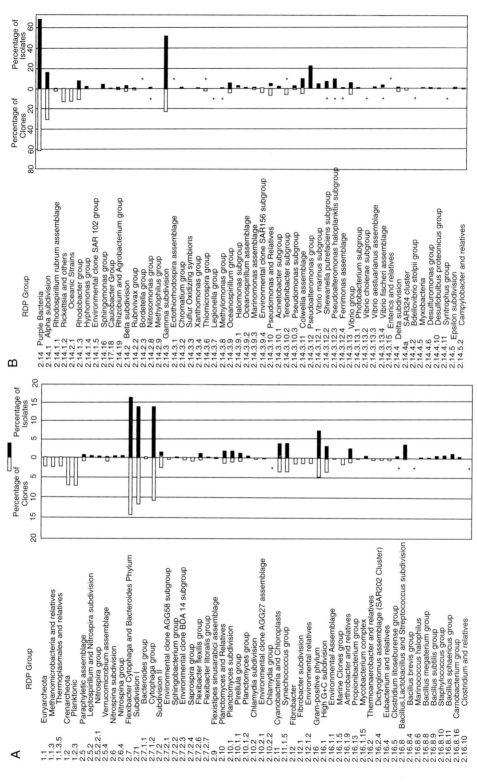

**Figure 12.2** Phylogenetic summary of more than 1,100 bacterial determination using 16S rDNA sequences. (From Suzuki MT, DeLong EF. Marine prokaryote diversity. In: Staley JT, Reysenbach AL (eds): *Biodiversity of Microbial Life*. New York, Wiley-Liss, 2002, pp. 209–234. This material is used by permission of Wiley-Liss, Inc., a subsidiary of John Wiley & Sons, Inc.)

**Table 12.1  Major Bacterial Divisions Isolated from Seawater between 1990 and 2002**

Verrucomicrobium
Cytophagales
Planctomycetes
Cyanobacteria
Actinobacteria
Low–Gram-positive cocci (G + C)
Proteobacteria

Adapted from Suzuki MT, DeLong EF. Marine prokaryote diversity. In: Staley JT, Reysenbach AL (eds): *Biodiversity of Microbial Life*. New York, Wiley-Liss, 2002, pp. 209–234. This material is used by permission of Wiley-Liss, Inc., a subsidiary of John Wiley & Sons, Inc.

every continent, and they are just a small subset of animals. The observation that members of such widely separated groups as the bacterial divisions are widely distributed should be expected. The correct comparison between the normal bacterial comparisons should be whether plants, animals, fungi, and microsporidia are found on every continent. The answer is yes. It is not the distribution of divisions that matter.

Taxonomic divisions are deeper than kingdoms. The really important questions concern what members of each division are present. How functionally and ecologically different are the species found within a site, and how do they compare with similar or dissimilar sites? How genetically similar are the same species found at increasing spatial scales? How temporally and spatially stable are the populations and communities found across large geographic areas such as oceans? Just because we have been able to measure something does not mean that the results are evolutionarily or ecologically important. The tools and techniques of molecular biology are powerful and have allowed insights into the ecology of microorganisms, and they will be continually improved and accessible to researchers. However, we must recognize the limitations of each method and the degree of resolution provided.

# Biofilm Communities

The growth of microbes on any surface results in the formation of a biofilm. All people have experienced biofilms quite intimately. Every night while sleeping, bacteria in the mouth attach to teeth and begin to replicate and grow, forming a biofilm on teeth and mouth tissues. Many people have walked in or along a rocky bottomed stream and felt the slippery stones. That slippery layer is a biofilm composed of bacteria and algae and inhabited by numerous microinvertebrates and macroinvertebrates.

Biofilms can potentially form on any hard or semi-hard surface, and they are extremely important in medicine and industry, although for different reasons. In medicine, bacteria can and do colonize various apparatuses that are inserted into patients. Pacemakers, catheters, and insulin pumps can be sites of colonization. These structures, once colonized, are extremely serious health hazards and very difficult to treat. This resistance to disturbance, including toxic or inhibitory substances such as antimicrobials or antibiotics, is a fundamental characteristic of biofilms. However, for patients who have the misfortune to have a biofilm form on an introduced structure or on internal surfaces of their bodies, the results can be debilitating or fatal.

In industry, scientists take advantage of microbial biofilms, and they work to create specific growth conditions suitable for a specific microbe or consortia of microbes that perform some desirable function. *Bioreactors* are engineered environments that promote the growth and production or decomposition of some compound or group of compounds. For example, many toxic chemicals have been released into the environment and found their way into groundwater supplies. Engineers have constructed bioreactors through which water is pumped. As water moves through the bioreactor, microbes growing in the bioreactor break down the toxic material, releasing carbon dioxide and water as waste products. The complete mineralization of an organic chemical pollutant is the desired goal, but other compounds often are formed during the degradation process.

Biofilms have been defined (Characklis and Marshall, 1990;Teske and Stahl, 2002) as microbial systems that have the following four elements:

1. A biofilm consists of cells immobilized at a substratum and frequently embedded in an organic polymer matrix of microbial origin.
2. A biofilm is a surface accumulation, which is not necessarily uniform in time or space.
3. A biofilm may be composed of a significant fraction of inorganic or abiotic substances held together by the biotic matrix.
4. Biofilms are distinguished from suspended growth microbial systems primarily by the critical role of transport and transfer processes, which generally are rate controlling in biofilm environments.

Biofilms may form on any surface exposed to microbes and that has sufficient water and nutrients to promote growth. Other surfaces may include leaves, roots, wooden debris, swimming pools, water delivery pipes and conduits, linings of the digestive tract, suspended particles in lakes, streams and oceans, hulls of ships, high-level radioactive waste containers, surfaces of hot springs, and many other surfaces. It should be clear from this incomplete list that microbes are capable of growth just about anywhere. Under natural conditions (i.e., habitats not impacted by human activities), biofilms form, are maintained, and perform many important ecosystem functions.

Let us consider the formation, maintenance, and performance of biofilms. Of all microbial communities, biofilms are perhaps among the easiest to observe, measure, and manipulate and therefore possibly are among the best understood communities. Even so, there is much to learn about biotic interactions occurring within biofilms. Biofilms may consist of a monospecies (e.g., *Staphylococcus aureus* on a pacemaker), or biofilms may be multispecies, multilevel structures that change over time such as the slippery surfaces of rocks in streams and rivers or the complex microbial mats found in hypersaline ponds.

Before we discuss natural biofilms and microbial mats, let us address the formation of biofilms under laboratory or contrived conditions. Microbes can colonize most surfaces. Some researchers think that most microbes have the ability to colonize. If a researcher places an uncolonized surface into medium containing bacteria, whether the medium is moving or stationary, colonization will proceed, and a biofilm will form. Why should this happen? What advantages do microbes get from living in a biofilm rather than living in microcolonies or as single cells? If we consider the simplest case (i.e., a biofilm made from a monospecies), at some point in the development of the biofilm, intraspecific competition will reach a maximum level. In a single-species biofilm, every individual is identical or nearly so to every other individual. Every individual is equally capable of taking up nutrients and obtaining suitable energy either from photosynthesis or from heterotrophic processes. Competition would be the most intense because there is presumably no best strategy because all individuals have the same strategies. Differences in uptake rates and energy acquisition may reflect point mutations that increase or decrease the ability to perform a function, but these differences would be minimal and of greatest importance under suboptimal conditions, when nutrients are scarce or not easily obtainable. Under these conditions, a single cell would be competing with others that are just as good at taking advantage of the resources available.

*Intraspecific competition* is defined as competition between individuals of the same species. With higher organisms, this type of competition can occur between siblings or between distantly or nonrelated individuals of the same species. In bacterial systems, there is one more level of intraspecific competition that is not experienced by sexually reproducing organisms: competition with yourself. All cells derived from a single cell through binary fission are essentially the same individual. As a colonized cell divides, the numbers of exactly identical individuals increases and so does the intensity of the competition because most of these cells will be in the immediate proximity. The evolutionary benefit obtained from surface growth must exceed the costs associated with competition.

Because all surfaces can be and are colonized given the opportunity, it seems likely that there are advantages to being attached to a surface over free living. What can a surface provide that a free-living cell does not have? Nutrient levels are often below the availability threshold of microbes; that is, although nutrients are present, their concentrations are too low for microbes to take them up. Morita (1997) pointed out that as far back as 1940, researchers had determined that the presence of surfaces increased the metabolic activity of microbes due to the concentration of nutrients and ions at the solid-liquid interface. Kjelleberg et al. (1982) showed that after surface attachment, starved bacteria enlarged, divided, and moved off the surfaces within 7 minutes after a minimal addition of nutrients. Attachment of bacteria to surfaces appears to be a strategy to obtain nutrients from oligotrophic environments. The advantage of reproducing in 7 minutes as opposed to 1 hour or longer offsets the disadvantages of competition.

Shapiro (1991, 1997a, 1997b) demonstrated in a number of elegant experiments that *Escherichia coli* has unique and repeatable behaviors during colony growth. These observational studies documented that *E. coli* had definite patterns of colony growth, with cells appearing to prefer growth under crowded conditions. Rather than growing away from the colony, individual cells remain associated with the colony and orient to increase contact with other members of the colony. Additional studies showed that the concentric rings that form during growth of the colony (Figure 12.3) can have different functions (i.e., division of labor) and that information and materials are shared or transported within the colony. Shapiro argues that colony growth habit is the rule and not the exception, and cites examples of free-living *E. coli* in chemostats adopting a biofilm (i.e., wall growth) growth form.

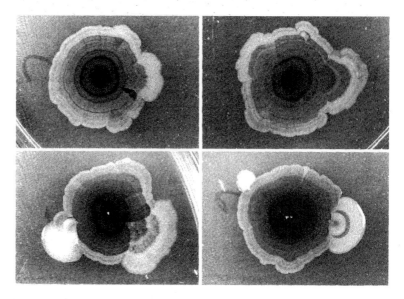

**Figure 12.3** Concentric rings of growth showing organizational patterns in *E. coli* colonies. (From Shapiro JA. Multicellularity: the rule, not the exception. In: Shaprio JA, Dworkin M (eds): *Bacteria as Multicellular Organisms*. Oxford, UK, Oxford University Press, p. 17. Copyright 1997 by Oxford University Press, Inc. Used by permission of Oxford University Press, Inc.)

The Shapiro studies are interesting and suggestive; however, colony growth on an agar plate containing required nutrients is similar in ways but fundamentally different in other ways from growth on a solid surface, where nutrients must come from the overlying media. The Shapiro model fits growth of bacteria on living surfaces such as the lining of the intestine or of growth on senescent organic matter such as leaves and fruits or perhaps accumulated sediments in lakes or oceans. It falls short of explaining growth on inorganic substrates such as rocks, plastics, glass, or other hard surfaces.

We must consider other aspects of colonization and growth particularly as they relate to life in communities. First, do all microbes have an equal probability of colonizing? The answer is no. Microbes closest to and those being wafted by air or water toward a newly exposed or created habitat have a higher probability of successfully colonizing than ones more distant or that have been carried past the area or are being transported in entirely the wrong direction. Second, the physiologic state of the microbe affects its colonization ability and subsequent growth.

Many microbes that attach to surfaces produce, often in copious amounts, extracellular material. The extracellular material can be primarily polysaccharide or a polysaccharide and protein mixture, and numerous exoenzymes are produced and released into the environment. The polysaccharide material is predominantly water (>95%) and forms the glue that attaches the microbes to the surface. However, this material does more than glue the microbe to the surface. These anionic polymers can act as ion-exchange materials that provide protection from metals, lysozymes, or other inhibitors such as bile salts (White, 1986). White (1986) further indicates that these polysaccharide capsules and slime layers can serve as diffusion barriers, molecular sieves, and

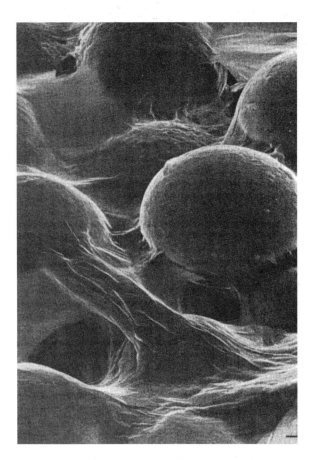

**Figure 12.4** Scanning electron micrograph of bacteria and glycocalyx formation on glass beads that mimic an open sandstone rock. (From Costerton JW. The role of bacterial exopolysaccharides in nature and disease. *J Industr Microbiol Biotechnol* 22:551–563, 1999, with kind permission from Springer Science and Business Media.)

adsorbants. They also can protect extracellular enzymes. It has been suggested (Hernandez and Newman, 2001) that there are mechanisms for extracellular electron transfer that results in various metabolic functions.

As microcolonies grow on some surfaces, they form a complete covering *glycocalyx* in which the cells become embedded (Figure 12.4). In a sense, these microbes form their own agar plate over and around each cell. The polymer acts to bring nutrients to the cell.

There are at least two difficulties with understanding the exocellular production of material. First, the extracellular polysaccharide (ECP) must be produced. This seems trivial until we consider that producing ECP must be done at the expense of producing more microbial cells. The energy required to make the polymer cannot be used to make new daughter cells. The benefit provided by producing the ECP must exceed the cost of not producing new cells. We can consider the problem as an equation in which the various costs and benefits are the variables. For example, if we let the benefit of adsorbance of nutrients be some value $A$ that is measured in some meaningful metric such as kilocalories, protection from metals be $B$, protection of exoenzymes be $C$, and so on for other "benefits" that may include adhesion, protection from antibiotics, diffusion barriers, and molecular sieve, we can estimate the total benefit provided by producing the material. In contrast, we can estimate the total cost to the microbe by summing the individual costs such as the adenosine triphosphate (ATP) required to synthesize the material, cost to take up carbon, cost to take up other nutrients, or cost of not reproducing. If the total benefits do not equal or exceed the total costs, we would expect that this production is not evolutionarily stable. In other words, cells that do not produce the substance would have an advantage over cells that do produce this material. All other things being equal, the cells that do not produce the ECP would be able to out-compete those that do in capturing resources and space. Because biofilm ECP is prevalent, we can assume that there is a net benefit to the cells that produce the material.

This observation on the prevalence of ECP production begs another question. What if a cell can become established in the ECP matrix, but it does not contribute to making or sustaining the matrix? Such a cell should be able to take advantage of all the good that comes from the matrix but have no costs in production or mainte-nance. These embedded freeloading cells could also take advantage of exoenzymes produced by other cells and save on metabolic costs. After an enzyme is released into the environment, the products catalyzed by the enzyme can be taken up by any cell that is physically close enough to the substrate or product. Theoretically, cheater cells should have an immediate advantage over the other cells. Why do they not become the dominant cells in the biofilm? If nonproducing cells dominate, no ECP would be produced, no exoenzymes would be released, and all cells would be at a disadvantage. Some mechanisms need to exist that prevent the increase of cheaters (discussed later).

We have been considering the formation and maintenance of biofilm communities that are composed of a single species. What happens when more than one species of bacteria, algae, diatom, and fungi are potential colonizers? Multispecies biofilms are prevalent in nature. These complex biofilms provide insight into many aspects of microbial ecology.

Colonization of biofilms and the subsequent development, maintenance, and per-formance of the biofilm are all affected by the ecology of the microorganisms forming the biofilm. Does the formation of a biofilm follow any predictable or repeatable pattern within a specific site or location? For example, if we placed sterile wooden dowels in a stream in South Carolina and similar dowels in streams in Virginia, Vermont, and Nova Scotia, would we expect to find similar colonization patterns? If we placed the dowels in different reaches of the same stream, would we expect similar colonization patterns? Would the same species of bacteria or algae colonize in the same order? Would the thickness of the biofilms be similar?

Scientists studying succession in communities of higher plants and animals have determined that various general characteristics apply to those organisms that colo-nize an area first and those that pervade during later stages of succession. Many com-munities follow a four-phase developmental cycle:

1. *Pioneer*: establishment and early growth of open patches by early colonizers–dominance of *r*-selected species
2. *Building*: maximum cover of colonizing species–few associated species and more species invading the area

3. *Mature*: complex associations of species and little new colonization–shift to
   *K*-selected species
4. *Degeneration*: patches open due to senescence of the biota or physical factors such
   as scraping and drying–colonization of patches increases species diversity and biological complexity

Pioneer communities depend on the available colonizers. This pool of potential
colonizers differs according to habitat. In *lentic* habitats such as ponds and lakes,
colonization occurs primarily through deposition or settling of species. Shlegel (1988)
found that certain vibrios can swim at up to 12 mm per minute, which corresponds to
approximately 3,000 times their body length. Even so, the effective distance a microbe
can swim excludes colonization through active processes for most species.

The sinking rate of free-living bacteria is only 0.5 to 3.2 mm per day (Jassby, 1975),
and this mode of transport of potential colonizers is also restrictive. However, passive
colonizers on particles settling out from the water column include algae, diatoms,
fungal spores, and bacteria. Other potential vectors of colonization in lentic habitats
may include invertebrates that can carry colonizers on their surface or through
deposition of their feces on surfaces.

The littoral zones of most lakes have well-developed biofilm communities. Wave
action in large lakes facilitates the transport of microbes into and out of the littoral
zones. Large-scale movement of free-floating materials can occur in response to internal currents generated by winds. These large-scale patterns can be easily seen during
algal blooms as visualized from satellite images (NASA web sites offer various images,
http://phytoplankton.gsfc.nasa.gov). Depending on the size of the lake, wave action
and currents may result in increased similarity among biofilms or decreased
similarity. Decreased similarity would occur if coves or certain shorelines are not
impacted by wave action or in the path of the currents delivering colonizers to an area.

In rivers and streams, the predominant colonization vector is downstream transport. Few microbes are capable of upstream migration except in association with vertebrates (i.e., fish) and some aquatic and semi-aquatic insects. As discussed in Chapter
4 on individuals, bacterial communities in lower reaches of a stream seem to be an
integration of types found upstream. This is not unexpected given the limited capacity of most microbes to move upstream. Stream biofilm communities can differ within
and between streams. Differences can occur over short distances. A river or stream
running through an open canopy that suddenly plunges into a closed canopy would
have very different members in the two biofilm communities. Algae and diatom species
would change rapidly with the decreased available light. Bacteria that obtain energy
from algal exudates would change accordingly as species of algae change or are eliminated. The thickness of the biofilm may also change depending on the availability of
nutrients and carbon.

Biofilm communities can change over much finer spatial scales. The biofilm community found on the top surface of a rock in a stream is fundamentally different from
that found on the sides or bottom. All of the rock can have a biofilm, but the nature
of each community will be driven by the source of energy, nutrients, and carbon available to the microorganisms.

Lock et al., in a series of papers (Lock, 1981, 1993; Lock et al., 1984; Lock and
Ford, 1985), developed an observational and theoretical framework of river biofilms.

The basis of this model was that extracellular polysaccharide produced by microbes formed a continuous matrix in which bacteria, cyanobacteria, and algae were fixed in place on solid surfaces. They (Lock, 1993) suggested this polysaccharide matrix functioned as a

1. Site for entrapment of soluble and particulate matter
2. Retaining and possibly protective mechanism for extracellular enzymes
3. Retention mechanism for any products of extracellular enzymatic hydrolyses
4. Framework or glue enabling a multistory of autotrophic and heterotrophic microorganisms to form under suitable conditions

Far from being a uniform assemblage of microorganisms glued into place, biofilm communities are complex and varied structures (Stock and Ward, 1989). In lentic and lotic systems, a firmly attached community can be overlain with a loosely attached biofilm component bound together by "mucilage" or polysaccharide. The thickness of these loose layers is considerably affected by the velocity of the overlying water. In streams and rivers, the layer is much smaller than in lakes or ponds. This loosely attached community has been shown to be the site of high microbial activity, and in some instances, it has been shown to inhibit microbial activity in the underlying stories of the biofilm (Lock, 1993, see references).

The three-dimensional structure of a biofilm is not haphazard. It has distinct organization. Through various microscopy techniques, this structure has been elucidated, showing the spatial arrangement of the different members of the community. In some cases, stalked diatoms attach to the substratum first and then bacteria and other algae or cyanobacteria attach to these diatoms. The three-dimensional structure is not restricted to stalked diatoms but occurs with bacteria and amorphous detritus (Rounik and Winterbourn, 1983). Others have shown that filamentous green algae and cyanobacteria formed multilayers on freshly exposed rock surfaces (Stock and Ward, 1989). Bacteria were found to be more prevalent between diatoms than above and much more prevalent on the attachment pads of the diatoms. Clearly, there is structure.

Most of the studies addressing biofilm structure have been conducted on biofilms exposed to light or on the surfaces of other structures like rocks or plants. Lock et al. (1984) were able to show that biofilms developed on the undersides of rocks in streams, and others have shown that biofilms exist on rocks found tens of centimeters below the surface of the stream in what is called the *hyporheic* zone. Terrestrially derived organic carbon that flows through these zones provides the carbon and energy necessary for metabolism to occur.

Rocks are not the only substratum that can be colonized by microbes and form biofilms. Any surface suspended in the water column will be colonized. In streams unaffected by human activity, tree limbs, roots, leaves, and macrophytes growing in the water all provide a solid surface and, to some degree, a nutrient source. However, tires, glass bottles, shoes, and plastic all have biofilm communities formed on them.

Are the biofilm communities associated with natural inorganic surfaces (e.g., rocks, cobble) different from natural organic surfaces? Are natural epilithic communities different from biofilms on glass or plastic? We would expect differences between epilithic and epiphytic communities because plants may release exudates that provide carbon or nutrients to the attached organisms, whereas microbes growing on a rock must obtain nutrients from the water column or from algae embedded in the biofilm. Lock (1993) states that although some differences may exist between biofilm communities, he expects fully that "biofilms from different surfaces will be convergent rather than divergent in their properties."

To this end, he proposed a generalized conceptual model of biofilm structure and functions. In this model, there is a close association between bacteria and algae, between bacteria and bacteria, and between the abiotic and the biotic. Algae, bacteria, and cyanobacteria are maintained in association with each other within the polysaccharide matrix. The amount of cyanobacteria and algae is controlled by the amount of light available. There is evidence that biofilms may have channels through which water, nutrients, metabolic products, and exudates may be exchanged within the biofilm and with the external water. Enzymes released from bacteria act on substrates brought into the polysaccharide matrix through these channels producing products that are used by neighboring bacteria. Sinsabaugh et al. (1999) have shown that exoenzymes are an integral part of biofilms. Several different hydrolytic enzymes have been identified. Algae release exudates that "feed" bacteria, which release more enzymes.

Newer techniques allow the visualization of biofilm structure during and after development. One such technique is confocal laser scanning microscopy (CLSM). Using CLSM, Battin et al. (2003) investigated the architecture of developing biofilms under two different water flow conditions in stream-side flumes. Water velocity was controlled in each flume such that there was a threefold to fourfold difference between the two treatments. Water was diverted from a natural stream into the flumes, maintaining the source of potential colonizers. Temperature, light levels, and flume substrata (i.e., cobble and gravel) were identical to those in the adjacent stream. Biofilm development was monitored on ceramic coupons ($1 \times 1.5 \times 0.3\,cm$). Different architectures were observed under the different flow conditions. In terms of biomass, the slow-flow treatment had a more than threefold higher bacterial abundance and more than twofold higher chlorophyll $a$ concentrations than did the fast-flow treatments. From these data and the increased contact between the overlying water and the biofilm in the slow-flow treatment, we might expect greater uptake of materials from the water column in the slow-flow treatment. Average uptake rates of dissolved organic carbon (DOC) were almost four times higher in the fast-flow treatment. Almost four times fewer bacteria but four times greater uptake rates were found in the fast-flow flumes. The exopolysaccharide measured as glucose were significantly higher in the fast-flow treatments. With lower bacterial densities and higher polysaccharide content, the exopolysaccharide-to-cell ratio was almost threefold higher in the fast-flow flumes.

The three-dimensional architecture of the biofilms differed by treatment (Figures 12.5). The initial surface (x,y plane) colonization was primarily of individual microcolonies and scattered bacterial cells. By day 4, ripple-like structures began to appear parallel with the flow. This was followed by development of long filamentous bacteria in the fast-flow treatment. After 8 days, clear spaces (*anastomoses*) began to form between the ripples and diatoms, and amorphous material became attached. After 21 days, a network of quasihexagonal structures formed, especially in the slow-flow biofilms. Pennate diatoms made up the edges of these structures.

Slow flow          Fast flow

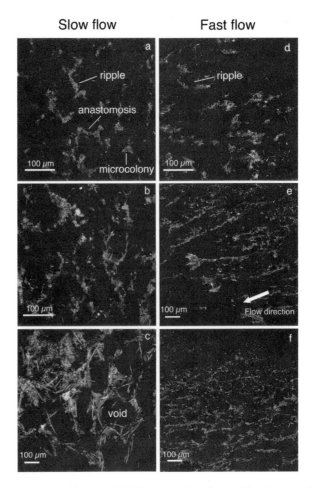

**Figure 12.5** Developmental architecture of biofilms under low-flow and fast-flow conditions as determined by confocal laser-scanning microscopy. (From Battin TJ, Kaplan LA, Newbold JD, Cheng X, Hansen C. Effects of current velocity on the nascent architecture of stream microbial biofilms. *Appl Environ Microbiol* 69:5443–5452, 2003.) (See color insert.)

Development upward from the surface of the biofilm (x,z plane) was noticeably different between the treatments (Figure 12.6). No algal cells were found in the 3- to 5-day biofilms. The base layer was composed primarily of bacterial cells, EPS, and other particles. As the biofilm grew outward, it formed finger-like projections, which were replaced with a canopy of diatoms. Extensive channel networks were formed with age of the biofilm. Stratification with distinct layers was observed between the EPS layers and diatoms.

How does this study compare with the conceptual model of Lock? Lock proposed that there were free-living bacteria embedded in matrix of polysaccharide. Battin et al. (2003) demonstrated that bacteria are the primary colonizers and that they can form distinct layers in the biofilm. Lock suggested that algae form the base layer. Battin et al. showed that the basal layer is primarily polysaccharide, nondescript particles and bacteria. Diatoms and algae attach later. Lock suggested that channels allow transport out of and into the biofilm. Battin et al. demonstrated the presence

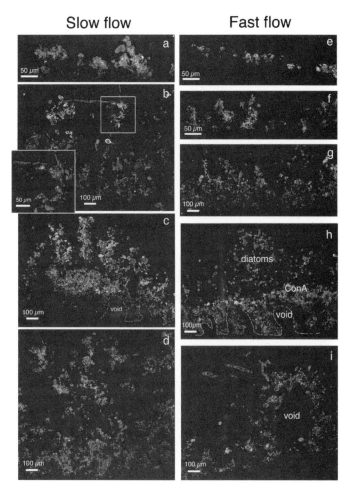

**Figure 12.6**  Three-dimensional development of river biofilms. Confocal laser-scanning microscopy micrographs of cryosections show the structural development of the biofilm x,z plane in slow flows (**a** to **d**) and fast flows (**e** to **i**). (From Battin TJ, Kaplan LA, Newbold JD, Cheng X, Hansen C. Effects of current velocity on the nascent architecture of stream microbial biofilms. *Appl Environ Microbiol* 69:5443–5452, 2003.) (See color insert.)

of numerous channels. The conceptual model based on observations and empirical data from the 1980s is fairly supportive of the observational data presented in 2003.

Neither the model presented by Lock nor the observations of Battin et al. give any indication of the species diversity and the complexity of species interactions that may be occurring within the biofilm. All observations and predictions are at the highest taxonomic levels (i.e., bacteria and algae). Even though our ability to visualize the minuteness of biofilm development has progressed, we are still restricted in knowing what organisms are present (with the exception of diatoms and algae) using these modern observational techniques. Differences in the morphology of some of the bacteria appear during development. However, we cannot determine the functional or other phenotypic differences between layers of the biofilm or between biofilms grown under different environmental conditions (i.e., flow rate in the Battin et al. study). We are limited to stating only observed physical differences.

Based on the uptake rates of DOC observed in the Battin study, we can make some suppositions on some relationships that may exist in these specific biofilms. The highest uptake of DOC occurred in the fast-flowing water. Although lower numbers of bacteria were present, more DOC was removed from the water than in the slow-flow treatment. Another significant observation was the ratio of polysaccharide to bacterial cells, which was three times higher in the fast-flow treatment. The canopy of diatoms was much more developed in the slow-flow treatment. It appears that the exopolysaccharide matrix facilitates the removal of DOC. Given the proximity and abundance of bacteria and diatoms or algae in the slow-flow treatment, some of the carbon or energy requirements of the bacteria must be being met by the diatoms and algae.

# Phylogenetics and Community Ecology

An understanding of how communities are created and maintained requires a synthesis of three major elements (Webb et al., 2002): phylogeny, community composition, and information on the traits of community members. In the previous discussion, we emphasized community composition, but the study of phylogenies, as applied to communities, greatly expands our understanding of the ecology and evolution of communities.

The construction of phylogenetic trees for communities allows hypotheses about the mechanisms that permit coexistence to be tested. The more genetically and phenotypically similar two individuals are, the higher probability of competition between them. The similarity of organisms is controlled by dispersal abilities, colonization successes, isolation events, and divergences that have occurred at various geographic scales. As Darwin observed, island faunas often produce significant divergence of characters within taxonomically close groups that seemingly originated from the same ancestral stock. The finches described by Darwin and studied by numerous others is a classic example of divergence. Each species of finch has exploited some aspect of the island habitat and consequently diverged phenotypically and genetically such that several species of finches are recognized.

Organisms have *intrinsic* and *extrinsic* traits. Intrinsic traits maintain meaning and can be measured whether the organism is removed from its natural environment or stays. Intrinsic traits include characteristics such as size, weight, shape, and amount of DNA, and for microbes, the organism is usually removed from its environment to measure these traits. Extrinsic traits include characteristics such as metabolic diversity, swimming speed, growth rates, metal tolerance, antibiotic tolerance, and drought tolerance. Extrinsic traits require that the organisms remain in their environment for the measurements. Artificial or constructed environments can be used to measure extrinsic traits, but the values obtained may be biased by known and unknown factors (e.g., temperature regime, lack of species interactions).

There are at least two ways in which species within a community can be phenotypically similar. First, the species are related to each other and have diverged in some characters but not in all. We would expect that species from the same genus to have more similar traits than species from the different genera but the same family. With each level of taxonomy, the amount of similarity should decrease. Organisms from unrelated taxa should have the highest level of dissimilarity. Traits change as the lineage diversifies (Webb et al., 2002).

Second, related or unrelated species have similar traits because of convergence. *Homoplasy*, or the independent evolution of similarity, results in traits being similar among independent lineages. The rate of homoplasy can increase as the number of potential traits decreases, and the number of ways species can be functionally similar increases (Webb et al., 2002). For example, there appears to be high levels of functional redundancy within many microbial communities (discussed later). Why should there be any functional redundancy? Would multiple species and or strains performing the same function in the same habitat not increase competition and negative interactions? Before we examine the consequences of functional redundancy, let us examine other microbial communities that have been well studied in nature.

# Soil Communities

Of all the communities and habitat types that we have discussed, the soil environment is perhaps the most complex and difficult to observe, measure, and understand. Soils are extremely heterogeneous matrices that vary over very small and very large scales. The size of particles, amount of organic matter, pH, concentration of nutrients, amount of clay, types of clay particles, amount of vegetation, types of vegetation, root mass, and amount of water all affect the ecology of the soil. Some of these variables, such as the role of interfaces, have been the subject of entire books.

The number of environmental variables found in soils can be reduced by grouping the factors based on mechanism or processes (Panikov, 1999). These groupings result in three major sets of independent environmental factors (Figure 12.7). The first group is the soil mineral resources. In many habitats, mineral resources are probably not limiting to microorganisms, even when they may be limiting to the plants growing in the same area. Most mineral resources are immobilized in various forms by both living and dead cells or by being insoluble, adsorbed, or chelated. The second group is sunlight and radiant energy. These determine most of an ecosystem's energy balance through changes in seasonal inputs. The primary variables are solar radiation and the physical state of the atmosphere. The physical state of the atmosphere is controlled by clouds, greenhouse gases, and the amount of aerosols present. The third group is the factors that control mass transfer. The soil solid phases control the rate of liquid and gas mass transfer. The amount of soil water, especially the soil pore water volume, is critical to soil biota because it controls gas transfer.

These factors interact with each other and with the soil microbes. The structure of soil communities is affected by the environmental factors and the species interactions that are present in a particular soil. Prokaryote diversity is affected by abiotic and biotic factors. Where the reservoir of species is high, we might not expect similarity among communities that appear to be physically similar, nor should we expect the communities to be stable. Communities that are similar and stable are usually observed where the diversity is low, even when the habitats are not physically similar. The source of diversity may be more important than the physical template of the environment.

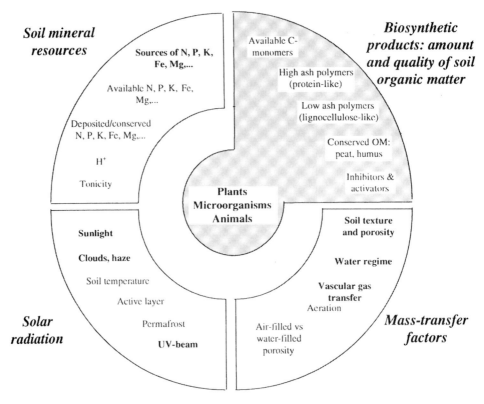

**Figure 12.7** Groupings of environmental factors that affect soil microbial communities. (From Figure 1 in Panikov A. Understanding and prediction of soil microbial community dynamics under global change. *Appl Soil Ecol* 11:161–176, 1999, with permission from Elsevier.)

## Oral Communities

Bacterial communities of the mouth are primarily associated with biofilms that are found on hard and soft surfaces. More than 500 different bacterial taxa have been isolated from oral surfaces. Differences have been observed between the communities on different teeth in the same mouth. However, the dominant species often are found on all similar surfaces (Figure 12.8). The identification of some taxa has significantly increased with the use of 16S rDNA probes. Some obligate anaerobic bacteria, such as *Fusobacterium nucleatum* and *Prevotella melaninogenica*, are frequently isolated. These obligate anaerobes cannot survive except in coaggregations with aerobic bacteria. *F. nucleatum* organisms often act as coaggregation bridges between species of bacteria that do not coaggregate, including aerobes and anaerobes. Additional studies have shown that *Actinomyces* organisms occur as a group or consortia of related taxa that cause tissue destruction. Oral microbial ecology is an expanding field of study, and new insights are being made with molecular biological techniques and with methods used to visualize colonization and possible species interactions through CLSM.

Oral microbes have been shown to share genetic information within and between taxa. For example, the genetics of *Streptococcus* is fairly well described, and other

**TOOTH SURFACE**

**Figure 12.8** Diagram of an oral biofilm on an enamel surface. The tooth surface is coated with a host-derived acquired pellicle (i.e., gold, curly meshwork at the bottom) to which accrete various colored, morphological shapes representing genetically distinct cell types. The *yellow arrows* indicate salivary flow, which provides a mechanism for individual bacteria, rosettes, corncob formations, and complex coaggregates to come in contact with the biofilm. Salivary flow also contributes to the detachment of cells from the biofilm. (From Figure 1 in Kolenbrander PE. Oral microbial communities: biofilms, interactions, and genetic systems. *Annu Rev Microbiol* 54:413–437, 2000.) (See color insert.)

taxa, including *Actinobacillus, Actinomyces, Porphyromonas,* and *Treponema,* have begun to be studied. Genetic communication between taxa has shown that the oral cavity is a dynamic and interactive habitat.

# Functional Diversity

Stability of communities has been suggested by some authorities to relate to the number of species occupying a location. Others suggest that stability is related more to the functional redundancy found in a location. Under this concept, communities are stable if the same functions prevail under varying environmental conditions. In other words, various species can be eliminated, and the community will remain stable if other species remain that perform the same function.

Various metabolic and genetic screenings exist that allow researchers to determine whether specific genes or traits are found in a particular sample. For example, by screening for various *nif* genes, it is possible to tell whether nitrogen fixation may occur at a certain location. The presence of the genes does not mean that fixation is being performed, only that the potential exists for that process. To determine whether nitrogen fixation is happening, we would have to perform specific tests, such as the acetylene reduction reaction.

The use of metabolic test strips allows characterization of communities based on the responses to the compounds and reactions tested. These test strips can also be used to characterize isolated microbes. It is from the database created from the responses of numerous microbes that the concept of functional redundancy was formulated. If species x, y, and z all can break down a specific carbon source and species x, y, and z were all isolated from a single sample, we could say that the ability to break down that carbon source has a level of redundancy.

At the ecosystem level, scientists often measure system level responses such as gross primary production or respiration. Sometimes, researchers may seek to determine the fate of specific materials as they move through an ecosystem. For example, transformations of various nitrogen compounds or carbon compounds (natural or anthropogenic) can be monitored. At the scale of the ecosystem, the scientists are not concerned about what organisms are performing the transformation but instead focus on whether the transformation is happening. Ecosystem scientists integrate across all lower scales of biological organization in measuring the important responses. When an ecosystem is chronically stressed, at some point, various processes are expected to be curtailed or completely interrupted.

Ecosystems are extremely complicated and complex systems. No aspect of the ecosystem is exactly similar to another contiguous or distant location. Microscale differences exist spatially and temporally based on environmental and biological conditions. The fact that numerous isolates of bacteria or other microbes isolated from a single environmental sample can be screened for the ability to break down or use a specific carbon resource does not mean that these organisms are necessarily found together under natural conditions. We have stressed repeatedly that the scale at which individual microbial processes are happening is below our ability to measure or observe (except molecular force microscopy). The very act of sampling disrupts the finely tuned relationships that exist between groups of microbes. Although several or many isolates may have the ability to break down a carbon resource, it does not necessarily mean that they have that ability under the environmental and biological conditions where they actually reside.

What does functional redundancy mean at the ecosystem level and at the individual species levels? Are they the same? It has been shown that stressed ecosystems have lower species and genetic diversity than do undisturbed ecosystems. Various thresholds exist along contamination or other stress gradients. These thresholds often appear as step functions. There appears to be no affect on a process with increasing stress levels until some set point is reached when the system abruptly changes (Figure 12.9). In the figure, there are three set points. At each point, the ecosystem process decreases, but between the set points, there does not appear to be anything happening. It is during these times between set points that careful examination of species and functional redundancy needs to be done. Is there loss of species during these times?

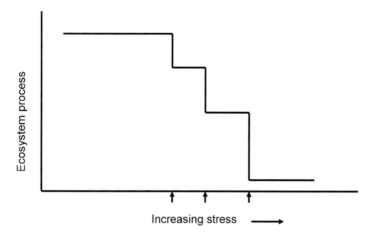

**Figure 12.9** Decreased ecosystem processes with increasing stress. There are three set points where significant changes in ecosystem processes occur. After each set point, the system functions at the same level until the next set point is reached, when there is an immediate decrease in function.

If species are lost, do the remaining species increase their processing through increased activity or do they increase in density?

At the individual species level, functional redundancy has a different meaning. Ecological theory states that two different species cannot occupy the same niche. Just because an organism has a metabolic capability does not mean that it uses that capability whenever it encounters a suitable substrate or compound. Two different organisms, both with the ability to break down some carbon compound, probably respond differently under the same environmental conditions.

Characterization of microbial (especially bacterial) communities using physiologic test strips must be based on an awareness of the bias implicit in these tests. Amazingly, characterization of different microbial communities have been done using this technology and with a certain degree of repeatability. Merkeley et al. (2004) used DNA and physiologic test strips to compare isolated wetland microbial communities in the Great Basin that had been impacted by cattle grazing or not. Unimpacted wetlands could be discriminated from impacted wetlands based on the carbon use patterns (Figure 12.10). Each symbol represents the mean of three replicate EcoPlates. Circles are impacted wetlands, and squares are unimpacted territories. There is little to no overlap between these wetland types, even though some of the wetlands (impacted and unimpacted) are hundreds of kilometers apart. However, even though the two wetland types can be discriminated using the physiologic test strips, there is considerable variation within each wetland type. The canonical scores have much spread along the first and second axes. These data suggest that inherent variability among wetlands exists but that the impact of cattle causes an overriding shift.

Functional characterization of communities gives some insight into similarities of differences between the communities being compared. How do these types of observations relate to functional redundancy? These types of studies can provide data on whether certain biochemical processes can occur, but they do not indicate that these processes do occur under the environmental conditions the samples were taken from.

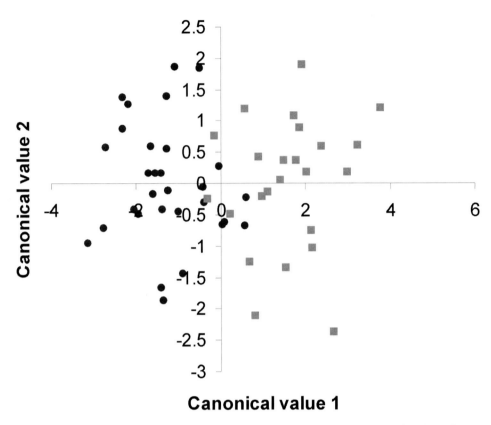

**Canonical value 1**

**Figure 12.10** Canonical discriminate plot of scores obtained from physiological test strips contrasting bacterial communities found in grazed and ungrazed springs. Most discrimination is obtained along the first axis. *Circles* are impacted wetlands, and *squares* are unimpacted territories There is very little overlap in test strip scores between grazed and ungrazed locations. (From Merkely M, Rader RB, McArthur JV, Eggett D. Bacteria as bioindicators in wetlands: bioassessment in the Bonneville Basin of Utah, USA. *Wetlands* 24:600–607, 2004.)

What does it mean that certain functions are found in very different communities or even fairly similar communities? The more organisms and types of organisms that can perform the physiological test seem to indicate greater health of the community. From our example, the highest metabolic diversity was found in the most disturbed locations. Based on metabolic diversity, we may conclude that the disturbed site is healthier than the undisturbed site. Care must be taken to interpret these results.

## Niche Constructionists

Organisms modify their environments, and over time, the environments modify the organisms. This is the fundamental basis of evolutionary theory and ecology. The very act of surviving causes changes in the abiotic and biotic components of an organism's environment. For example, all aerobic organisms consume oxygen. If oxygen is not replenished, the environment becomes more and more inhospitable for aerobes and more and more suitable for anaerobes.

Beavers are classic examples of organisms that modify their environment to make it more suitable for themselves. Beaver cut trees and build dams, which flood the adjacent riparian zones. Lodge-building beavers then construct a retreat that is protected from predation in the middle of the reservoir created by damming.

Organisms must obtain carbon from some source. Organisms that mineralize organic compounds to derive carbon have, at least seasonally, a limited source of the organic matter. The act of breaking this material down results in a decreasing abundance of the material and a subsequent increase in the breakdown products and waste products of the process. The accumulation of these waste and breakdown products further modifies the environment. Sometimes, these modifications can be extreme and result in an almost exclusive environment in which only the modifier can survive. Examples are bacteria that produce $H_2SO_4$. The production of the acid keeps the pH at such a low level that no or relatively few other bacteria are able to survive. These organisms therefore eliminate potential competitors and predators. The downside is that they are restricted to the harsh environments they help to create. For example, the Tinto River in Spain is an extreme environment that has a mean pH of 2.2 and fairly high concentrations of heavy metals (López-Archilla et al., 2001). The metals and the pH are products of the metabolic activity of chemolithotrophic microorganisms, specifically sulfur- and iron-oxidizing bacteria. The very act of surviving by these microorganisms alters their environment such that no other organisms are able to live there.

Some organisms modify the environment in ways that benefits themselves and other organisms. Often, the end products of one metabolic process are the starting elements of another. This topic is discussed later, especially as it relates to nutrient cycling in the environment.

To what extent are organisms able to construct a suitable habitat for themselves and their posterity? *Niche construction* is the process whereby organisms are able to construct, modify, and select important components of their environment (Day et al., 2003). In the process of altering their environment, these niche constructionists alter the selection pressures that will act on their posterity, and they do this in a nonrandom manner. Theoretical work by Laland et al. (1996, 1999) suggests that niche constructionists can have very strong evolutionary effects that can lead to evolutionary inertia and momentum, fixation of otherwise deleterious alleles, support of stable polymorphisms where none is expected, elimination of other stable polymorphisms that would have been stable, and influence on disequilibrium.

These studies are intriguing and suggestive. How extensive in nature are these sorts of interactions and modifications? Clearly, there are limitations. For example, anaerobic photoautotrophic bacteria have a very limited niche in which they can survive. Their ability to alter or maintain that environment is similarly restricted because of the limitations imposed on their metabolism and physiology. Any organism or physical condition which alters the light regime, especially light penetration, will affect the distribution of these organisms.

The earth's atmosphere and physical conditions are very different from those 4 billion years ago. The alterations to the atmosphere and physical conditions have been affected by the origin of life and life processes. There was initially no free oxygen available. With the advent of photosynthesis, oxygen was produced faster than it was bound up by physical or chemical processes, beginning the trajectory of environmental changes that favored the evolution of life that is familiar to most of us and resulted in the form of life capable of studying and investigating the earth. Humans, over the short haul, are the greatest niche constructionists, but over geologic time, microbes have done more to alter the environment that any other group of organisms, including the green plants.

ccurred in the watershed of rivers and in the sediments and water column of the
ivers provide various forms of nitrogen to the oceans hundreds to thousands of kilo-
neters downstream of the source waters.

To avoid confusion about cycle linkages, we approach nutrient cycling from a different perspective, that of evolutionary ecology. Although the unique and singular
mechanisms and chemistry involved in the transformation of the various nutrients are
beautiful and complex, they are discussed only briefly. Instead, we seek to understand
how these mechanisms interact within and between organisms to promote survival
and reproduction.

All living organisms need carbon for almost every aspect of metabolism, especially
anabolism. Only organisms that can perform photosynthesis are able to take inorganic
carbon dioxide and convert (i.e., fix) this gas to organic molecules. Photosynthesis in
green plants is limited by the availability of water and carbon dioxide gas. Carbon
dioxide ($CO_2$) makes up only about 0.03% of the atmosphere. However, plants annually produce approximately 75 billion metric tons of organic carbon through photosynthesis from an estimated 500 billion metric tons stored as dissolved carbon dioxide
in the oceans and 700 billion metric tons in the atmosphere. Although there seems to
be plenty of carbon dioxide, plants can be limited by the local distribution of the gas.

All green plants take up carbon dioxide gas through *stomata* located on their leaf
and stem surfaces. Once inside the plant, differences in biochemistry (e.g., C3 versus
C4 plants) determine the pathways for organic carbon synthesis. However, the movement of carbon dioxide into the plants and the subsequent biosynthesis are fairly
similar among the wide diversity of plant types. Differences in leaf and plant morphologies in part result from interactions among the needs for gas exchange, light
capture, and prevention of water loss. All green plants must take up carbon dioxide.
None is exempt. Organisms that obtain carbon from sources other than photosynthesis are not plants.

All living organisms require nitrogen to make proteins, enzymes, and other cellular
components. Nitrogen gas makes up 78% of the gases in the atmosphere. Most organisms are unable to use elemental atmospheric nitrogen, and nitrogen is therefore often
a *limiting* nutrient because its availability restricts the growth and reproduction of
organisms. Seventy-eight percent of the atmosphere contains a nutrient that is required
by all living things, but this nutrient is unavailable for most organisms. Those that can
take up nitrogen gas do so through a process called *nitrogen fixation*. Carbon dioxide
can sometimes be a limiting factor, but it is hard to conceive of a situation in which a
nutrient that constitutes 78% of the total volume of atmospheric gases would be limiting. Although many species can fix carbon through photosynthesis (i.e., all green
plants), relatively few organisms can fix nitrogen. Why are there relatively few organisms that are able to fix elemental nitrogen? Why has evolution not favored more, if
not all, organisms to have the ability to capture this abundant and plentiful nutrient?

## Nitrogen Cycle

Before we can understand the nitrogen cycle, we must first examine the nitrogen cycle
of much earlier times. The earth's early atmosphere was different from that of today
in a number of ways. It was a reducing atmosphere, and there was no free oxygen gas

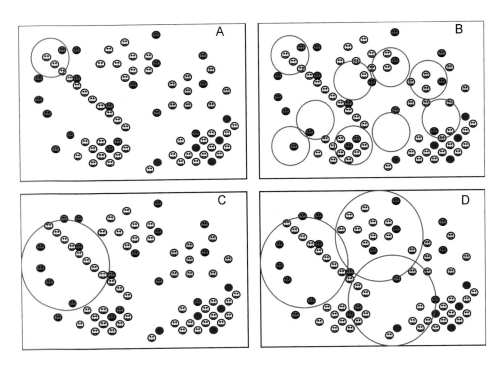

**Figure 11.1** Hypothetical populations of different smiley faces are used to show the effect of different-sized sampling devices on capturing the true level of microbial diversity.

Slow flow    Fast flow

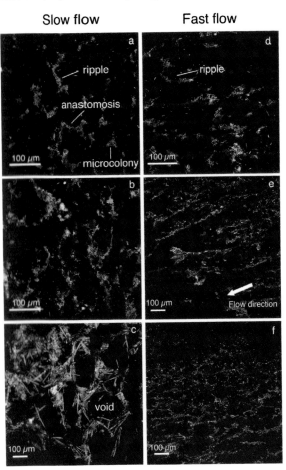

**Figure 12.5** Developmental architecture of biofilms under low-flow and fast-flow conditions as determined by confocal laser-scanning microscopy. (From Battin TJ, Kaplan LA, Newbold JD, Cheng X, Hansen C. Effects of current velocity on the nascent architecture of stream microbial biofilms. *Appl Environ Microbiol* 69:5443–5452, 2003.)

# Microbes and Processing of Nutrients

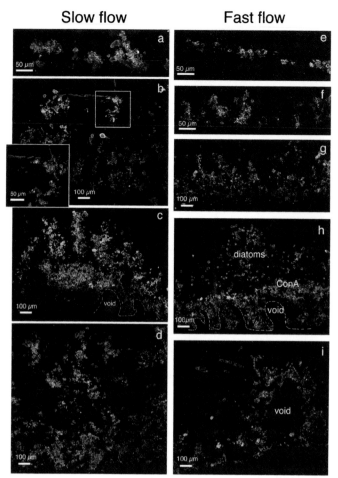

Slow flow        Fast flow

**Figure 12.6**  Three-dimensional development of river biofilms. Confocal laser-scanning microscopy micrographs of cryosections show the structural development of the biofilm x,z plane in slow flows (**a** to **d**) and fast flows (**e** to **i**). (From Battin TJ, Kaplan LA, Newbold JD, Cheng X, Hansen C. Effects of current velocity on the nascent architecture of stream microbial biofilms. *Appl Environ Microbiol* 69:5443–5452, 2003.)

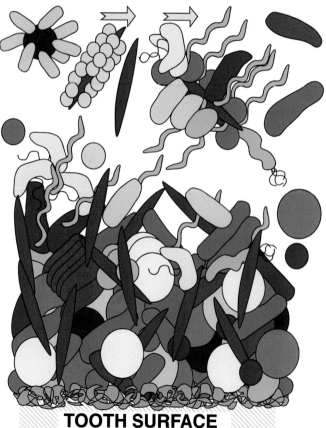

**TOOTH SURFACE**

**Figure 12.8**  Diagram of an oral biofilm on an enamel surface. The tooth surface is coated with a host-derived acquired pellicle (i.e., gold, curly meshwork at the bottom) to which accrete various colored, morphological shapes representing genetically distinct cell types. The *yellow arrows* indicate salivary flow, which provides a mechanism for individual bacteria, rosettes, corncob formations, and complex coaggregates to come in contact with the biofilm. Salivary flow also contributes to the detachment of cells from the biofilm. (From Figure 1 in Kolenbrander PE. Oral microbial communities: biofilms, interactions, and genetic systems. *Annu Rev Microbiol* 2000.)

## Nutrient Cycling

All living organisms require some basic substances to grow an things require some form of carbon, nitrogen, and phosphorous ture, including protein composition, nucleic acid compleme enzymes, and other information and structural molecules. Many needed in greater or lesser amounts and in various molecular fo *ent* has meaning only in the context of some living organism. organism, the chemical modifications of these substances are a fu or organic chemistry. It is the intimate coupling of the chemical c by living organisms to maintain growth and reproduction that def

Most ecology textbooks discuss microbes in the chapters deal cycling. These discussions graphically and verbally demonstrate tl linkages that serve to move nutrients from abiotic to biotic to abioti An understanding of these intricacies is important. The diagrams cussions of complex nutrient cycles appear as seamless logical flows o energy. It is true that the metabolic products of one set of organism the necessary materials for another set of organisms, but each of t evolved independently or nearly so. Microbes that produce a product ing it as a service to other organisms; release is a function of the bioche organism. The fact that some other organism evolved to take advantage resource is totally independent of the first process.

One of the dangers of the compact, often circular, diagrams that grapl nutrient cycles is that these processes appear to be tightly linked in spac even though they may be separated by significant geographic distances ar cases, temporally by millennia of years (e.g., materials sedimented to oce Although tight spatial and temporal linkages may be true for some aspe cycles, it is not necessarily true for all components. For example, much of gen load in coastal waters is imported from rivers. Transformations of nitro

278

The primary gases were nitrogen, ammonia, hydrogen, methane, and water vapor. Any free oxygen released by geochemical processes was rapidly bound up in oxides of various metals. For more than 2 billion years, the levels of oxygen gas in the atmosphere were less than 0.01% of today's atmosphere (Loomis, 1988).

Plenty of energy existed in the conditions of early earth. Solar energy in the form of ultraviolet light and high-energy particles provided continuous inputs of energy. Heat energy derived from geologic and solar energy maintained relatively hot oceans and baked the land. Water vapor collected into clouds and precipitated back to the surface after cooling. Lightning from thunderstorms and from static electricity generated during volcanic activity provided electrical energy. Numerous studies have shown that any or all of these potential energy sources are capable under laboratory conditions of converting the gases in the early atmosphere into complex organic molecules, including amino acids and nucleotides. Volcanism was a primary source of large amounts of ammonia needed to make the amino acids.

Because the origin of life is discussed in Chapter 2, we can omit the development of the earliest living organisms here. In the primordial mixture of the earth's oceans, there was a ready supply of amino acids and other molecules of life derived from the abiotic processes previously described. Biological synthesis of these building block molecules was not needed, and we may suppose the cycling of nitrogen or any other element by living organisms was not needed. However, at some point, the biosynthetic requirements and, ultimately, the reproduction of new organisms exceeded the abiotic capacity of the earth to provide ready-made molecules. Even before the abiotic source of nitrogen compounds was exceeded, early microbiota probably began to take advantage of the nitrogen-based organic compounds released after senescence and cell lysis.

Nitrogen in the early earth was probably fixed abiotically as nitric oxide during lightening discharges. Based on this probable scenario, can we surmise which aspects of the nitrogen cycle should have evolved first? Was this evolution *monophyletic* (i.e., single evolutionary event) or *polyphyletic* (i.e., originated in more than one type of organism)? How can we determine the answers to these and similar questions about processes that occurred millions of years ago? Based on our answers to these questions, can we predict where certain links in nitrogen cycling should occur today? Can we predict something about the metabolic and species diversity associated with these processes?

Before we proceed much further, it is important that we understand something about the chemistry of nitrogen. Nitrogen was discovered in the 18th century by French chemists who named it *azote* or *gaz nitrogene* (Aulie, 1970). *Azote* means "without life." Organisms that can fix nitrogen are collectively called *diazotrophs*, which means "$N_2$ eaters" (*diazo* means "di-nitrogen," *troph* means "eat").

Because of the different valence states of nitrogen, natural selection can favor organisms or enzymes that can facilitate alterations in the chemical state. Nitrogen, like carbon, can be completely oxidized or completely reduced. Ammonia ($NH_3$) is the reduced form, and nitrate ($NO_3^-$) is the oxidized form of nitrogen. Various intermediates form, such as nitrite ($NO_2^-$).

In the early earth's environment, organic molecules containing nitrogen were formed through various abiotic processes. These fundamental building blocks of life were formed without any metabolic reactions associated with a living organism. Every amino acid is made up three components: an amino group ($NH_2$), a carboxyl group (COOH), and a central carbon atom. From this basic arrangement, other functional groups can be appended to form the different amino acids. Amino acids contain carbon, oxygen, hydrogen, and nitrogen. The process of breaking down organic nitrogen compounds is called *ammonification*. In ammonification, organic molecules containing nitrogen are hydrolyzed and *deaminated* (i.e., $NH_2$ group removed), and carbon dioxide is respired with concomitant assimilation of needed nitrogen compounds into the *ammonifying* organism and the release of excess nitrogen in the form of ammonia or ammonium ions.

Most microorganisms, plants, and animals are capable of ammonification. Paerl (1993) lists photoautotrophs (oxygenic and anoxygenic), heterotrophic bacteria (including most genera of aerobes and facultative anaerobes), and most genera of chemolithotrophs as ammonifiers. In freshwater lakes, significant increases in ammonia release occur when organic matter sediments into the anoxic hypolimnion of stratified lakes (Wetzel, 1983). Given the antiquity of amino acids, it seems probable that ammonification was among the first links of the nitrogen cycle.

At some point, the density of organisms that relied on ammonia produced through volcanism and ammonification exceeded the production of ammonia by these processes, and the supply of nitrogen became a limiting factor. At this juncture, organisms capable of removing nitrogen gas dissolved in water or directly from the atmosphere began to have a selective advantage. Organisms capable of capturing nitrogen gas were no longer limited to finding nitrogen in compounds released by other organisms.

The evolution of this process is incredible from a biological and an energetic point of view for at least two reasons. First, nitrogen fixation is an energy-expensive reaction. The primary reaction is the formation of two ammonia molecules from elemental nitrogen.

$$N_2 + 3H_2 \rightarrow 2NH_3 \, (\Delta G = +150 \, kcal/mole), \, 6 - 15 \, ATP$$

This reaction requires significant inputs of energy, because $N_2$ is a very stable molecule that contains a triple bond and the activation energy required to break these bonds is extremely high. The reduction of ammonia in cell-free extracts of nitrogenase is coupled to the hydrolysis of up to 15 ATP molecules, depending on the conditions. However, in vivo, the number of ATP molecules required is probably closer to six (Gottschalk, 1979). This energy demand is met through autotrophic or heterotrophic processes.

Nitrogen-fixing organisms can be free living or symbiotic. Free-living nitrogen fixers can be found among the cyanobacteria and bacterial species belonging to most of the known orders and families of bacteria. The symbiotic nitrogen fixers are often associated with various types of plants. For example, all species of the bacterial genus *Rhizobium* are associated with legumes, *Frankia* species with early successional plants like alders, and *Spirillum lipoferum* with certain species of tropical grasses. These symbiotic associations developed after the evolution of terrestrial green plants, an event that occurred long after nitrogen fixation as a biological process evolved. Because plants evolved much later than bacteria and cyanobacteria, these associations are of comparatively recent origin. However, the free-living nitrogen fixers are among the most ancient of organisms. Fossil remnants of cyanobacteria-like organisms have been dated to the earliest of times, which is extremely interesting from evolutionary and ecological perspectives. It implies that photosynthesis and perhaps nitrogen fixation evolved fairly early.

Both processes have had considerable impact on the nature of the earth, including the physical and chemical environment of much of the earth's inhabitable places. The chemical and physical changes have brought about conditions that favored the evolution of most other organisms. Nitrogen fixation and photosynthesis evolved in microbial taxa and altered the evolution of other microbial taxa long before higher plants or animals evolved.

The second reason more organisms are not nitrogen fixers is that oxygen inhibits the enzyme (*nitrogenase*) that catalyzes this reaction. Oxygen irreversibly inactivates nitrogenase. In the early earth's atmosphere, this was not a problem because oxygen was absent or existed at very low levels. Any available oxygen was bound quickly to various metals. Cyanobacteria are the only diazotrophs (i.e., nitrogen-fixing bacteria) that produce oxygen during photosynthesis. This creates a unique problem in that the by-product of one metabolic process inhibits or prevents the other metabolic process. To overcome this problem some cyanobacteria have evolved unique structures known as *heterocysts* (Figure 13.1). Heterocysts are specialized cells with thick walls, and they do not produce oxygen. Nitrogen fixers that do not have heterocysts rely on other adaptations of physiology, behavior, or morphology (Paerl, 1993) to create microzones of oxygen depletion. For example, aggregations of nitrogen-fixing cells can deplete oxygen through metabolism. Similarly, various consortia of heterotrophic non–nitrogen-fixing bacteria associated with the nitrogen fixers can result in oxygen being consumed to levels that allow nitrogen fixation to occur.

Nitrogen fixation requires primordial conditions. These conditions exist now only in microenvironments and various biologically controlled situations, but all life depends on the fixation of gaseous nitrogen so that structural, informational, and

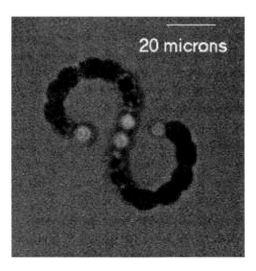

**Figure 13.1** *Anabaenopsis circularis.* The central pair of heterocysts is visible (magnification ×200). (From Roger Burks [University of Californial at Riverside], Mark Schneegurt [Wichita State University], and from Cyanobacterial Image Gallery at http://www-cyanosite.bio.purdue.edu.)

metabolic molecules can be formed and allowed to function. It is incredibly interesting how nitrogen fixers continue to capture nitrogen from the atmosphere for their own use and how all other organisms have evolved to take advantage of transformations of this nitrogen.

The nitrogenase gene is a complex gene made up of several subunits. Analyses of the nitrogenase subunits using a variety of molecular biological tools have revealed interesting patterns and suggestions of evolutionary development. One analysis of 30 nitrogenase DNAs deduced amino acid sequences obtained from five major taxonomic subdivisions of nitrogen fixers and demonstrated great variation in phylogenetic distances (Zehr et al., 1997). These 30 samples were from heterocystous and nonheterocystous cyanobacteria. These two major groupings (with or without heterocysts) formed distinct clusters. Within the heterocystous group, there was not as much variation as that found within the nonheterocystous group. For example, branching forms of heterocystous cyanobacteria were not different from nonbranching forms. In contrast, representatives of the filamentous nonheterocystous cyanobacteria did not form tight clusters within the group, but rather formed deep branching patterns between the heterocystous forms.

Using similar techniques on another group of nitrogen-fixing bacteria, the *nifH* genes were sequenced in five genera of methanotrophs: *Methylococcus*, *Methylosinus*, *Methylocystis*, *Methylomonas*, and *Methylobacter*. Phylogenetic analyses revealed that the type I and type II methanotroph nitrogenase genes were very different from each other but that the type I nitrogenase gene was similar to that of the alpha-Proteobacteria, whereas the nitrogenase gene from type II methanotrophs were more similar to the gamma-Proteobacteria (Boulygina et al., 2002).

Did the nitrogenase gene arise in the proteobacteria, and was it subsequently passed horizontally to other groups such as the cyanobacteria and methanotrophs or vice versa? The answer seems to depend on which component of the nitrogenase gene is examined. Phylogenetic analysis of *nifH* has suggested that this gene has been horizontally transferred from a proteobacterium to the cyanobacteria. Some analyses based on the *nifK* sequences suggest a vertical descent (i.e., from cell to daughter cells during binary fission), whereas others suggest horizontal transfer (i.e., between related or unrelated species through conjugation) (Hirsch et al., 1995). If the *nifK* gene evolved by vertical descent, it suggests that the Proteobacteria and the Cyanobacteria are sister phyla. Extant eubacterial *nif* genes appear to have at least three distinct origins from ancient gene duplications. Gene duplication is an effective mechanism to increase the metabolic diversity of an organism. As long as one copy of the gene continues to function within the bounds of the metabolic process it codes for, all other copies are free to undergo mutation and change.

What is the ecology of nitrogen fixation? Nitrogen fixation occurs in myriad environments, and by an amazing diversity of organisms, it produces an incredible amount of fixed nitrogen. It is estimated that approximately $20 \times 10^6$ tons year$^{-1}$ of molecular nitrogen is fixed in the oceans (Rheinheimer, 1992). However, most of the fixed nitrogen in the oceans remains in the oceans. Although this immense amount of nitrogen is converted from gas to organic and inorganic forms of nitrogen and drives or limits the productivity of the oceans, it is not available for terrestrial or freshwater organisms. Without nitrogen fixation in terrestrial and freshwater environments, life as we know it would not exist.

An understanding of the ecology of nitrogen fixation is based on an understanding of the biochemistry of nitrogenase and nitrogen fixation. The activity of the nitrogenase enzyme depends on certain environmental conditions, including anaerobic conditions, and a source of low-potential electrons, ATP, and trace metals, especially molybdenum and iron (Benson, 1985). Discovery of new techniques to measure nitrogen fixation (i.e., acetylene reduction, allowed the assessment of almost every habitat). $N_2$ fixation has been found in many different habitats.

## Fixation in Soils

Free-living and symbiotic bacteria and cyanobacteria carry out nitrogen fixation in soils. Early investigations into factors controlling the fertility of soil resulted in awareness that certain plants increased fertility. Crops such as clover began to be rotated with other crops in England and France during the 17th century, because the land was more fertile after the clover than before. Legumes such as peas, beans, and soybeans increase soil fertility. This observation was made centuries before farmers began to rotate their crops. Benson (1985) reports that Neolithic and Bronze Age farmers used legumes as founder crops or crops that increase the fertility. Ancient Greeks and Romans recognized that beans and alfalfa in some way "manured" the soil. Rice farmers knew that the water fern *Azolla* increased their harvest. Given the close association of ancient farmers and their environment, it might be expected that they would notice differences in harvest efficiencies and over time subscribe the effect to one plant over another. However, other nonleguminous plants that had little or no agricultural use had similar affects on associated plants. For example, alder (*Alnus glutinosa*) was observed in the 1600s to "nourish" plants set near by (Benson, 1985).

Free-living bacteria, including *Azotobacter* and *Clostridium pasteurianum*, were isolated from soils and in pure culture shown to fix nitrogen. After the determination that free-living nitrogen fixers could be isolated from soils, many different soils were examined to determine whether these organisms were present and whether nitrogen fixation was occurring. The broad range in soils examined included dune soils, beach soils, silt loam, paddy soil, forest soils, arctic soils, and desert soils. Although there is considerable variation in the amounts of nitrogen fixed within these different habitats, significant amounts could be detected in all habitats. Rates of nitrogen fixation were affected by the type, acidity, temperature, moisture content, element content, and origin of the soil (Benson, 1985). Despite the different rates among soil types, every soil had nitrogen fixers, and they were fixing nitrogen. Considerable amounts of nitrogen are fixed in climax vegetation communities. In most of the soils, cyanobacteria were easily isolated and probably contributed to most of the nitrogen fixation.

The deep phylogenetic branching of heterocystous and nonheterocystous nitrogen fixers is indicative of the ancient origin of these types (Figure 13.2). The evolution of the heterocyst allowed nitrogen to be fixed in the presence of oxygen being produced by photosynthesis in the other cells. This creates a dilemma of sorts, because the nitrogen is being produced in specialized cells that are not dividing. The actively growing cells, where photosynthesis is occurring, must obtain the fixed nitrogen. What delivery system allows the nitrogen to travel from the heterocyst to the other cells? Under anaerobic conditions, the vegetative cells have been shown to fix nitrogen, which indicates that the initial condition is nonheterocystous.

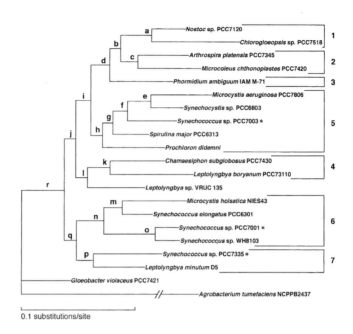

**Figure 13.2**  Maximum likelihood tree of selected 20 cyanobacteria and one outgroup. The horizontal length of each branch is proportional to the estimated number of substitutions. *Asterisks* indicate the strains for which 16S rRNA gene sequences were determined in this study. Groups 1–7 are phylogenetically distinct groups. Group 1 represents the heterocyst forming taxa. (From Honda D, Yokota A, Sugiyama J. Detection of seven major evolutionary lineages in cyanobacteria based on the 16S rRNA gene sequence analysis with new sequences of five marine *Synechococcus* strains. *J Mol Evol* 48:723–739, 1999, with kind permission of Springer Science and Business Media.)

Symbiotic associations between species of nitrogen-fixing bacteria and green plants result in considerable amounts of nitrogen being fixed. Among the best studied of these symbiotic associations are those between various species of *Rhizobium* and other species of legumes. Because considerable details of the formation and maintenance of these associations exists in the primary and review literature, we offer only a cursory description. These associations are often species specific.

Free-living *Rhizobium* organisms living in the *rhizosphere* of the plants are stimulated by exudates from the roots of the plants and increase in density within the membranous layer around the outside of the roots. By growing in close association with the roots, the bacteria prevent the root exudates from escaping into the soil and thereby making the exudates unavailable for other soil bacteria. Further modifications occur in the plants and the bacteria. Enzymes secreted by the bacteria break down the cellulose fibers that surround the root hairs and allow the bacteria access to the root. The bacteria change their morphology to spherical highly flagellated cells from rod-shaped cells. These new cells are called *swarmer cells*, which enter the root hair and increase in number, and the plant produces a hyphal-like infection thread that grows and branches. After the infection thread has formed, the bacteria move within this thread and infect root cells. All root cells are susceptible to infection, and most infections results in cell death. If the root cell happens to be *tetraploid* (i.e., four sets of chromosomes), nodulation may develop. The bacteria reproduce within these cells and become enveloped in a membrane that is made by the host cell. The bacteria then undergo another morphological change into swollen and irregularly shaped cells called *bacteroids*. The bacteroids are involved in nitrogen fixation. Nodulated plants continue to produce exudates.

This relationship between the bacteria and the plants are usually described as a positive symbiosis in which both species benefit from the association. The plant benefits by obtaining a supply of usable nitrogen, with the cost of some metabolic products being released into the rhizosphere. In this case, the benefits in terms of energy and nutrition are clearly advantageous for the plants. How about the bacteria? The usual answer is that the bacteria benefit by having the plant exudates as a readily usable carbon and energy source. However, the bacteroids are terminally differentiated cells that are incapable of reproduction in many associations. Exactly what are the benefits to the bacteria? In evolutionary terms, the bacteria continue to benefit the plants by allowing the plant to have sufficient nitrogen to grow and reproduce, but they themselves have become an evolutionary dead end. Natural selection can favor only individuals that reproduce, and the bacteroids do not reproduce.

Free-living rhizobia can reproduce. Swarmer cells that penetrate and infect the plant are destined to not reproduce, but their nodulation and nitrogen fixation cause the plant to continue to produce and release exudates that benefit the free-living cells that remain outside the host. What is the degree of relatedness between the cells that do not infect and those that do? The evolution of symbiotic nitrogen fixation may rely on kin selection (Bever and Simms, 2000), but the release of exudates can and probably does benefit totally unrelated nonmutualists.

Nitrogen-fixing populations of *Rhizobium* may be susceptible to invasions and competitive exclusion by non–nitrogen-fixing species, especially saprophytic rhizobia. However, these symbiotic relationships between plants and bacteria continue to exist. Some bacteroids produce special compounds called *rhizopines* from plant metabolites. Rhizopines are compounds that cannot be re-assimilated by the plant, nor can they be used by unrelated free-living bacteria. The continued release of root exudates and rhizopines assumes certain spatial distributions of the free-living rhizobia to be evolutionarily meaningful. If the free-living bacteria are too diverse or in a large mixture of unrelated bacteria, there would be little benefit to the free-living forms that would not be available to other nonrelated bacteria.

Phylogenetic analysis of the rhizobia based on the 16S rRNA subunit places the species into three genera: *Rhizobium*, *Bradyrhizobium*, and *Azorhizobium*. All three of these genera are within the alpha subdivision of the Proteobacteria but on distinct branches (Young and Haukka, 1996). Young and Haukka suggest that the ability to form nodules may not have been present in the ancestral rhizobia but that the nodulation genes were transferred between phylogenetically distinct bacteria. The phylogeny of the nodulation genes is more than likely different from the phylogeny of the bacteria. Although nitrogen-fixation genes can be linked to nodulation genes, they may have different evolutionary histories. Observations such as these are exciting and suggest the diverse ecologies of bacteria and their genes.

The almost global distribution of nitrogen-fixing organisms and the universal dependence on fixation to provide nitrogen to microorganisms indicate the age of this process. Given the diversity of habitats and environmental conditions that have been examined, it should be fairly obvious that levels of abiotically fixed nitrogen are not sufficient and must be replenished through biological fixation. Loss of nitrogen fixers through senescence and death is continuous, and nitrogen-fixing bacteria must reproduce at rates higher than the rate of senescence. Nitrogen fixation did not evolve to benefit other microorganisms or higher organisms; it instead evolved to benefit nitrogen-fixing bacteria. The worldwide distribution of these organisms suggests that the nitrogen-fixing way of life is robust and successful and that even if all other components of the nitrogen cycle were disrupted, $N_2$ would continue to be fixed.

## Denitrification

Combined nitrogen can be removed from terrestrial and aquatic systems through the process of denitrification. Denitrification is the reduction of oxidized nitrogen (nitrite and nitrate) by bacteria during the oxidation of organic matter under anaerobic conditions. The following reactions from Wetzel (1983) are examples of denitrification and concomitant glucose oxidation using nitrate and nitrite, respectively.

$$C_6H_{12}O_6 + 12NO_3^- = 12NO_2^- + 6CO_2 + 6H_2O$$
$$\text{(free energy} = -460 \text{ kcal mol}^{-1})$$

$$C_6H_{12}O_6 + 8NO_2^- = 4N_2 + 2CO_2 + 4CO_3^- + 6H_2O$$
$$\text{(free energy} = 720 \text{ kcal mol}^{-1})$$

The amount of free energy obtained from the anaerobic oxidation of glucose with concomitant denitrification is almost equal to that obtained aerobically ($-720$ kcal mol$^{-1}$) (Wetzel, 1983). From these reactions, the complete reduction of nitrate and nitrite results in nitrogen gas being released, which may be lost from the system if not refixed by nitrogen-fixing bacteria.

As with nitrogen-fixing bacteria, bacteria that perform denitrification have evolved to take advantage of conditions and resources that ensure that these organisms leave copies of themselves in subsequent generations. The fact that the ending product of denitrification is the starting product of nitrogen fixation is not destiny but rather biochemistry. Both processes evolved under anaerobic conditions, and both require anaerobic conditions to complete the reactions. The optimum temperature for denitrification is, as Wetzel (1983) states, "well above the temperature of most natural fresh waters. At high temperatures the primary product is $N_2$ . . ." At cold temperatures or under acidic conditions, denitrification occurs very slowly.

Nitrogen fixation and denitrification are found among diverse groups of ancient anaerobic bacteria (autotrophic and heterotrophic), which include sulfur and sulfate reducers, cyanobacteria, methanogens, and other photosynthetic bacteria. Both processes are highly conserved and widespread. Many facultative anaerobes are capable of denitrifying. These facultative organisms have completely developed oxygen respiration, but in the absence of oxygen, they are able to switch to nitrate respiration. The switch from oxygen respiration to nitrogen respiration requires time before the enzymes used in the process (i.e., membrane-bound nitrate reductase and nitrite reductase) can be formed. However, changing back to oxygen respiration is nearly instantaneous, indicating that oxygen respiration is constitutive (Rheinheimer, 1992).

Because of the requirement of anaerobic conditions, denitrification is restricted to habitats that meet the oxygen-depletion requirement and to locations where there are sources of oxidized substrates (i.e., $NO_3^-$ and $NO_2^-$). This confinement of the process results in a few habitats with very lively denitrification and others with little to no denitrifying activity. Paerl (1993) suggests that, at least for the oceans, biogeochemical impacts over large spatial and temporal scales might have influenced the flux and limitation of nitrogen at various times and places by affecting the source and sinks involved in nitrogen transformation processes.

In the earth's early biotic conditions, denitrification might have been very widespread and resulted in the replenishment of $N_2$ fixed through biological and abiotic processes. With the advent of photosynthesis and the production of free oxygen, denitrification would have been relegated to the limited habitats mentioned previously. Respiration based on nitrate or nitrite is only about 10% less efficient than respiration based on oxygen. However, a 10% higher efficiency over many generations is significant and would result in many more copies of bacteria able to use oxygen than those relying solely on nitrogen respiration. Bacteria that obtained a complete oxygen respiration system and maintained the nitrogen respiration capability would clearly have advantages over obligate aerobes or obligate anaerobes. Over ecological and evolutionary time scales, changes in oxygen concentrations have occurred in many habitats. Facultative organisms have advantages over the long and short hauls during which variation in resources may change. An organism that is obligate for some resource probably has an advantage in habitats where the chance of change in that resource is rare. These organisms presumably are more efficient at reproduction than organisms that must maintain the genes or the genes and the products of alternate strategies.

Denitrification is also known as dissimilatory nitrate reduction, as opposed to assimilatory nitrate reduction. In assimilatory nitrate reduction, the reduced nitrate or nitrite is incorporated into bacterial biomass. Studies done that measured denitrification have shown that up to 90% of labeled ($^{15}N$) nitrate was reduced in the first 2 hours after introduction, with most of the nitrogen going to the formation of $N_2$. However, other studies (Keeney et al., 1971) have shown that almost 40% of introduced labeled ($^{15}N$) nitrate was incorporated into bacterial biomass by assimilatory nitrate reduction.

## Nitrification

The oxidation of ammonium ion ($NH_4^+$), which is the most reduced form of nitrogen, can be performed by bacteria with the release of energy. This energy is used by the bacteria to reduce carbon dioxide and make organic substances. The complete oxidation of ammonium is accomplished in two steps by a number of different bacteria. This process is collectively called *nitrification*. The first step involves the oxidation of ammonium to nitrite ($NO_2^-$), and the second step involves the oxidation of nitrite to nitrate ($NO_3^-$). The reactions involved in nitrification can be written as follows:

$$NH_4^+ + 1\tfrac{1}{2}O_2 \rightarrow NO_2^- + H_2O + 2H^+ + energy$$
$$NO_2^- + 1\tfrac{1}{2}O_2 \rightarrow NO_3^- + energy$$

Both reactions require oxygen, and organisms that are involved in nitrification therefore must be aerobic. The requirement of oxygen suggests that nitrification evolved much later than the other nitrogen transformation processes. This idea is strengthened by the observations that relatively few bacteria are capable of performing nitrification.

Formerly, all nitrifiers were placed in a single family (Nitrobacteriaceae), but based on recent molecular analyses, the bacteria that perform the first step and the bacteria involved in the second step are not closely related. All nitrifiers are obligate chemoautotrophs (Rheinheimer, 1992), using carbon dioxide as the sole carbon source. The nitrite bacteria include species in the genera *Nitrosomonas*, *Nitrosococcus*, and *Nitrosospira*. The nitrate bacteria fall into the genera *Nitrobacter*, *Nitrospina*, *Nitrococcus*, and *Nitrospira*. Species of nitrifying bacteria found in soils and aquatic environments appear to be the same.

Because nitrite and nitrate bacteria are inhibited by light, we should expect higher numbers of these organisms associated with sediments. In open water, where light is most abundant, other planktonic organisms, especially phytoplankton, out-compete nitrifiers for available ammonium. However, nitrification in aerobic sediments is usually higher than in the open water.

## Nitrogen Transformation Summary

If we consider the major nitrogen transformation processes together, these processes seem to form an integrated whole, or a cycle. The ending products of one transformation are the starting products of another process. However, the simplistic view that the products of one process are immediately used by organisms in another process is at best misleading. Each of these processes has environmental constraints, such as temperature and pH, that affect the process positively or negatively. These environmental constraints are not the same for each process. Ideal conditions for denitrification may be very much different from the ideal conditions for nitrogen fixation, and these differences may increase or decrease the available pool of starting or ending products. Some forms of nitrogen are much more mobile than others, and they may be transported considerable distances before additional transformations are performed. The availability of various carbon sources affect the rates of some of these transformations. Walvoord et al. (2003) found high concentrations of nitrate a few meters below the surface of many desert environments. This accumulation, or sink, for nitrogen demonstrates that the presence of a particular form of nitrogen does not mean that transformation of that product will necessarily occur.

# Sulfur Biogeochemical Transformations

We have spent considerable time on aspects of nitrogen transformations. Microorganisms have evolved to take advantage of other minerals and nutrients that are necessary for growth. Many of these substances have fewer *valence* states and therefore fewer possible transformations that can occur biotically or abiotically.

Sulfur is an essential mineral required by all organisms for the synthesis of proteins and enzymes, especially those made with the amino acids methionine, cystine, and cysteine. Transformations of sulfur are somewhat similar to those of nitrogen. The release of sulfur as a component of proteins is accomplished by a wide number of microorganisms during the degradation of the protein into its component amino acids. Sulfur is taken up by plants chiefly in the oxidized form of sulfate ($SO_4^{2-}$) and subsequently converted into plant protein. In a process analogous to ammonification, sulfur is converted into $H_2S$, which is highly toxic and unstable under aerobic conditions. The $H_2S$ can be oxidized to sulfur by some bacteria and fungi or spontaneously in the presence of oxygen. The sulfur is then oxidized to sulfate by sulfur bacteria.

Various chemoautotrophic bacteria, primarily members of the genera *Thiobacillus*, *Beggiatoa*, and *Thiothrix*, oxidize sulfur, thiosulfate, and sulfite and obtain energy used to reduce carbon dioxide. In addition to these organisms, other heterotrophic bacteria and fungi can oxidize sulfur compounds.

Only a small amount of $H_2S$ is released during the decomposition of protein, but significant quantities can be produced anaerobically by bacterial reduction of sulfate. *Sulfate reduction* requires the absence of oxygen and a source of hydrogen for reduction. The hydrogen is usually supplied through an organic acid or alcohol. Under some conditions molecular hydrogen can be used. The process of sulfate reduction is analogous to denitrification, with the primary difference being that sulfate-reducing bacteria are obligate anaerobes and entirely dependent on sulfate reduction. Denitrifiers can switch between anaerobic and aerobic respiration. Sulfate reduction appears to be an ancient process because of the obligate anaerobic conditions.

*Sulfur oxidation* can be performed by two different groups of bacteria. The first group is the colorless chemosynthetic bacteria. These bacteria obtain energy by the oxidation of $H_2S$ and elemental sulfur.

$$2H_2S + O_2 \rightarrow 2H_2O + 2S$$
$$2S + 2H_2O + 2O_2 \rightarrow 2H_2SO_4$$

The end product of this oxidation is sulfuric acid. These organisms modify their environment by making it much more acidic. The genus *Thiobacillus* is among the best studied of the sulfur-oxidizing bacteria. Different species of *Thiobacillus* have very different pH optima. Wetzel (1983) states that *T. thioxidans* requires an acidic pH between 1 and 5, whereas *T. thioparus* has a pH optimum near neutrality, and *T. denitrificans* performs best at pH values that are alkaline. *T. thioxidans* organisms are some of the bacterial niche constructionists. The end product of their sulfur oxidation (i.e., sulfuric acid) contributes to environmental conditions that favor *T. thioxidans* by maintaining the pH at acidic levels.

# Carbon Cycling

Nitrogen and sulfur are only two of the nutrients needed to sustain life, and the transformations of many of these other nutrients have been described in detail. We will consider only aspects of the global carbon cycle as it relates to microbes. Carbon is a major basic constituent of all living things. The transformations of carbon are essential to the sustaining of life. Every beginning biology student knows the main equation of the carbon cycle:

$$CO_2 + 2H_2A + light \rightarrow (CH_2O) + H_2O + 2A$$

where $H_2A$ represents any substance from which electrons can be removed. For all higher plants, this is usually water, but for many prokaryotes, it can be substances such as hydrogen sulfide. Carbon dioxide in the atmosphere and dissolved in water is the source of the carbon needed for all living things. However, most of the carbon on earth is not gaseous carbon dioxide in the atmosphere, but bicarbonate and carbonate ions stored in the oceans. Almost 71% of the carbon available for cycling is found in the oceans, and most of this carbon is in the deep oceans. Terrestrial environments have approximately 3% of the carbon stored in living and dead biomass. Fossil fuels have most of the remaining carbon. The atmosphere has very little (about 1%) of the available carbon.

Carbon cycling in the oceans is essentially a closed system with a little exchange at the air-water interface. However, given the huge volume of the oceans, this accounts for very little in terms of net flux of carbon. Oceans are the primary sink of carbon.

Carbon dioxide fixed into organic molecules by photosynthesis is released back into the environment through the degradation of these organic substances, primarily through the feeding of heterotrophic microorganisms. Bacteria and fungi are able to break down nearly every organic substance. If conditions are suitable (i.e., oxygen is present), microorganisms can break this material down to carbon dioxide and water. Some substances can be degraded only when other easily degraded substances are present, such as amino acids or simple carbohydrates.

Degradation of organic matter takes place at different rates, and these rates are influenced by the composition of the material and the environmental conditions. In the absence of oxygen, fermentations are usually the rule and indicate incomplete oxidations. Under anaerobic conditions, various ending substances can be produced, such as different alcohols, lactic acid, acetic acid, butyric acid, propionic acid, and formic acid. Carbon dioxide and hydrogen also can be produced.

Primary producers often make important structural carbohydrates, such as starch, cellulose, hemicellulose, and lignin. Starch is a reserve substance that is easily degraded by a number of bacteria and fungi by means of exoenzymes. By these enzymes, the starch is hydrolyzed to a disaccharide, which is further hydrolyzed to glucose. Cellulose is the primary skeletal substance of most plants. Cellulose must also be hydrolyzed by the action of exoenzymes known as cellulases. Anaerobic breakdown of cellulose can occur and results in the formation of various fermentation products. Xylans or hemicellulose act as a strengthening material for plants and as a reserve of food. More microorganisms are able to degrade xylans than are able to degrade cellulose. Lignin is an important constituent of wood. Because lignin is degraded at much slower rates than the other plant compounds, it can accumulate in the environment. Lignin is not a single compound; it is composed of various units that can be linked together with C–C bridges. These bridges are resistant to microbial degradative enzymes. Fungi are responsible for most of the breakdown of lignin in water.

All other naturally occurring and most man-made organic substances, including fats, proteins, pectins, chitin, fatty acids, and various hydrocarbons, can eventually be degraded by microorganisms. Short-chained aliphatic hydrocarbons such as ethane and butane can be oxidized by several different bacterial types, including *Pseudomonas*, *Flavobacterium*, and *Nocardia*. Many more types of bacteria are able to degrade long-chain aliphatic hydrocarbons. The more carbon atoms there are in the chain, the more types of microorganisms there are that can break down the substances. Aromatic hydrocarbons can be degraded by bacteria, yeasts, and other fungi. Complete breakdown of aromatic hydrocarbons requires oxygen. Some hydrocarbons are broken down by microorganisms that do not obtain any nutrition or energy in the process. This is called *co-metabolism*, which is misleading because the name implies that the organisms are obtaining something in the process. Instead, certain microbial enzymes can be used to break down these hydrocarbons, but the end products are not used by the organism, and the energy released remains unavailable to the organism. However, knowing what substrates stimulate these organisms to perform the co-metabolism can aid in the clean-up of toxic wastes from the environment. Many studies have shown that additions of suitable substrates can result in nearly complete clean-up of various toxic compounds.

# Information Spiraling

The naturalist Aldo Leopold described the transport of a single molecule from a tree through many biological and abiotic transformations until the molecule eventually migrated into the ocean, where it was essentially lost from the terrestrial world. Jerry Elwood, Dennis Newbold, and others (Newbold et al., 1981) considered the movement and cycling of nutrients in streams and rivers. The end products of one chemical transformation may be transported some distance away from the source. In streams, this movement away from the source is assisted by the nearly unidirectional flow of water. Elwood and Newbold considered what factors affected the transport distance and turnover of a molecule of some nutrient, such as nitrogen. In their concept, the distance a molecule moves in a river depends on the biology and geology or hydrology of the system. All other things being equal, rivers or streams with high levels of retentive devices such as debris dams have higher abundances of attached microbial communities. These communities presumably take up materials flowing past them through passive or active means. Once retained, the molecule can be subjected to assimilation by the biota or additional chemical transformation before being released again into the water column and subsequent transport downstream.

The movement and transformation of stream nutrients can be modeled and the parameters of the models estimated from empirical studies. Based on these models, a greater understanding of stream nutrient dynamics has been developed. These new models predict the physical stream length and time the nutrient will remain within a particular compartment. For example, Figure 13.3 predicts the fate of a nutrient under various natural and altered stream conditions. Because microbes control many of the biological processes that trap, transform, and transport nutrients in streams, it is important to understand factors that determine the distribution and abundance of the microbes. Community-level interactions are particularly important because the coupling between transformations greatly affects the residence time of the nutrient within a certain reach of stream.

Inorganic nutrients that become embedded within the extracellular polysaccharide matrix may be retained in an inorganic form or become assimilated into the biota. After the nutrient has become assimilated by a microbe, it may be transformed and

| Mechanism | | Effect on Nutrient Cycling | | Ecosystem Response to Nutrient Addition | Ecosystem Stability | Categorization of study Streams | |
|---|---|---|---|---|---|---|---|
| Retention | Biological Activity | Rate of Recycling | Distance Between Spiral Loops | | | | |
| A. HIGH | HIGH | FAST | SHORT 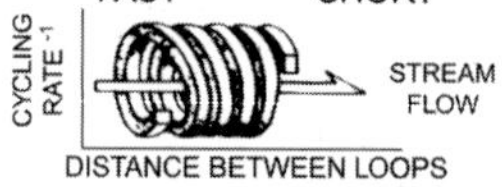 | CONSERVATIVE (I>E) | HIGH | MI 2,3 PA 1,2,3 | |
| B. HIGH | LOW | SLOW | SHORT 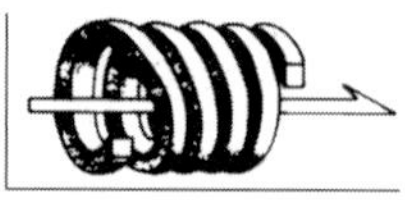 | STORING (I>E) | HIGH | OR 1,2 ID 1 MI 1 | |
| C. LOW | HIGH | FAST | LONG  | INTERMEDIATELY CONSERVATIVE < A but > D | LOW | ID 3 MI 4 PA 4 | |
| D. LOW | LOW | SLOW | LONG  | EXPORTING (I=E) | LOW | OR 3,4 ID 2,4 | |

Figure 13.3 Fate of a nutrient under various natural and altered stream conditions. (From Figure 18 in Minshall GW, Peterson RC, Cummins KW, Bott TL, Sedell JR, Cushing CE, Vannote RL. Interbiome comparison of stream ecosystem dynamics. *Ecol Monogr* 53:1–25, 1983.)

released in a different form, or the microbe may become assimilated into higher trophic levels through predation.

Let us consider the distribution of stream bacteria. This is a specific example of the "everything is everywhere" paradigm. With the strong longitudinal vector of downstream water movement, it seems obvious that bacteria from the headwaters of a stream should be found in lower reaches of that same stream. Unfortunately, very few studies have examined the community composition of stream microbes along any stream continuum. The classic stream concept paper, "The River Continuum Concept" (Vannote et al., 1980), predicts changes in stream biota along a hypothetical stream continuum. The article accurately predicts changes in stream macroinvertebrates, algae, and fishes functionally and specifically. However, although microbes are found at all stream locations, there is no prediction of changes within the microbes functionally or by species. Are the same bacteria found at all stream locations?

Rivers and streams usually have significant changes in physicochemical characteristics along their continua. Headwater streams in many locations are heavily canopied and derive much of their nutrients and organic carbon from the terrestrial vegetation that forms the canopy. The input of organic carbon in temperate climates usually comes in a large pulse associated with leaf fall in the autumn. This input may be dominated by a single species or composed of many different tree species. Different species of trees have leaves that vary greatly in decomposition rates. Some species of leaves such as oaks may be very recalcitrant to decomposition because of waxy cuticles or because of increased levels of toxic plant secondary compounds such as tannins or because of very low levels of nitrogen relative to carbon. Other species are decomposed rapidly because they have high levels of easily degraded compounds such as sugars.

Some streams pass through very different types of vegetation, often over relatively short distances. For example, streams that originate in tall grass prairie have headwaters that are dominated by different species of grass with few or no trees. Further along the stream course, there is often an abrupt change from grassland to a gallery forest. Gallery forests are heavily canopied forests dominated by various hardwood deciduous trees.

Should we expect the bacteria and microbes found in the headwater grassland-dominated reaches to process carbon from plants they have never been exposed to? Similarly, are the same microbial functional genes needed within the headwater as are needed in the lower reaches under the forest canopy? In other words, are the stream microbial communities at least functionally if not taxonomically different between the headwater and gallery forest reaches of the same stream? If the same functional genes are found within the headwater as in the lower reaches, we can predict that the forms and delivery of carbon and nutrients within these reaches are very similar.

Leff et al. (1992) presented a new concept in stream microbial ecology called *information spiraling*. This concept sought to link the movement of bacteria or their genes in lotic ecosystems with nutrient spiraling and carbon turnover, and it suggested that the genetic information of bacteria should be tightly linked to spatial or longitudinal position in the stream. Given that material differences in the quality and quantity of nutrients and organic matter are seen among stream reaches separated by changes in vegetation, slope, and nutrient source pools (Minshall et al., 1985), it was hypothesized that the genetic information necessary to take up and process this material under ambient environmental conditions should have equally distinct distributions. Bacteria in one reach should not necessarily be identical to bacteria found in another reach. However, because various materials are integrated along stream continua and because lower-order tributaries may contribute materials similar to upstream reaches, certain phenotypic characteristics of bacteria may be maintained among stream reaches.

Entire bacterial genomes are not necessarily conserved among stream locations; instead, various genes may be maintained independent of the whole bacterial cell through movement into or out of other taxa. The length of stream over which certain gene functions are adaptive may vary for each gene and may be defined as the *information length* ($L_I$). Information length is a distance measure that characterizes genetic-level phenomena. Information length is a measure of the influence over stream reach distance of genes that affect the catabolism of specific organic molecules or facilitate the transformation of some other nutrient. It is based on the "mix" or assemblage of microorganisms that collectively determine these processes. The information length is the total downstream distance over which some gene is maintained in the community. Because a gene may be shared among different strains or species of bacteria, the information length may be much longer than the distribution of any one species of bacteria.

It is reasonable to assume that the expression of a given gene in a stream ecosystem is environmentally determined. A range of biotic and abiotic conditions within the stream establishes the selection pressures to preserve or eliminate non-negligible gene frequencies in the population. Such conditions include the strength and nature of seasonality, meteorological norms, local edaphic and geochemical regimes, and regional topographic and watershed features. All of these conditions set the magnitude and rates of energy and material inputs to the stream ecosystem and select for the facultative ability to match energy use rates to prevailing energy levels.

The only physical constraint or maximum to information length is the total length of the stream, which is the distance from headwaters to the ocean, to a brackish water interface, or to the mouth of an inland sea. It is possible that certain genes are distributed along the entire length of a river system. However, due to changes inherent along most stream continua (i.e., changes in vegetation, temperature, discharge, light), information length is normally less than the stream length.

Natural selection favors increasing efficiency of use up to the biological limit. The biological limit is the maximum amount of energy or nutrients an organism can take up or process as set by the energetics and biochemistry of the organism. Because bacteria are the oldest creatures on earth, it is reasonable to expect these rates to be near their biological limits and to correspond to the temporal average energy or nutrient load. With increasing temporal variance, new strains of bacteria with wider ranges of tolerance come to dominate in the community. In other words, evolution constantly exerts pressure to plug leaks in stream ecosystems. However, because of hydrological and other abiotic constraints, plugging those leaks entirely is unlikely.

Genetic differences among spatially distinct bacterial communities measured on a single date are an integration of all genetic or selection events before sampling. Communities at a specific location similarly represent an integration of all stream events (e.g., floods, litter import, changes in temperature, chemical perturbations) that have occurred before sampling. Sufficiently long inter-disturbance time periods may allow the establishment of "resident" floras that propagate clonally. Are these resident floras randomly distributed? Are bacteria that colonize first after a disturbance the ones that predominated beforehand, or are specific bacterial forms selected because of their ability to survive under the newly established environmental conditions?

The information length is a property of specific genes. Because bacteria can and do share genes among unrelated taxa, the information length may vary for each gene. The actual length is determined by whether the gene function is required at a specific location. This concept is not concerned with the genes of central metabolism, but rather with genes required to process complex plant compounds or transform available nutrients.

Changes in information length due to anthropogenic activities may make the overall system more inefficient, with more material exported downstream. Alterations to bacterial assemblages that affect genetic exchange or species removal should affect the information length. Disturbances that alter the energy or nutrient load directly or alter the temporal energy or nutrient load variance subsequently affect information length and ecosystem processes.

Leff et al. (1993) surveyed bacterial assemblage DNA collected from three stream habitats—leaves, mid-channel sediments, and bank sediments—for the relative abundance of the neomycin phosphotransferase gene relative to total eubacterial DNA as assessed by hybridization. Relative gene abundance was significantly different among the habitat types. Bacterial DNA isolated from stream bank sediments had higher relative amounts of the gene than did the other two habitat types independent of stream location. These data suggest that this particular gene (or genes carried with it) is more advantageous in bank sediments than on leaves or mid-channel sediments. Given the extremely high potential of inoculation and cross-inoculation of all stream habitats over time, this observation is not trivial. Specific genes have unique distributions in streams.

# Geostatistics and the Spatial Patterns of Microbes

The concept of spatially explicit distributions of specific bacterial genes or traits is not restricted to rivers and streams. We can estimate the distribution of specific bacteria or genes spatially using a technique known as geostatistics. *Geostatistics* is a subdiscipline of statistics that uses the correlation between pairs of samples to describe the spatial patterns of the traits of interest. It was developed in the mining industry, in which knowledge about the amount of ore present at one location could be used to predict amount of ore at another location in two-dimensional spaces. The spatial correlation among sampling locations for a particular bacterial trait should decline with distance from a source for which that trait is adaptive.

The coupling of spatial statistics and molecular biological techniques allows us to know how widespread a particular adaptive gene or function was and whether that gene or function was restricted to one or many bacterial types or species. This approach is applicable to determining patterns in oceans, in soils, on leaves, in epidemiology—in any system that has a spatial component.

Knowing whether the distribution of certain microorganisms is spatially correlated with the distribution of contaminants can be valuable information in the remediation of contaminated sites. Because an organism has been isolated from the environment where a contaminant is found, it does not necessarily mean that the organism is always found with the contaminant. Recall our discussion on the spatial distribution of microbes and the difficulty in sampling. Spatial statistics could help us fine-tune our understanding of microbial distributions and explain some of the variability found in the natural environment.

Some large-scale practices such as logging have been shown to impact ecosystem processes. It is not clear whether such practices affect the distribution of bacteria or their genes across the landscape. One study (Shaffer et al., 2000) examined the temporal and spatial distribution of genes used in nitrogen fixation across a landscape affected by logging. Certain genes could not be found in the clearcut areas over a 200-km, east-west direction from clearcut areas ranging in age from 5 to 10 years. However, these genes were consistently found in the litter below intact forests. It appears that logging can greatly affect the distribution of certain genes and that it reduces genetic diversity and removes certain genes from the gene pool.

# Species Interactions and Processes

14

## Species Interactions

Within a community, species may or may not interact. Species interactions can occur between related taxa or between totally unrelated species. For example, some animals interact with various species of plants. In these instances, the plants can provide food, shelter, or other services such as chemical protection from predators. Consider the well-known example of ants and acacia plants. In Central America, there are several species of acacia. Some acacia have especially large thorns in which the ants live. The ants hollow out the bases of the thorns. There are several related species of ants that do not live in the swollen thorns, and there are other related species of acacia that do not have the swollen thorn bases and therefore do not harbor ant colonies.

The ants raise their young within the thorn bases. The acacia has *nectaries*, which provide the adult ants food. These nectaries are located on the leaves and not on the flowers. The plants have specialized leaf tip structures known as *Beltian bodies* on which the larval ants are fed. This association benefits the ants, but what about the plants?

The ants patrol the acacia 24 hours each day, attacking and stinging any insect or vertebrate that comes in contact or lands on the plants. The ants macerate any plants that grow against the acacia or that sprout nearby. Acacia plants are shade and fire intolerant. The ants help the plant by keeping other plants from shading or producing fuels near the acacia.

This example, often cited and well documented, appears to benefit both the animal and the plant, but there are other types of species interactions. Many interactions occur that do not even require proximity of the interacting organisms; the interactions occur over various spatial and temporal scales. Substances produced by one species may affect behavioral or other changes in another species. Changes to the environment caused or created by one species may alter the growth or reproduction of unrelated species.

From the perspective of one species, there are a number of different ways other species can interact. First, the other species may be considered a food source or prey. Second, the other species may provide habitat. Third, a species may be the predator, or it may be a predator on competing species. Fourth, the other species may compete directly for resources such as food, habitat, or space. Fifth, the species may provide services for each other, or one species may benefit and the other not, or vice versa.

Species interactions often are complex, with the intensity of the interaction changing both temporally and spatially. Futuyma (1998) is careful to warn that most ecologists measure and classify species interactions based on their effect on population growth. However, the effect of an interaction must be measured in the fitness of individuals, not populations. The effect on a population and the effect on an individual need not be the same. The interaction between a predator population P and a prey population V may be positive if P also feeds on a close competitor of V for food, say population W, and keeps this population at levels that minimize competition. In this example, the overall population V may benefit from the predator population P, but any individual of V that is consumed by the predator would not. Selection favors individuals of the prey population V that are able to escape predation. Selection likewise favors individuals of the other prey population W that are able to escape predation and out-compete individuals from population V. Population W is not providing a service to population V by being eaten by the predator. The consequence is ecological and not evolutionary.

Microbial species interact with other microbial and macrobial species. Microbes can be prey or predator; they compete for resources, space, and habitat. Microbes enter into what appear to be mutualistic interactions in which all species involved benefit. Some of the biology's interesting secrets may be linked to ancient mutualisms. Mitochondria and various plastids such as chloroplasts are now considered to be relics of past mutualisms or commensal interactions. Mitochondrial DNA is haploid and circular, and it has high homology with bacterial DNA. The evidence suggests that some bacterium was "captured" or parasitized a primitive eukaryote and established a permanent interaction. The host cell provided basic biochemical building blocks, and the captured cell provided adenosine triphosphate (ATP). Both types of cells benefit. Both sets of genes (mitochondrial and genomic) get replicated and passed on to future generations. Other interactions, although not as elegant, have remarkable properties and deserve attention.

Darwin formulated his theory based on the selection of individuals through interactions with the environment that result in the "most fit" individuals surviving and less fit or unfit individuals not passing genes on into future generations. Subsequent modifications and enhancements to the theory continue to support the notion that natural selection operates only on individuals. Considerable controversy has been generated by those who feel that selection can operate at levels higher than an individual and those who maintain the Darwinian view of evolution.

## Proliferation Hypothesis

Microbiology developed as a science almost independently from evolution and ecology and certainly independent of any coherent concept of microbial communities. Some of the most sacred concepts in microbiology are based on the assumption that pure cultures are needed to understand processes and disease. In the following discussion, we consider a conceptual framework that views microbial *communities* as being subject to selection. This new conceptualization offers significant insights into patterns and processes that occur in the microbial world that are not easily explained by standard evolutionary theory. These ideas run counter to the cogent and thoughtful arguments about group selection that have been proffered by Williams and Maynard Smith. The concept is called the *proliferation hypothesis*, and the following discussion is largely based on the thoughts and observations of Caldwell et al. (1997).

Biology is a science of hierarchies. Every introductory biology textbook has a section on the molecules of life. These molecules include inorganic and organic substances. Amino acids, proteins, enzymes, nucleic acids, DNA, RNA, lipids, fats, water, oxygen, carbon dioxide, and numerous other compounds and molecules are discussed. Being part of a cell, tissue, organ, organ system, or individual facilitates the replication or synthesis of these molecules in biological systems. For some organisms, the complexity stops at the cell level. In microbiology, we are aware of certain genetic elements that reproduce independently of the cell, such as plasmids and transposons. However, biology does not end with the individual. Groups of taxonomically similar organisms are called populations; the association of populations in space and time is known as a community, assemblage, or consortium; and the interactions among communities form the ecosystem.

The proliferation hypothesis assumes that self-replicating molecules can proliferate more effectively if they associate through chemical bonding (ionic and covalent) and then propagate as macromolecules. Macromolecules through association with the cell membranes and walls of prokaryotes can replicate more effectively than unassociated macromolecules. Based on our understanding of the origins of various eukaryotic organelles, it appears that some prokaryotes are sometimes more effective at propagating as part of eukaryotes through endosymbiosis or after attachment. Similarly, prokaryotes and eukaryotes sometimes proliferate best when associated with and propagating as communities through behavior. Communities themselves sometimes proliferate when they associate to form ecosystems.

The more time that has elapsed in the process of self-organization, the more likely that more effective associations have arisen. Based on this hypothesis, we can look back into the beginnings of life and expect anaerobic processes to have higher levels of organization, greater interactions among participants, and more diversification than processes that are aerobic. Consider that in the biogeochemical cycles discussed earlier, the most complex cycles are those based on anaerobic metabolism. Plants and animals have been interacting for only a fraction of the time that microbes have been. Although there are intricate and wonderful associations between animals and plants (e.g., pollination systems), these associations are not as complex as some microbial associations.

Biological diversification and complexity are hypothesized to be the products of proliferation through biological adaptation and association at each level of biological organization: molecular through ecosystem categories. Community ecology and evolutionary ecology have their roots in macroscopic organisms, which have especially long life spans, large sizes (relatively), and slow growth rates. These characteristics make studies of community-level evolution in higher plants and animals difficult to accomplish within the lifetime of an investigator. However, there are numerous microbial examples in which the interactions or networks among interdependent species result in the success or failure of the entire community. Examples include degradative consortia, lichens, dental plaque, and rumen communities. We first consider the curious taxonomy of lichens as an example of community-level evolution.

Some lichens are interesting associations of fungi, algae, and cyanobacteria. Taxonomists have given each unique lichen association a binomial name (i.e., genus and species). Are lichens individual organisms or simple communities? It appears from the taxonomy that because they function as an integrative whole, they are considered an organism. Why would separate species come together in an association such as a lichen? Consider that the ecological range of the free-living members of the lichen community is very much different from that of the lichen. Cyanobacteria and algae are primarily aquatic organisms, and fungi are soil dwellers. Lichens have been successful in exploiting novel niches that are primarily *lithospheric* (i.e., associated with rocks) or *phyllospheric* (i.e., associated with plants). Lichens have extended the habitat range of the free-living organisms. It would have been nearly impossible to predict the habitat range of the association from the preferred habitats of the free-living organisms. Caldwell et al. (1997) suggest that this lack of overlap between the range of the free-living organisms and the extended range of the association is critical in establishing whether communities come into existence as discrete units. Heterotrophic bacteria are excluded from the association by antibiotics produced and released by the fungi that in effect maintain the composition of the group, allowing the group to proliferate.

The lichen example may be at the end of a continuum of possible associations. Few other associations result in species designations, but all such associations result in increased habitat ranges and increased or optimal use of environmental resources, including space, nutrients, and food, and these associations often modify the environment such that suboptimal microenvironments become favorable.

Natural selection is presumed to produce new species through selection acting on individual organisms, but this view is too restrictive and does not include much of the microbial world. For instance, eukaryote evolution is directly related to the endosymbiosis of microbes (e.g., organelles such as mitochondria and chloroplasts). Speciation of "higher" organisms occurs through community-level evolution of two or more separate organisms contributing to increased proliferation of the group. In most instances involving microbes, members of the community remain physically separate but achieve an increased synergism through the association.

For evolution to act on communities, there must be some genetic resources that when changed affect the overall functioning and survival of the group. Any genetically altering event in one member of the community, such as mutations or recombination, affects the individual and the entire community. The genetic resources of the association change through individual mutation and recombination but also through movement into and out of the group through immigration and emigration. Success of the group depends on the collective genetic resources, and the community can propagate, adapt to new environmental conditions, and evolve because of this communal genetic resource. Evolution in microbes occurs at the community level, the species level, and the subspecies level (i.e., genetic elements).

Many microbial associations can evolve through the lateral transfer of various degradative genes. Higher organisms are constrained in their evolution by sexual mechanism and by sexual recombination. Recombination in higher organisms occurs usually between members of the same species because various sexual reproductive mechanisms prevent sharing of genes between unrelated species. Prokaryotes are not so constrained. Genes and/or genetic elements can be and are transferred between distantly related microbes. Evolution in bacteria can be very rapid, but because they do not have complex sexual mechanisms, speciation as defined by higher organisms does not occur. Speciation is not necessarily the only evolutionary force acting on living things, and this is especially true for microbial communities.

Bacteria are small. Because of their extremely small size, an individual bacterium can contain only about 0.1% of the DNA found in vertebrate animals, and this DNA takes up almost 50% of the cell volume. There is not much room for anything else in a bacterial cell. The volume limitation makes maintenance of essential DNA of major importance. If genes for a specific trait were not needed, elimination of those genes would be advantageous. This is exactly what happens to many bacteria when they are placed in conditions where a once important trait is no longer required. For example, traits such as gas vacuoles in planktonic bacteria, heterocysts in nitrogen-fixing cyanobacteria, or the growth of filaments or aggregation behavior in predator-resistant bacteria have all been shown to be lost when the selective pressure was removed or reduced.

Plasmids carrying genes for the degradation of various complex xenobiotic compounds can move between genetically unrelated bacteria under strong selection imposed by the presence of the xenobiotic compound. Sometimes, the strain of bacteria that originally had the plasmid is not able to survive under the selection, but the catabolic plasmid proliferates through transfer to other members of the community. Numerous studies have shown that biodegradative communities of bacteria respond to chlorinated hydrocarbons by forming consortia of bacteria that are unable to degrade the compound as isolated populations. These consortia mediate only the most thermodynamically favorable reactions of all possible reactions. The formation of specific bacterial communities is not a random process; the coming together of specific members results in limited environmental resources being used optimally.

A polymer can be completely mineralized to carbon dioxide and water (see Chapter 5, Figure 5.9) through the concerted efforts of, for example, four different microbial species. Species A is unable to break down the initial polymer completely and produces products that become the "food" of other species. Under the constraints of natural selection, we might envision a condition in which a single organism is selected that is capable of degrading a compound and its degradative substrates. In the laboratory, some strains have been found that can do exactly that. However, when these strains are released into the complexities of the environment, they inevitably die out, although the mobile genetic elements they contain may survive in members of the natural community. In nature, diverse communities are the result of evolution selecting differentiation of roles performed by various group members. For example, some members may cause a compound to adsorb to particles, whereas others are able to degrade part of the molecule, and other members protect the community from predators through antibiotics or release of extracellular polysaccharide. Each member performs a role and provides nutrition or protection for members of other groups.

Our understanding of microbial communities is far from complete. New technologies have helped researchers uncover previously unknown relationships, but much more information is needed. Caldwell et al. (1997) suggest 10 areas that need to be determined in microbial community ecology:

1. The relative stability of various community networks under defined laboratory conditions
2. Whether there is only one community network that can "lock into" and occupy a specific environment (or several environments)
3. Whether the presence of one community precludes the development or encroachment of another
4. The habitat range of organisms alone and in association with their community
5. The *ecotones* between communities and whether they represent sharp boundaries or gradual transitions
6. Whether the communities occupying a specific environment have been optimized through evolution in terms of their ability to proliferate and to make most efficient use of environmental resources and habitat (i.e., optimize the conversion of abiotic to the biotic)
7. Whether spatial or temporal pathways are necessary for the development of specific microbial communities or associations
8. Whether communities create favorable microenvironments within unfavorable microenvironments and homeostasis
9. Whether the composition of the community depends on the timing of immigration by specific species
10. Whether some traits of organisms specifically affect the proliferation of the community directly, while affecting the proliferation of the individual only indirectly (habitat range of parental strains and adaptation-negative mutants alone versus their respective range in combination with other members of their community)

Microbial communities appear to behave differently from communities of higher organisms because of increased linkages, networking, novel genetic mechanisms, and extension of habitat ranges brought about by association. However, these community characteristics need to be examined in many more environments. The concept of niche construction suggests that there may be multiple end points for microbial communities that are a function of the characteristics of the starting members of the community. Is this true in nature? Under laboratory-defined conditions, do microbial communities always come up with the same or very similar species compositions and abundances (e.g., predators, producers, heterotrophs), and if not, how much variation is there in the ending communities?

The stability of a community relates to how much stress can be applied before the community begins to break down through the loss of community members. At the microbial scale, can there be more than one community occupying the same microenvironment? Consider a habitat in which some chlorinated hydrocarbon has been released and a consortium of bacteria has developed that is capable of degrading this compound. Would this consortium supplant other consortia or communities that existed before the release of the hydrocarbon? Under pristine conditions, there was a preexisting community that was able to survive under the physical, nutrient, and carbon conditions then present. Would this community continue to exist and interact in the presence of the hydrocarbon-degrading community, or would their activities be interrupted?

In the lichen example, it was clear that the habitat ranges of the fungi, algae, and cyanobacteria were greatly increased because of the association between the different organisms. Is this generally true for all microbial associations whether tightly associated through endosymbiosis or more loosely connected through linked metabolic processes? Unfortunately, observing microbes in situ is difficult, although new techniques (e.g., confocal laser microscopy) allow scientists windows that look into the lives of microbes. To answer whether habitat ranges have been expanded, we would need to capture and characterize free-living microbes that can be compared with the same microbes that are found only in an association. The very act of sampling may disturb associations, making the members appear as free-living forms.

There is much to be done in the area of microbial community ecology. The development and maintenance of communities, whether they behave and are selected as units or not, is controlled by the species interactions that occur among members. Species interactions can have negative effects on one or both of the interacting species, they can have positive effects on one or both parties, and they can have no or little effect on one or both groups. The interactions discussed can occur between totally unrelated organisms, related organisms, or in the case of clonal organisms, between the same individual. Interactions between distantly or unrelated organisms are collectively called *interspecific interactions*. The same interactions between closely related organisms are called *intraspecific interactions*.

# Negative Relationships

Organisms can negatively impact each other without killing or damaging cells. Some interactions are only chemical, and these chemicals illicit various responses or suppress activities of potential competitors. These interactions can be subtle and require carefully planned experiments to detect effects. For example, Kearns and Hunter (2002) were able to show that extracellular products of the green alga, *Chlamydomonas reinhardtii* can actually suppress heterocyst formation in co-occurring cyanobacteria and presumably the amount of nitrogen fixation. To what extent are chemical interactions occurring between and among microbes?

Predation in the largest sense includes organisms that eat other organisms from the outside in and those that eat all or parts of other organisms from the inside out (i.e., parasites), as well as herbivores or grazers that consume all or part of various autotrophs (i.e., photoautotrophs and chemoautotrophs). Predation in most instances has a negative impact on the individuals that are the resource for the predator. For example, grazers, even though they may not consume all of a primary producer, probably reduce the overall productivity of the grazed prey and reduce the individual fitness of that organism. This is intuitive but often confused with population benefits. Many green plants set more seed after grazing than before. However, if an individual plant was continually grazed such that it had to put more resources into making new tissue than a plant that did not have to expend that energy, we would expect the grazed plant to produce fewer offspring, all other things being equal.

Microbes can be prey and predators. The term *predacious microbes* needs qualification. Because microbes may include prokaryotes and eukaryotes and because many microscopic organisms (e.g., Protozoa) feed almost exclusively on bacteria and fungi, this topic is broad in scope. If we narrow the perspective to include only the evolutionary ecology of predatory bacteria, the problem is still daunting but manageable. I therefore restrict the discussion to predacious and parasitic bacteria, fungi, and viruses.

## Parasitism

Certain aspects of life history are important for understanding parasites and their hosts. Many microorganisms live in or on other organisms and obtain nutrients and suitable environmental conditions for growth and reproduction. Microbial parasites can be *obligate* or *facultative*. According to Rheinheimer (1992), obligate microbial parasites include viruses, whereas facultative microbial parasites include most bacterial and fungal parasites. Facultative parasites are able in the absence of a living host to obtain nutrients and carbon from saprophytic activities. Obligate parasites require living cells to survive and reproduce.

In a sense, novel DNA that is incorporated into cells through transformation, transduction, and conjugation is a form of parasitism. The new DNA uses the cell's replication machinery to ensure that it is copied into future generations. Although some of this new DNA can and does confer selective advantage to the host cell (e.g., antibiotic resistance or metal resistance), the ultimate benefit is that the DNA, all of it, is replicated in new environments. Cells without such traits may not divide, are killed, or have greatly reduced rates of reproduction. The distribution and abundance of specific genes in a community is independent of the species make-up of the community. Instead, the distribution of specific genes depends on the selective pressure on the hosts. If the entire genetic background of a particular cell is not adaptive to a specific location, but certain genes found within that cell confers an advantage, we might expect that those genes would have a distribution different from the host cells and could be found in related or unrelated taxa. A more detailed discussion of information length and spiraling is available in Chapter 13.

There are few organisms that cannot be infected by viruses. From this observation, it seems fairly obvious that the virus strategy is incredibly successful. Regardless of whether we consider viruses alive or not, their impact on living things can be great. Free-living viruses and or virus-like particles have been detected in most aquatic environments with densities of about $10^7$ to $10^8 \, \text{mL}^{-1}$. Often, the density of these virus-like particles is orders of magnitude higher than bacterial densities, suggesting that most bacteria are at risk for viral infections. Such high densities may impact various events or processes. For example, viruses may control cyanobacterial blooms.

Do specific bacteria have specific viruses? Host-specific phages can be isolated from the environment. However, there is also evidence that some phage types collected from very distant and different environments are able to lyse the same species of bacteria (Wolf et al., 2003). In this study the researchers isolated 44 phages from 13 very different habitats that included freshwater and brackish locations from distant geographic locations. Some of the phage isolates were found at all locations, but others were site specific, and there was evidence of coexistence of different phage isolates within a location. Based on these observations, phages probably shape the structure of most microbial communities and help in maintaining high bacterial diversity.

Not all virus infections are virulent, and they do not result in the death or lysis of the host cell. This fact is extremely important in the genetic modification of bacterial and other cells through transduction. Novel DNA can be incorporated into a host cell through viruses and be passed on to future generations.

Bacteria that are infected with viruses but that are not lysed may survive, but they are at an evolutionary disadvantage compared with cells that are not infected. Viruses capture the cell machinery of an infected cell and use it to produce multiple copies of the virus genome. Infected bacteria in competition with uninfected bacteria of the same species may have a reduced replication rate because of having to replicate the virus and bacterial genes. This is the same problem associated with the carrying of any extrachromosomal material, including plasmids and transposons. Why has selection not favored strains that are completely resistant to virus attack or to plasmid exchange?

An answer lies in the potential benefit the introduced DNA may confer on the host cells. Transduction and conjugation may provide the cell with novel DNA and genes that allow tolerance or resistance to various compounds and metals or that confer abilities to process or derive energy from new sources such as anthropogenically produced materials.

Plasmids and viral DNA appear to behave as parasites. Is this true? The answer is not simple. Some plasmids may differ in size, transferability, genes, copy number, and probably many other traits. Based on observations about the increase in antibiotic resistance among bacteria and the coupled observations showing antibiotic resistance on mobile genetic elements such as plasmids, it is clear that extrachromosomal DNA exchange can and does occur. The effect of such an exchange can be worldwide.

In 1978, Reanney suggested that we needed to rethink the classic view of a chromosome as a static molecule and use a new model in which chromosomes are viewed as dynamic and subject to frequent rearrangement. Microbial chromosomes may be broken apart into smaller subunits. Because of the interactions between viruses, plasmids, and chromosomal DNA, what in eukaryotes may take millions of years to generate a qualitatively new trait requires single generation through the combination of unrelated replicons that contain separate genes that together ensure the survival of the newly naturally engineered organisms.

The evolution of viruses, plasmids, and cells is one of coevolution. Chromosomal genes are very resistant to evolutionary change. Conservation of genetic information is a hallmark of important genes, and differences between organisms in these genes are the basis of determining relatedness among organisms. Organisms that have more similar gene sequences of these conserved genes are more closely related than organisms that diverge significantly from each other because of the time required to bring about the large cumulative differences. Extrachromosomal material is not under the same constraints of conservation. It is free to change and therefore becomes experimental DNA (Reanney, 1978). Under various environmental stresses, the adaptive response of microbes is not chromosomal mutation and subsequent selection, but rather the uptake of extra DNA from the environment with the required genes necessary to handle the stress. Extrachromosomal DNA is free to change. Although the genes found on these subunits do confer advantages under various stresses, the genes are not essential, and any changes will not affect the basic housekeeping traits that are required for life.

## Predation

Many ecology textbooks discuss predator-prey interactions in sections devoted to population biology or ecology. Although predators often feed exclusively on one or a few prey types, most organisms are themselves the prey of some other organism. A specific predator may affect the abundance and distribution of only a few species, but the combined affect of all predators may be very important in shaping community structure. Bacteria are the prey of numerous other organisms. However, there are also bacterial predators that eat other bacteria. In this section, we consider predation by higher organisms and predation by other bacteria.

Every living thing will eventually be eaten by something. For most, the eating takes place after death without sharp teeth being used and involves microbes and inverte-brates. One of the most frightening considerations for humans is the thought of being eaten by some big mean animal. This fear is played on by the makers of many movies where some large predator is terrorizing cities and eating the inhabitants. In ecology, the fear is tempered by the choice of terms. Trophic interactions and trophic levels are used to indicate that something is eating something else.

In many people's eyes, "survival of the fittest" conjurers up the image of some cute herbivore (think bunny) being chased down and killed by a vicious predator (think coyote or bobcat), and every now and then, one of the prey escapes, and the people watching the documentary applaud. On the other hand, many people are impressed with the strength and cunning of many large vertebrate predators.

Predation is an integral aspect of many organisms' ecology. Evolution has selected organisms capable of consuming other organisms; it has selected and counter-selected traits for predator avoidance and more efficient prey capture. Plants, which are incap-able of escaping predation through movement, have been selected to escape or ame-liorate the effects of predation by producing compounds that act as deterrents to the herbivores that feed on them, by increasing their growth rates to replace the consumed parts, or by the production of structures such as thorns or stingers that deter preda-tors. There is evidence that natural selection has favored a coevolution between some plants and their grazers such that the saliva of the grazers actually stimulates regrowth of the plants. The maintenance of the tallgrass ecosystem seems to be based on large ungulate and arthropod grazers consuming wide expanses of prairie and then moving on. The roots were not eaten and the animals left behind fertilizer that aided the regrowth of the prairies.

Bacteria are consumed by many different protozoan and metazoan predators under wide environmental conditions. Bacteria are the first link in many trophic transfers, including photoautotrophy and chemoautotrophy. Bacterial autotrophs fix inorganic carbon and convert that carbon into bacterial biomass. Bacterial heterotrophs and saprotrophs convert organic molecules previously fixed by other bacteria, animals, or plants into bacterial biomass. Even though the importance of bacteria and their prod-ucts is recognized in biogeochemical cycles and secondary production, little is known about the regulation of their abundance, biomass, or productivity in any system.

In most environments, there are always bacteria and other microbes present. The density of these organisms may be site specific, but in general, regardless of season, microbes are present. If we survey the predator densities at the same times and places, we usually will find thriving populations of predators. Why do the predators not eat the prey into extinction? How do the prey escape predation? What environmental or ecological factors control the predator density other than the prey density? These and other questions about predation have been the focus of numerous studies using higher organisms. From such studies, it has become apparent that predation is a complex ecological process involving many components of the environment. The complexity of the environment provides refugia for prey species that allow them to escape predation. For example, the more shrubs and woody debris a woodlot has, the higher probability a rabbit can escape being caught by a fox. However, the reproduction of the rabbits is also linked to the availability of food and suitable nesting sites, which may be directly related to the shrubbery and wood debris. The effects of reproduction and predator refugia are confounded. Are there more rabbits because they are able to escape predation, or are there more rabbits because there are more places for rabbits to feed, nest, and reproduce?

Because of the extremely small size of most microbes and their habitat, the complexity of their environment is probably beyond our ability to measure and describe it. Although the complexity of the environment may increase or decrease, the probability of being consumed is independent of habitat complexity. For example, suspended particulate matter in the pelagic zones of oceans and lakes provides habitat for many microbes. Aggregates may limit the ability of some predators to feed but at the same time increase the probability of being consumed by some other predator. Free-living microbes may escape predation by not being large enough of a particle to be filtered out by a predator. Some bacteria and fungi in biofilms may be consumed by grazing invertebrates or scavenging protozoa, but many others are preserved because of the extracellular polysaccharide matrix they are embedded in satiates the predators.

Predation of bacteria by their predators is perhaps the oldest predator-prey system on the planet (Posch et al., 2001), and we would expect evolution and coevolution to have occurred and resulted in adaptations in the prey and the predator. Posch et al. (2001) identify three theoretical aspects of bacteria-predator interaction that need to be considered:

1. Protistan feeding can directly influence the size structure, community composition, and productivity of a bacterial community. Selective grazing of protists on specific bacteria can consequently have an impact on the whole bacterial community. Characteristics of bacterial cells for a protistan predator to use as criteria to prefer or avoid include motility, cell surface properties, size, nutritional content, and activity state (growth or division). Not only does the uptake of distinct prey size classes change the size distribution of a bacterial community, but feeding on larger (i.e., more active or dividing) cells affects the productivity of the whole community. Grazing controls the bacterial production rather than the standing stock, but size-selective feeding can change the bacterial community composition if the mean sizes of some particular populations lie in the optimal food size range of the predator. As a consequence, grazing can modify interspecific competition within a bacterial community.

2. Bacteria have several grazing-defense strategies to decrease vulnerability or to avoid grazing-induced mortality. Some bacterial strains show a high phenotypic plasticity and can grow larger than the optimal food size range of the predator.

Another strategy depends on an increase of growth rate to compensate for cell losses. We can also speculate about a decrease in growth rate and consequently cell size to minimize the grazing-induced cell losses. The formation of inedible colonies yields another efficient strategy against being grazed.

3. Flagellates and ciliates must adapt to more grazing protected bacterial communities. In theory, this reaction can start a cyclic relationship whereby predators affect the main characteristics of a prey and vice versa. In nature, it is very difficult to observe this phenomenon for microbial predator-prey interaction. This relationship is complex because protistan predators reduce bacterial biomass, and feeding results in remineralization of nutrients whereby protist themselves support their bacterial prey.

Can bacterial predators change the composition and size structure of bacterial communities as suggested? Are bacterial predators indiscriminate feeders that consume any and all bacteria? These two conditions are not necessarily mutually exclusive. An indiscriminate predator may impact the composition and size structure if the bacteria are not randomly distributed. For example, if all bacteria of species A, which are of mean size X, are clumped together, and all individuals of species B, which are of mean size X − 1, are clumped together some distance from species A, a predator that finds the patch with species A first will reduce the diversity and the size of the community. If the bacterial species were randomly distributed in space, an indiscriminate predator would not be expected to alter the composition or the size structure of the community because every patch would be, on average, identical with respect to species and size classes.

## Control of Community Structure

Before discussing specific examples of predation on microbes, let us consider the effect of predation as a general concept.

There are two schools of thought on the structure of communities: predation, or what is called top-down control, and production, or bottom-up control. Much research has gone into trying to determine which processes structure the patterns we can observe and measure in nature. In general, top-down theory holds that predation controls the biomass, numbers, and composition of their prey (e.g., herbivores), which affects the biomass, numbers, and composition of the primary producers. The bottom-up concept holds that availability of nutrients and light control the distribution and abundance of primary producers, which controls how much biomass is available for herbivores, which determines how much is available for predators. Both approaches probably affect the composition and functioning of a particular ecosystem.

Predators or predation control the numbers of microorganisms and the composition of microbial communities. Predators of microbes are a diverse group of organisms, including metazoans, protozoans, and other bacteria. In this section, we consider the affect of metazoan and protozoan predators on microbial communities and populations.

Protozoa are often predacious on bacteria. Several studies have shown that these organisms require bacteria in their diet. However, all bacteria are not equal when it comes to feeding by protozoa. Some species of protozoa grow best when fed a diet composed of several species of bacteria rather than a single species. It has been shown that protozoa fed an exclusive diet of some bacteria resulted in death of the protozoan presumably from build up of toxic substances. The size and shape of bacterial cells also determines the level of predation. Large, chunky cells are easier to detect and capture. However, there are numerous strategies used by protozoa to capture bacterial cells that will not be enumerated here. The ultimate question of concern is whether bacterial communities are affected adversely by predation. For predation to affect bacterial or microbial species, the intensity must be great enough to become a limiting factor. Predators must reduce the density of microbes sufficient to affect overall growth and reproduction or be responsible for a high proportion of microbial mortality.

Predation is often discussed in the setting of population ecology. In microbial ecology, it is often too difficult to follow a single population of bacterial species. Instead, researchers lump all bacteria (see the earlier discussion about the inherent problems of this approach) together and monitor the affects of predation on the entire bacterial community. In some examples, what appears as a predator effect can be seen in the annual cycle of total bacteria and heterotrophic nanoflagellates (Figure 14.1). Increases in flagellate density follow increases in bacteria density. However, it is not clear whether the oscillations in bacterial density result from increases and decreases in predator density. These changes may be caused by other factors not measured in this study. For example, the major peak in flagellate density occurs after a major drop in bacterial density. There is some relationship, but these data are not sufficient to tease apart the interactions among the various species. A much more informative approach would require monitoring specific bacterial species over time and not the entire community. Predation has been shown to affect the species composition of bacterial communities, and these changes in composition may affect various nutrient cycles.

The fact that the microbial populations fluctuate on an annual basis is of interest. These data suggest that even if predation is not responsible for the decreases in microbial density, something is. In other words, the microbes do not have unlimited growth, and subsequent population crashes are seen in some laboratory experiments. Density is regulated spatially and temporally.

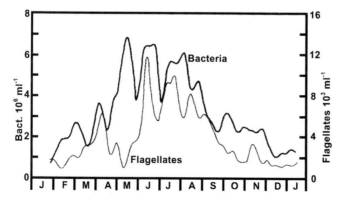

**Figure 14.1** Temporal relationship between bacteria and heterotrophic nanoflagellates that prey on the bacteria. (From Rheinheimer G. *Aquatic Microbiology*, 4th ed. West Sussex, UK, John Wiley & Sons, 1992, © John Wiley & Sons Limited. Reproduced with permission.)

There are numerous species of metazoa that feed on microbes. Many of these are filter-feeding organisms, such as sponges, mussels, oysters, and many species of arthropods. Some of the feeding apparatuses of these organisms are restrictive to certain sizes and shapes of microbes as discussed in Chapter 5. Other feeding strategies include scraping and grazing. Over localized areas, predators can significantly lower the microbial biomass. However, microbes continue to exist everywhere. Given the fact that most microbes are incapable of rapid escape, what mechanisms do microbes have that lower the risk of predation? What controls over these mechanisms do microbes have? In other words, can they initiate a predator defense in response to some environmental cue or stimuli? Alternatively, must it be an all or none response?

Many organisms escape or minimize the probability of predation through several mechanisms, including behaviors, secondary or poisonous compounds, camouflage, and warning coloration. Much attention has been given to these modes of predator escape. All of these methods require energy, which is then not available for reproduction. This means there must be a trade-off between selection favoring the fastest, most armored, worst tasting, best concealed, and so on organism and the ability to reproduce and leave offspring. In other words, an organism cannot put too much energy into predator avoidance and defense. To do so results in less energy being available to make babies or in the case of many microbes, more copies of oneself.

Most ecology textbooks consider plant and higher animal predator avoidance, and no attention is given to microbes. This is curious considering that microbes and bacteria have been in existence longer than all other organisms combined. As with most things microbial, the general mindset is that if you cannot see it, it probably does not exist. However, microbes are consumed, they are targeted as a preferred food by numerous other organisms, and they are subject to viral attack. Given their exceedingly long existence, it seems highly likely that microbes have evolved ways to reduce the effects of predation. They have.

When an asexual organism reproduces, are the daughter cells new individuals or copies of the original? The answer to this philosophical question is germane to understanding one mechanism of dealing with predators. The whole purpose for the existence of any organism is to leave copies of its genes in subsequent generations. Having multiple copies of oneself that are distributed both spatially and temporally is an effective mechanism to ensure the survival of one's genes. If we consider offspring of a single cell as multiple copies of the same individual, it does not matter if some of the cells get eaten or die of any other means, because the individual survives. Predators can eat some copies, and there are still plenty of copies left. This strategy can work only if some of the cells disperse away from the original cell. Any mutation that favors dispersal, detachment, migrations, or motility should increase under predation. Cells that stay together may run the risk of being consumed en masse. Alternatively, colony formation may sometimes be an effective predator avoidance mechanism. Colony formation can be an effective strategy for obtaining or processing food, so there must be trade-offs between staying and immigrating between efficient feeding and predator avoidance.

Let us consider the decomposition and mineralization of some leaf that falls into a stream as an example of strategies to escape predation. The interaction of invertebrates, fungi, and bacteria in the breakdown of leaf material in streams has been conceptualized in a model. In this model, the leaf material is attacked first by bacteria, followed by fungi. Macroinvertebrates, usually aquatic insects or crustaceans, called *shredders* begin to macerate the leaf tissue to obtain the high-quality fungal and bacterial biomass growing on and in the leaf material. As the shredders open the leaf matrix, bacteria are able to colonize inner surfaces and contribute to the breakdown. Microbial growth rates must be faster than the decomposition rates and especially faster than the feeding rates of the invertebrates, or they will be consumed before being able to reproduce and disperse.

Many shredding invertebrates are able to selectively feed on specific species of fungi that have colonized the leaves. In feeding choice experiments, Arsuffi and Suberkropp (1989) have shown that shredders selectively feed on certain species of fungi over other species. Aquatic hyphomycete fungi reproduce by producing spores that in the absence of shredding invertebrates can completely cover the surface of leaves. When shredders are present, the fungi must penetrate the leaf matrix and produce spores at rates faster than the shredders are consuming the leaf material. Although it is known that the numbers and types of shredding invertebrates affect the rate of decomposition, it is not known whether the numbers and types of invertebrates can affect the reproductive capacity of the microbial species involved in decomposition.

Many bacteria suspended in air or water may be the excess production originating from terrestrial environments. For example, Edwards et al. (1990) have shown that most of the bacteria being transported down a river are dormant. Are the bacteria in transport inactive because they have been washed into a new (in this case, aquatic) environment through storm or rainfall events? Alternatively, are they dormant aquatic bacteria that are being transported downstream after being dislodged from some surface or after actively entering the overlying water? Let us consider the fate of bacteria originating from terrestrial environments first.

Can bacteria of terrestrial origin colonize aquatic surfaces? If these bacteria are in a dormant state, it seems unlikely that they would be able to actively attach and colonize surfaces. Successful colonization is an active process and involves attachment and subsequent competition for space. We have discussed biofilm development and maintenance. Most bacteria in transport are probably dormant. The bacteria being transported downstream are inactive and incapable of changing their ultimate fate. Transport from freshwater to marine or estuarine waters would ultimately result in the death of these bacteria due to chemical and physical differences. In contrast, native aquatic bacteria are quite capable of colonizing surfaces.

Terrestrial and aquatic strains of the some species of bacteria are genetically different from the aquatic strains of these same species. In one study, isolates of *Burkholderia cepacia* were taken from several terrestrial and aquatic locations within a drainage basin. Isolates of *Burkholderia pickettii* were collected from the stream locations. Based on multilocus enzyme electrophoresis (MLEE) the two different species of aquatic *Burkholderia* (*B. cepacia* and *B. pickettii*) were more similar genetically than were the isolates of *B. cepacia* collected from terrestrial and aquatic habitats (Figure 14.2). Terrestrially derived bacteria differ genetically from aquatic bacteria of the same species. The origin of a particular strain can be identified based on the genetic make-up of the specimen. It appears that selection has favored unique combinations of genes in one habitat over another. Selection within a habitat is stabilizing with respect to the genes examined.

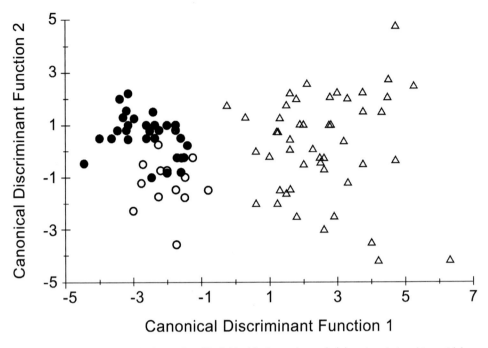

**Figure 14.2** Canonical discriminant plot of individual isolates of aquatic (*closed circles*) and terrestrial (*triangles*) *Burkholderia cepacia* and aquatic (*open circles*) *Burkholderia pickettii* based on the frequency of electromorph types of various enzymes. (From Figure 3 in McArthur JV, Leff LG, Smith MH. Genetic diversity of bacteria along a stream continuum. *J North Am Benthol Soc* 11:269–277, 1992.)

What do these observations have to do with predation and predator avoidance? Stream filter-feeding invertebrates such as blackflies (Diptera: Simuliidae) and freshwater bivalves that feed on suspended bacteria may be harvesting only introduced or nonnative bacteria. *Autochthonous* bacteria may not be consumed by filter-feeding organisms because they rarely or never enter the water column; instead, *allochthonous* bacteria washed in from terrestrial environments may be providing most of the bacterial biomass in transport and the food for these invertebrates. Terrestrially derived carbon input as microbial biomass may form the trophic base of some aquatic food webs.

## Satiating the Predator

Swamping the predators with an abundance of prey is an effective predator avoidance strategy. Some species of plants produce huge amounts of seeds in 1 year and then have subsequent years when little or no seed production occurs (Shibata et al., 1998). This strategy is called *masting*, and it has two main effects on predators. First, the predators are satiated. So many seeds are produced that the predators cannot consume them all, and some of the seeds survive to germinate. The second effect is temporal. Having asynchronous production of seeds predators cannot key in on when the next mast year will occur. Predators need resources to make predator babies. In essence, masting plants avoid predation by preventing the predator from becoming a specialist on them. Selection cannot favor a predator to specialize on a resource that is not predictable. Predators that feed on mast production do so by chance (i.e., they are lucky enough to be alive when a mast year occurs).

Individual microbes do not have mast years or anything like them. Microbial production of a specific species varies spatially and temporally but can be explained on the basis of availability of nutrients or physical factors such as temperature. Although production of any one species of microbe can be affected by the availability of a specific resource, not all bacteria or microbes are able to respond to every resource.

Microbes as a group have been successful in swamping their predators with more biomass than the predators can consume. It is not clear whether specific groups of microbes are at more risk of predation than other groups. What is obvious is that microbial reproduction in every environment is sufficient to offset the effects of predation and disease (discussed later). In many instances, the reduction due to predation may be up to one or more orders of magnitude, but this is minimal when we consider the density of microbes, especially bacteria, in most habitats and the potential for relatively rapid growth.

Microbes fall within a fairly narrow size range, but predators can specialize in specific sizes or shapes, and there is evidence that certain predators cannot grow on certain microbes. Only a small amount of bacterial carbon being transported is consumed by other heterotrophic organisms. Total microbial production is greater than the ability of even the best predators of microbes to consume.

Microbial predators have varied mechanisms for capturing and harvesting their prey. Predators that feed on microbial biofilms can rake the biofilm (e.g., ostracods), scrape the biofilm (e.g., snails with their radula), or gather from the biofilm (e.g., some mayflies use their setae). In one study, representatives of each of these predator strategies were used to determine their effect on a natural river biofilm assemblage (Lawrence et al., 2002). All feeding strategies reduced aspects of the biofilm (Figure 14.3). Snails reduced algal biomass, biofilm volume, and the amount of extracellular polymer by more than 90% of the controls, whereas mayflies reduced these

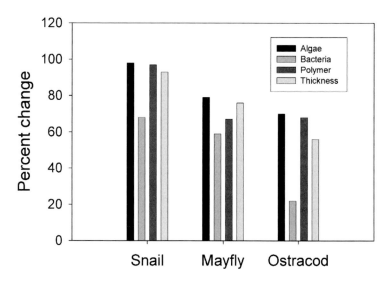

**Figure 14.3** Percent change in components of a river biofilm by three different macroinvertebrate grazers. (Data from Table 1 in Lawrence JR, Scharf B, Packroff G, Neu TR. Microscale evaluation of the effects of grazing by invertebrates with contrasting feeding modes on river biofilm architecture and composition. *Microb Ecol* 44:199–207, 2002.)

components to approximately 70% and the ostracods to between 60% and 70%. Both mayflies and snails reduced bacterial biomass by >60%, whereas the ostracods reduced bacterial biomass by only 20%.

In this laboratory experiment, the predators were unable to completely consume the microbial biomass or products. Other predators feed on microbes through filtration. For example, blackflies (Diptera: Simuliidae), net-building caddisflies (Trichoptera: Hydropsychidae), bivalves, and bees (Lighthart et al., 2000) filter bacteria from freshwater or air. The filtering apparatus can be so specialized (i.e., selective) that filtering species can be identified based on the net pore sizes or space between filtering combs. The size of the filtering structure would be as selective of bacteria-sized particles as machined laboratory filters.

Bivalves and other marine invertebrates filter feed in marine and freshwater ecosystems. However, there are billions of bacteria or microbes per milliliter of water, per gram of soil, or per liter of air, indicating that the effect of predation on microbes as a group is minimal.

The densities of marine flagellates and ciliates have been shown to increase with a concomitant decrease in bacteria (Figure 14.4). In one study (Sime-Ngando et al., 1999), bacterial densities in ice brine decreased from $1.2 \times 10^9 \, L^{-1}$ to $0.2 \times 10^9 \, L^{-1}$ over a 30-day experiment, a reduction of 1 trillion cells, whereas the densities of bacteriovorus flagellates and ciliates increased from $0.1 \times 10^6$ and $2 \times 10^3$ to $10 \times 10^6$ and $15 \times 10^3$, respectively. As the density of bacteria found in ice brine and open water decreases over a 30-day period, the numbers of bacterial predators increases more than tenfold. Figure 14.4 shows the importance of carefully examining the scale used on the axes to know the level of an effect. Something is affecting the bacterial density. The reduction in bacterial cells shows the effect of predation and represents an 83% reduction in bacteria. However, there are more than $10^8$ cells/mL remaining! Predator densities increased and supposedly would continue to exert an effect on bacterial numbers, especially in this closed system. However, it would be unlikely that the predators could ever cause the population to crash. The correlated increase in predator density is interesting. However, Sime-Ngando et al. (1999) estimate the average number of potential bacteria these predators can consume is between 3 and 6 cells protozoan$^{-1}$ hour$^{-1}$. Assuming an average of 1,000,000 protozoa per liter are consuming 6 cells every hour for 30 days, these protozoa would ingest more than $4.2 \times 10^9$ cells from every liter of water. This number is four times higher than the average density of the bacteria. Although predation may account for a significant reduction in bacterial biomass, the bacteria are able to reproduce faster than the effects of predation and therefore keep from being completely consumed.

Besides the ability to swamp their predators with high densities, microbes have additional mechanisms and strategies to avoid predation. These include changes in morphology and colony size, production of toxins or inhibitory compounds, and production of extracellular polymers.

We have considered only a few groups of organisms that obtain all or part of their nutrition from microbes. Many other taxa feed directly or indirectly on microbes. Many ciliates and flagellated protozoa feed on bacteria and so do the slime molds (class Eumycetozoea). Sponges are indiscriminate feeders and as such capture free-living bacteria or organisms attached to detrital particles. Gastrotricha, members of the Entoprocta, and some polychaete and oligochaete worms do not feed directly on

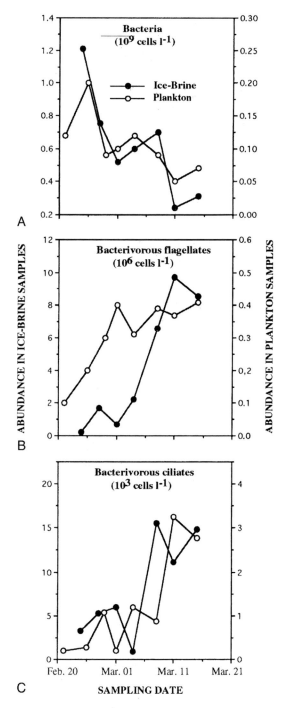

**Figure 14.4** Abundance of different bacteria and their predators found in ice brine. Notice how bacterial populations decrease as predator populations increase. (From Figure 3 in Sime-Ngando T, Demers S, Juniper SK. Protozoan bacterivory in the ice and water column of a cold temperate lagoon. *Microb Ecol* 37:95–106, 1999.)

bacteria, but they do consume detritus that is usually colonized by bacteria. All detritus-feeding organisms obtain most of their nutrition from bacteria and fungi that have colonized the detritus and not from the recalcitrant plant-derived carbon. Free-living nematodes feed on bacteria, yeasts, fungal hyphae, and algae. Each of these groups of organisms have unique feeding mechanisms and clearly each have been and continue to be successful in capturing sufficient food to produce or reproduce themselves. How is it that microbes have and continue to escape being overpredated?

**Predation-Induced Changes in Morphology: Filament Formation**

Some bacteria have the ability to alter their morphology in response to predation. This ability appears to be taxon specific with different taxa having greater phenotypic flexibility than others. One microbial mechanism that has been related to predation is the formation of filaments by certain species of bacteria. We have discussed filament formation in response to various environmental conditions (e.g., temperature, salinity). Several field observations have reported the occurrence of filamentous bacteria under strong predation by protozoa. Based on these observations, it has been suggested that these filaments are a defense mechanism against the predators.

What evidence would be required to determine whether this suggestion was supported? First, it must be shown that the filamentous morphology develops in response to predatory pressure. Several studies have demonstrated this. Often, these changes in bacterial communities occur quite rapidly and appear to be a response of normal sized bacteria.

In the laboratory, experiments that examined the effects of contrasting feeding modes of predacious protists on bacterial morphology and community composition produced some remarkable results. When mixed bacterial assemblages were subjected to predation by the flagellate protist *Bodo saltans*, the composition of the assemblage shifted from beta- to alpha-Proteobacteria with a concurrent increase in the number of threadlike cells in the beta-Proteobacteria (>3 μm) (Pernthaler et al., 1997; Simek et al., 1997). *Bodo saltans* feeds almost exclusively on bacteria smaller than 3 μm in diameter. Three days after introducing the flagellate into the experimental vessels, the protist-inedible bacteria constituted more than 60% of the assemblage. No threadlike cells were present before predation. There was a rapid increase in the inedible cells in the first 2 days, after which the biomass of these cells remained constant suggesting a low growth rate.

Pernthaler et al. (1997) have shown that a medium-sized bacteria (beta-Proteobacteria) developed significant filaments after exposure to a flagellate in response to what the authors thought was a chemical stimulus. Hahn et al. (1999) observed a similar response in a different species of bacteria, but the authors found that filament development in the absence of predation occurred as a function of growth rate. In additional studies, Hahn et al. (1999) demonstrated for some species of bacteria that filament formation was a response to increased growth rates. In their experiments, filaments were not a direct response to the presence of the flagellate, but rather the reduction in bacterial biomass by the predator resulted in an increased growth rate and filaments (Figure 14.5). Filamentous growth has also been reported as a possible phage resistant mechanism or as an aid in gliding (motility).

Second, the growth of the filaments makes the bacterium larger than the normal size. Evidence that shows an increase in actual size makes it more difficult for the

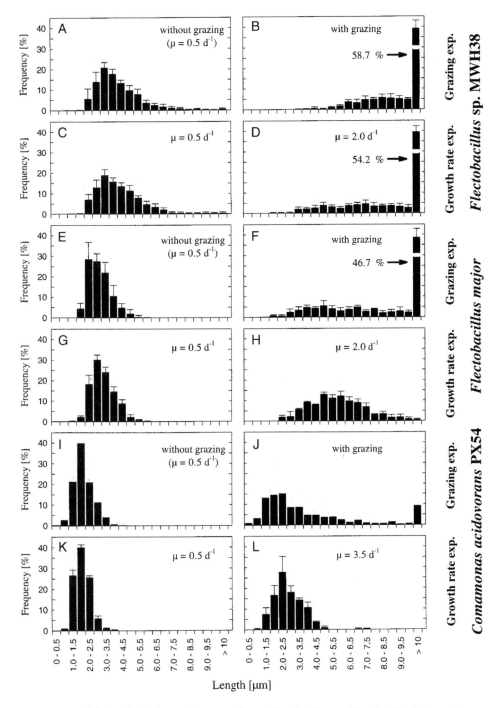

**Figure 14.5** Cell size distributions of chemostat-grown *Flectobacillus* sp., *Flectobacillus major*, and *Comomonas acidovorans*, with and without flagellate predation and at different growth rates in flagellate-free culture. (From Figure 5 in Hahn MW, Moore ERB, Höfle MG. Bacterial filament formation a defense mechanism against flagellate grazing, is growth rate controlled in bacteria of different phyla. *Appl Environ Microbiol* 65:25–35, 1999.)

predator to feed on bacteria would be strong evidence in favor of supporting the suggestion. This response would be effective only against metazoa predators that consume the bacterium, not against viruses or bacterial predators. Wu et al. (2004) demonstrated that the size of bacteria does increase in the presence of a predator (Figure 14.6), but they did not find that this increase affected the ability of the predator to consume the bacteria. In every experimental condition, the authors established that the predator showed an exponential increase, but the efficiency of ingestion was reduced as the size of the bacteria increased (Figure 14.7).

A shift to threadlike morphology had an apparent evolutionary cost that prevented a continuous increased growth rate. The cells escaped predation, but they were unable to reproduce. The study did not show the effect when the predation pressure was eliminated. It would be interesting to see how long it took the population to shift back to the original composition and size structure if it did shift.

### Predation-Induced Changes in Morphology: Change in Cell Volume

Some bacteria escape predation by forming micro-colonies that are larger than the size preferred by predators. Bacteria can also ameliorate the effects of predation by changing other aspects of their morphology or by growing together into micro-colonies that are larger than the preferred size of the predator. The micro-colonies produce extracellular materials that make their effective size even larger and protect some of the embedded cells from being consumed. Most bacterial predators have a restricted or optimal size range on which they can feed. Any particle that is larger or smaller than this size range would not be used as prey. By altering their size, microbes can avoid predation, at least by a predator that feeds on their normal size. Changes in size can be accomplished by forming long, threadlike strings. Bacteria can also become larger by changing their cell volume.

It appears that bacteria can alter their cell volume in response to predation. Posch et al. (2001) used a multistage, continuous culture experiment to test the effect of predation on bacterial community cell volumes. In their experiment, the first stage contained an alga (*Cryptomonas* sp.) and its associated bacterial community. This association appears to be obligate in that the algae dies without the bacteria being present and the bacteria feed on algal exudates. No predators were present in the first stage. This stage flowed into the second stage, which was the control for the experiment. This stage contained an established algae/bacterial community. The third stage was fed from the second but included a predator (*Cyclidium glaucoma*). *Cyclidium glaucoma*, a ciliate bacterial predator, has an estimated optimal food size of about $0.1\,\mu m^3$. When *C. glaucoma* was allowed to feed on a mixed bacterial community, the mean cell volume of the bacteria increased relative to controls that had no predators. This increase in cell volume reflected a change in width of the cell, and not elongation. Community composition of the mixed bacteria changed under predation. Components of the bacterial community were negatively impacted by predation, whereas others appeared to be less affected. No other morphological changes (i.e., clumping, threadlike cells, and cilia) in the bacteria were observed during the 65 days the experiment ran. The results are intriguing (Figure 14.8). The cell volume distribution in stage one quickly equilibrated to a mean of around $0.05\,\mu m^3$ and then maintained that distribution throughout the experiment. The second stage vessel had basically the

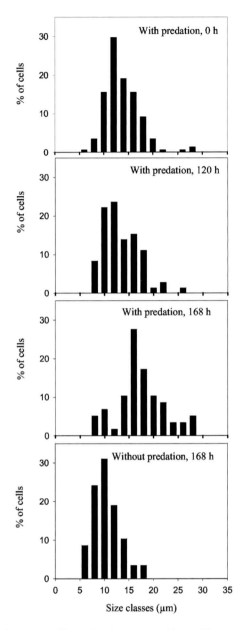

**Figure 14.6** Change is size classes of bacteria when grown with or without a predator. (From Figure 2 in Wu QL, Boenigk J, Hahn MW. Successful predation of filamentous bacteria by a nanoflagellate challenges current models of flagellate bacterivory. *Appl Environ Microbiol* 70:332–339, 2004.)

same size distribution as the first stage. However, when the predator was introduced, the cell volume of the bacterial community shifted to a much larger size. The preferred size of prey for the predator is marked with a dashed line in Figure 14.8. Predation caused the cell volume to shift such that the average cell size was larger than the predator could ingest. How long would it take the bacterial community to reduce its cell size when the predator was removed? Larger cell volumes mean more resources must be obtained and then sequestered at the expense of not reproducing as frequently.

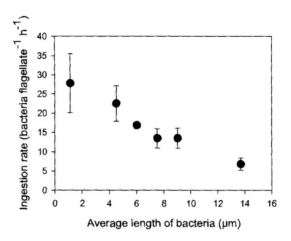

**Figure 14.7** Flagellate ingestion rate as a function of the average length of bacteria. (From Figure 5 in Wu QL, Boenigk J, Hahn MW. Successful predation of filamentous bacteria by a nanoflagellate challenges current models of flagellate bacterivory. *Appl Environ Microbiol* 70:332–339, 2004.)

*Pseudomonas* species formed micro-colonies of more than 1,000 cells to prevent grazing. The individual bacterial cells did not change their size or their growth rates. However, under the pressure of predation, they began to form micro-colonies. These colonies were larger than the optimal size for the predators, and the bacteria were able to escape the predator.

In these examples of bacteria avoiding predation, it appears that the bacteria are able to detect and then respond to predation pressure. We may expect that some chemical cue is being detected by the bacteria. In the *Pseudomonas* micro-colony example, the researchers placed the bacteria into media where the predator had been but was currently removed. No micro-colonies were formed. If a chemical cue was present in the medium, it was not detected, it was too dilute, or it was broken down after predator removal. Alternatively, the physical presence of the predator may be needed to initiate this response.

### Other Effects of Predation

Predator feeding, although directly impacting the microbes that are being eaten, may stimulate the growth of the remaining cells. Grazing of higher plants does not usually result in the ingestion of all of the plant by the grazer. Various amounts of biomass are consumed, but the plant remains. Grazing of microbes usually results in all of an individual organism being consumed. However, if we consider that the progeny of asexual binary fission are copies of the original organism, if a few or many cells are consumed, the organism still remains. Grazing may promote survival. In one study, regeneration of nutrients by a grazer resulted in more than 70% of the nutrients (including nitrogen) being released back into the immediate proximity of the initial grazing. These regenerated nutrients presumably enhanced bacterial growth. The consumption of some bacteria promoted the growth of the remaining bacteria, and because the nutrients were released near the site of grazing, we may conjecture that these nutrients were taken up by closely related, if not exactly related, bacteria. Survival of a species depends on the transmittal of genes from one generation to the next.

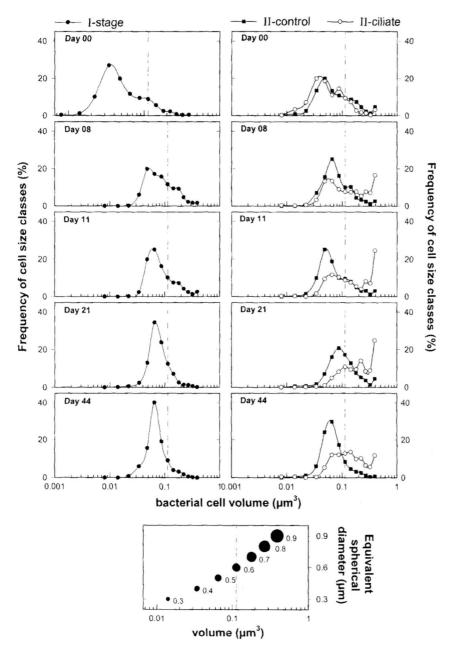

**Figure 14.8** Bacterial cell size distributions in all stages of the continuous cultivation system on 5 selected days. Volumes of spheres with a diameter between 0.25 and 0.85 μm (0.05-μm steps) were used as size classes to generate frequency distributions. The lowest panel shows volumes that would correspond to 0.3- to 0.9-μm spheres. The *dashed line* indicates the probable optimal food size of *Cyclidium glaucoma*, a ciliate protozoan. (From Figure 6 in Posch T, Jezbera J, Vrba J, Šimek K, Pernthaler J, Andreatta S, Sonntag B. Size selective feeding in *Cyclidium glaucoma* (Ciliophora, Scuticociliatida) and its effects on bacterial community structure: a study from a continuous cultivation system. *Microb Ecol* 42:217–227, 2001, with kind permission of Springer Science and Business Media.)

In this scenario, predation may enhance the survival of the species by providing limiting nutrients. Zubkov and Sleigh (1999) demonstrated that grazing by three different predators caused an increase in the release of radiolabeled molecules from labeled bacteria (Figure 14.9). When grazers were absent, very little label was found in the medium. Maximum release occurred after 6 days with most grazers.

### Predation and Growth Rates

In contrast to the formation of filaments by the beta-Proteobacteria, the alpha-Proteobacteria cells found in the same experiments, all of which were in the preferred size range of the flagellate predator, were grazed much faster than their density would indicate. To compensate for losses due to predation, these bacteria increased their growth rate. These rapidly reproducing cells were able to balance the effect of predation and maintain fairly constant densities. Increased growth rates compensated for increased predation. Both of these strategies (i.e., morphological change and increased growth rates) are effective against predation. Both have costs associated with them.

Increased growth rates would only be adaptive under favorable nutrient and temperature conditions. If the supply of nutrients failed or decreased these cells would be unable to reproduce at the higher rates and would presumably decrease in numbers by being consumed. The change to threadlike morphology kept the bacteria from being consumed but reduced their reproduction.

### Preferential Feeding

Are there preferred food items? Preference can be given to size and shape and to the quality of the food item. Predators that have the ability to discern or detect differences in the nutritive quality of a prey item should preferentially select the food item with the higher nutritional content all other things being constant. In other words, two food items that are approximately the same size and shape but that differ in nutritive content should not be predated at the same rate if the predator has some mechanism for evaluating the food items.

Dini and Nyberg (1999) observed the response of 31 stocks of *Euplotes*, a cosmopolitan marine protist on six different food species that included three species of algae and three species of bacteria. The 31 stocks of *Euplotes* came from nine reproductively isolated groups and included three morphospecies (i.e., morphologically defined species), *E. vannus*, *E. crassus*, and *E. minuta*, each with autogamous (i.e., self-fertilizing) and cross-breeding breeding groups. We would predict that the largest morphospecies should be better at handling and translating large food items (e.g., algae) into predator biomass than the smaller morphospecies. Similarly the smaller morphospecies should be better at handling and processing small food items than the larger morphospecies. From the experimental design, we cannot predict if there will be differences in the "quality" of the food items as perceived by the predators.

The results indicated that the largest morphospecies was more efficient at turning algal biomass into predator biomass than bacterial biomass, and the reverse was true of the smallest morphospecies. Mean number of predator fissions (i.e., reproduction) completed in 5 days after feeding on one of the six food items ranged from 0 to more than 17, with a strong statistical interaction between morphospecies and food items. Two of the predator groups were unable to grow on *Escherichia coli*. No choice test was given (i.e., the predator had only one food item to eat and could choose only to

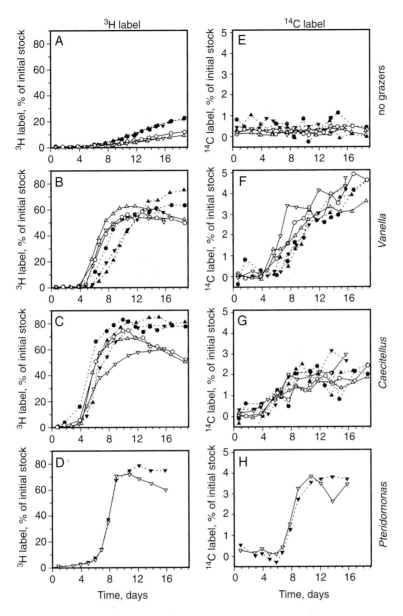

**Figure 14.9** ³H and ¹⁴C label released in the water in various incubation experiments is expressed as a percentage of the amount initially present in the stock of microwave-treated, dual-radioactive–labeled *Vibrio* (DRLV) deposited on insert filters. In control experiments (**A**, **E**), bacteria were incubated without protozoan predators. In other experiments, the bacteria were incubated with the amoeba *Vanella* (**B**, **F**), or the flagellates *Caecitellus* (**C**, **G**) or *Pteridomonas* (**D**, **H**). In most cases, three experimental series (*circles*, *up triangles*, and *down triangles*), each with two different amounts of bacteria—0.3 pg (*solid symbols*, *dotted lines*) and 0.9 pg (*open symbols*, *solid lines*)—of bacterial protein (cm) were monitored. One series with two different amounts of bacteria is shown for *Pteridomonas*. (From Figure 1 in Zubkov MV, Sleigh MA. Growth of amoebae and flagellates on bacteria deposited on filters. *Microb Ecol* 37:107–115, 1999.)

eat or not). However, because two of the groups could not grow on one of the bacterial species, we might expect optimal feeding by the predator and preferred selection of food items that do translate into predator biomass.

## Bacteria and Viral Interactions

Bacteria and their predators have been interacting for a very long time. Similarly, bacteria and viruses have been interacting for probably as long a period. Viruses that attack bacteria are numerous and diverse. We will discuss in more detail the interactions between bacteria and viruses in a later section. Here, it is important to understand that individual bacteria that are infected can be killed or otherwise incapacitated by the infecting virus. At the individual cell level, the affects of infection are essentially a binary response (i.e., effect or no effect). At the population and community levels, the affect of viral infections may be very large and affect the dynamics of bacterial growth and therefore the rates of various ecosystem processes.

We have previously discussed individual bacterial responses to predation, which can be morphological, behavioral, and act through production of various secondary compounds. Nevertheless many different organisms consume bacteria. One question of interest is whether the level of predation is such that it controls the population density and distribution of microbes or whether the reproductive rate of the microbes is faster than the predation rate. If the reproduction exceeds predation, the density of the microbes will increase, all other things being equal.

## Microbial Loop

An insightful contribution to understanding microbial involvement in the trophic structure of ecosystems is the *microbial loop*. This conceptual model was developed based on observations in pelagic systems. Organisms found in plankton or the suspended microbial communities of the water column of freshwater and marine aquatic ecosystems are of three main types: *phytoplankton* or the photosynthetic/autotrophic organisms, *zooplankton* or the protozoans, and the *bacterioplankton* or heterotrophic bacteria. Predation of bacteria by bactivorous flagellates can be sufficient to affect the abundance and species composition of bacterial communities. If predators of bacteria can control the numbers and types of bacteria present, they may exert an impact on microbial processes. Some predators are capable of feeding exclusively on microbes. The importance of microbes as prey is somewhat dependent on the environment. In eutrophic environments such as many rivers, lakes, and coastal waters, there are usually many microbes and a rich *meiofauna* that feed directly on the microbes. In oligotrophic environments such as some lakes and the open ocean, microbes may provide little nutrition because of their relatively low numbers. Soils and sediments usually have well-developed microbial communities that are eaten and used as food by numerous soil inhabitants.

In some productive aquatic ecosystems, dissolved organic carbon (DOC) has been shown to accumulate. If DOC is the primary food source for many bacterial species, why would this material accumulate? Why do some species not key in on the excess DOC and degrade this material? Perhaps the DOC is of poor quality and is preferentially avoided. An alternative explanation involves species interactions that control the consumption of DOC. Bacterial consumption is controlled by food-web mechanisms that control growth and biomass. Growth rate may be kept low by competition between bacteria and phytoplankton for essential mineral nutrients. Biomass is kept low by bacterial predators. A model based on these food-web interactions has shown that the nature of what limits bacterial growth rates is a key aspect for understanding the release or accumulation of DOC (Thingstad et al., 1997).

## Bacteria as Predators

Because of their enormous population sizes (actual densities), high nitrogen content, and relative inability to escape, microbes have been exploited as a food source by diverse groups of organisms. Species interactions within microbes are also varied, diverse, and sometimes intense. Competitive ability and mutualisms have been selected and honed over the 3.8 billion years of microbial interactions. Predation, as an ecological and evolutionary concept, has been developed and studied most thoroughly in higher animals. When microbes have been studied as part of trying to understand the consequences of predation, they have been primarily considered as prey. Microbes are the prey of numerous other organisms, and some microbes are predatory on other microbes. This is especially true of bacteria.

Bacterial predators can be obligate or nonobligate. Nonobligate predacious bacteria can obtain nutrition from sources other than live bacteria. However, under certain circumstances nonobligate bacterial predators are effective predators on other bacteria. Studies have revealed that some bacterial predators can be predators of other bacterial predators. These predators can be Gram-positive or Gram-negative organisms.

Casida (1992) isolated a copper-resistant, nonobligate bacterial predator that exhibited a number of curious behaviors. First, it was found to be predacious on most other nonobligate bacterial predators, and it therefore exerted control on the level of predation in the soil. Second, it could inhibit fungi. Third, after attachment to the prey cell it produced a toxic growth initiation factor. Fourth, it produced a novel compound that was antibacterial and antifungal. Under the study conditions, these characteristics allowed the strain to quickly become dominant and persist over extended periods. This rapid growth obviously requires sufficient nutrients supplied by prey. However, this strain did not have to search out prey; instead, other predatory bacteria did this, and as they multiplied, these predacious bacteria became prey.

Other bacteria are obligate predators. Examples include species of *Bdellovibrio* and related organisms. These bacteria are Gram-negative cells that possess one sheathed polar flagellum. Extremely fast swimming rates have been recorded for these organisms, with rates approaching 100 body lengths per second, which make them among the fastest bacteria. They are the cheetahs of the microbial world.

*Bdellovibrio* organisms are obligate predators of other Gram-negative bacteria. Special attack cells attach to and then penetrate prey cells. After penetration, the prey cell's metabolism is inactivated, and the cell is turned into a *bdelloplast*. Bdelloplasts serve several functions, including offering protection against phage attack, protection against photo-oxidation, and increasing resistance to pollutants. This does not have many of the characteristics of classic predator-prey systems. Instead, it appears to be more parasitic and less predacious. However, these predators do not commandeer the prey's cell machinery. Using the cytoplasmic contents of the prey, the predator produces filamentous growth and DNA replication. After exhaustion of the resource, the predator divides through multiple fissions into attack cells complete with a flagellum that bursts the bdelloplast, releasing the predator into the surrounding environment. *Bdellovibrio*-like organisms are fairly common in both natural and man-made environments. Yair et al. (2003) reported that they are easily isolated from soils, the rhizosphere, fresh and saline habitats, and wastewater treatment facilities. They are often associated with surfaces and biofilms. These organisms have also been isolated from the gills of crabs and from oysters, and they have been found in the feces of chickens and some mammals. The latter observations are interesting given that *Bdellovibrio*-like organisms are aerobic, indicating that for at least the time necessary to clear the gut, they are able to withstand anoxic conditions. However, some marine forms, especially the attack cells, are able to survive anoxic conditions outside of a host.

Normal estimates of *Bdellovibrio*-like organism densities based on plating techniques are between tens and tens of thousands per gram or milliliter of sample. These densities are what we would expect for a predator population. Predator populations generally do not exceed densities of their prey. A general rule of thumb in trophic transfer is that the energy available to the next higher trophic level in a food web or chain is less than 10% of the level just below. Although this generalization has not been definitively shown for most ecosystems, the concept is based the transfer of energy between trophic levels. Whatever the real relationship is, it is clear that predator biomass or numbers cannot exceed their food base. Exceptions are found among those organisms that feed on rapidly reproducing species. At any one point, the number of prey appear to be below what is needed to provide energy to the predators. However, if we examine total reproductive output over some extended time period, it can be seen that the predators are able through consumption to maintain the prey population at reduced levels. For *Bdellovibrio*, the prey populations are at much higher density. Each species of *Bdellovibrio*-like organism exhibits different prey ranges. In other words, the predators are selective. Many species are able to use a number of prey species, whereas others have been shown to attack only one type of prey species.

Not all bacterial predators use the same strategy as the *Bdellovibrio* organisms. Yair et al. (2003) describe several other strategies. Whereas the *Bdellovibrio* organisms attack and then penetrate the prey cell, others have no periplasmic stage but rather remain attached to the surface of the prey. These predators feed on the cytoplasm of the prey and divide while attached, and the emptied prey cell maintains its original shape. Other micropredators include *Ensifer*, *Vampirovibrio*, and *Vampirococcus*, which are extracellular epibiotic predators. The gliding bacteria (i.e., *Myxobacteria*, *Cytophaga*, and *Herpetsiphon*) have the ability to lyse living bacteria and then use them for food. Some bacteria feed in groups in what has been described as wolf pack predation (e.g., *Myxococcus*) (Martin, 2002). This is an area open for more research. Many questions remain. How prevalent are bacterial predators? Most bacterial predators attack Gram-negative bacteria, but are there bacterial predators of Gram-positive bacteria? Do bacterial predators exert any control over the distribution and abundance of bacteria? Does bacterial predation affect community composition? What are the interactions between bacterial predators and protozoan or metazoan predators? There is some evidence that bacterial and protozoan predators do not interact except in competition for available prey (Casida, 1989).

Bacterial predators may also affect large-scale algal blooms. In particular, lytic bacteria were isolated from a lake where dense cyanobacterial blooms occur (Rashidan and Bird, 2001). These isolates were incubated with 12 cyanobacteria and 6 heterotrophic bacteria. Some of the isolates could lyse only specific species of cyanobacteria. These results suggest that these predatory bacteria can control and potentially eliminate cyanobacterial blooms in this lake system.

## Competition

Competition among species usually is a negative interaction in which one species outcompetes the other species for limiting resources. The results of competition are the elimination of one or more species that have niches that overlap too much with the winning species. We have discussed the effects of competition in other chapters. Here, we discuss the antagonism between competing species through the production of antibiological substances.

Some microbes can avoid competition by producing compounds that negatively impact their competitors. Often, these compounds (e.g., antibiotics) are considered in discussions on competition. It is thought that antibiotic-producing microbes do so in response to negative interactions with unrelated species for space or other resources. Antibiotic production in the environment is difficult to detect. However, the production of antibiotics would be most important at the scale of an individual cell or microcolony. At that scale, our ability to detect production is probably exceeded. Intuitively, large-scale production of complex molecules such as antibiotics is metabolically very expensive and must be done as a trade-off for reproduction. The profit from the production of an antibiotic must exceed the cost. In other words, the bacteria that produce the antibiotic must produce more offspring over time than the same species that does not produce the antibiotic. Interestingly, some of the slowest-growing culturable bacteria, the actinomycetes, produce antibiotics.

## Actinomycetes and Antibiotics

In one of the few studies on the ecology of actinomycetes, one of the slow-growing, antibiotic-producing bacterial groups, Wohl and McArthur (2001) observe that growth rates of actinomycetes are typically slower than most fungi and bacteria. However, these organisms produce secondary metabolites, which include a suite of enzymes and antibiotic substances that makes them resistant to many other microorganisms. These metabolites aid in their ability to degrade and assimilate a broad range of recalcitrant compounds, including chitin, cellulose, hemicellulose, keratin, pectin, and lignin. The ability to break down recalcitrant compounds requires time. Slow-growing bacteria would be inferior competitors for space if they did not have some mechanism that gave them a slight advantage. Antibiotics provide such an advantage. In the case of actinomycetes, competition is not for the organic resource, because most of the fast-growing species are incapable of breaking down these complex molecules. Competition is instead for space to grow or for access to other limiting nutrients. Presumably without the aid of the antibiotics, the faster-growing species could become established.

# Neutral Relationships

A neutral relationship is one in which neither of the interacting species obtains a benefit or a negative effect. Some relationships may at first glance seem to be neutral, but on closer examination, it is discovered that there is or may be a negative effect on one of the species. For example, many species of bacteria and fungi use plants and animals (externally and internally, such as the intestinal tract) as a solid surface on which to grow and reproduce without producing any harm or advantage to the colonizing species.

In aquatic systems, bacteria, fungi, and some algae may settle out on plants and animals. These layers of organisms may be fairly dense. Microbes that settle onto animals may aid the animal by creating a camouflaged or cryptic cover that allows them to escape predation or be a more effective predator. If there is an advantage to the host, the interaction is not neutral. These same microbes settling out onto aquatic plant species may inhibit light penetration to the plant. Some plants and cyanobacterial mats are never colonized by microbes, which suggests that there is a negative impact of passively settling bacteria and fungi on some species.

Some plant species produce antimicrobial compounds or alter the surface pH, which keeps their surfaces clear of bacteria and fungi. Rheinheimer (1992) reports that in some north German lakes, bacteria were rarely found on the surfaces of cyanobacteria or diatoms. In contrast, others have shown continuous covering of planktonic algae by bacteria. In general, the differences between these studies and numerous others may be summarized as follows. Vigorously growing algae are usually free from bacterial growth on their surface, whereas stagnating or declining algae are increasingly colonized by bacteria. Declining or stagnating algae release cell products, presumably decreasing production of antibiotics or inhibitory compounds. If the settling organisms obtain some sort of benefit from being on the surface of the plant or animal, the interaction is not neutral because the benefit (i.e., increased growth rate of the commensal) is a net drain on the resources obtained by the host, which subsequently cannot be used to produce more copies of the host. Algae release dissolved organic substances, and many animals shed tissue and excreta that serve as nutrients for microbes. Some bacteria and fungi do not obtain any nutrition from the plants or animals; they are attached to but use the other organisms as physical supports and obtain their nutrition from the surrounding media. These relationships are neutral only if there are no negative or positive effects that are derived from the association or interaction. Besides decreasing light penetration to leaves or photosynthetic cells, aufwuchs (i.e., settled microbial material) may be harmful to plants by attracting grazers, scrapers, or other predators that may damage leaf tissue in the process of feeding.

There are numerous microbes, including many bacteria and yeasts that grow as commensals on the surface and gut linings of many higher animals. These organisms in some instances provide nutrients to their hosts. Others use the surface as an attachment site but provide nothing in return to the host. Because they do not negatively impact the host, these relationships are commensal.

# Positive Relationships

Much ecological theory has been based on models of competition and predation, both of which are negative interactions. Many ecological textbooks cite a few examples of mutualisms but give little attention to these interactions. Profitable cooperation by and between different organisms may play a very important role in many ecosystems. We will compare two positive interactions. First, if one organism creates conditions that favor or are required by succeeding unrelated organisms, the relationship is called *metabiosis*. Second, if two or more contemporary organisms form an interaction that provides an advantage to all members of the interaction, the relationship is called *symbiosis* or *mutualism*.

## Metabiosis

There is some ambiguity about metabioses. For example, is the interaction always one direction (i.e., is one species always benefited and the other gets no benefit out of the interaction)? Can the interacting species coexist side by side, or is there always a succession in direct response to modifications of the environment by one species that is exploited by the next?

Many degradative consortia may be metabioses. If a potential nutrient cannot be exploited except in a mixed culture, we are dealing with metabiosis. For example, *E. coli* and *Proteus vulgaris* together can survive on a lactose-urea medium. In this instance, *E. coli* can cleave lactose, and *P. vulgaris* cannot, and only *P. vulgaris* can cleave urea. The breakdown products from these cleavages can, however, be used by the other organism as a carbon and nitrogen source (Rheinheimer, 1992). Other examples include the degradation of lignin or cellulose.

Microorganisms can also change the physical conditions of an environment and make the environment more suitable for other microorganisms. Any change in $E_h$; pH; gases such as oxygen, carbon dioxide, carbon monoxide, $H_2S$, and $CH_4$; and any number of other substances or conditions by one microorganism or group of microorganisms may produce conditions that are suitable for the development, growth, and reproduction of other microbes. Increases in methane by methanogens would create the substrate for methanotrophs.

It is not always clear that the interaction is successional and not concurrent. In the *E. coli* and *P. vulgaris* example, each microorganism created suitable conditions for the other. If the medium had been only lactose or urea but not both, one or the other species would not have been able to survive until the former species had produced sufficient product. When both substrates were present, each species could feed on a select component of the medium, and each could feed on the breakdown products of the other species. In this instance, the two species coexisted, and the species were successional to each other.

## Symbiosis

Symbiosis has been observed between microorganisms and many plant species and between microorganisms and most animals. Microbes live in close association with many organisms, but when that association is permanent, at least during some part of the life cycle, we consider the association to be symbiotic. Symbioses are many and diverse and include associations between hosts and *symbiotes* (i.e., smaller partners in the association) that occur external to each other (i.e., *ectosymbiosis* or *exosymbiosis*) or internal to the host (i.e., *endosymbiosis*) but extracellular. Some symbiotes reside in the cells of the host (i.e., *endocytobiosis*). According to Herre et al. (1999), mutualisms are ubiquitous and ecologically dominant, and they exert a profound influence at all levels of biological organization, from cells to ecosystems. Overmann and Schubert (2002) list several functions provided by microbes in a number of different symbioses. These include acting as a hydrogen sink in transfer of hydrogen between eukaryotes and prokaryotes in anaerobic environments, oxidation of sulfide and fixation of carbon dioxide in interactions with nematodes, producing exoenzymes for the breakdown of cellulose and keratene in association with shipworms, providing vitamins to tsetse flies, recycling amino acids in aphids, producing toxins that are used in defense by various protozoa, and increasing the cytopathogenicity of amoeba. This list is far from being exhaustive, but it is representative of the diversity of interactions.

Symbiotic relationships between eukaryotes and prokaryotes are between 20 and 250 million years old. However, most bacterial and archaeal symbiotic partners have yet to be cultured, and their associations must be inferred from molecular detection. Overmann and Schubert (2002), in one search of molecular databases, determined that there were at least 937 symbiotic prokaryotes identified. Moran (2002) states that ecological interactions between large organisms may be controlled in part by associations with microorganisms.

For something that has that much influence on the biology of this planet, it seems interesting that little attention was paid to the phenomenon in many ecology textbooks. A notable exception has been the symbiosis between nitrogen-fixing bacteria and legumes.

Mutualisms can be diffuse indirect interactions, or they can be extremely highly evolved relationships where there is a direct exchange of goods or services between the partners. There may be highly evolved modifications in behavior or structure in either of both of the partners. The following discussion relies heavily on the observations of Herre et al. (1999) and Nardon and Hubert (2001) as they relate to microbial symbiotic relationships.

### Theoretical Considerations

Selection acts to maximize the fitness of individuals. When there are limited resources, selection should favor organisms that can effectively harvest or use these resources. Cooperation may seem antithetical to increasing fitness because the organism must give up something during their interactions. This something has a metabolic or reproductive cost associated with it. Cooperation must provide fitness enhancements that exceed those lost if the organisms remained outside an association. This is not a chicken or egg argument. Organisms evolved before associations. Because each organism is trying to maximize its fitness, mutualisms should be viewed as reciprocal exploitations that provide benefits to each member of the association. That sounds like an oxymoron. How can exploitation provide a benefit? What are the conflicts of interest between the partners, and how are these mediated in both space and time as a function of the life histories of the partners? As with many things ecological, our understanding of mutualisms will increase as we examine the effects of symbioses at multiple evolutionary and ecological scales.

Some of these relationships extend over long time frames, sufficient to allow speciation events to occur. What phylogenetic patterns exist between and within partners? In other words, is speciation in the host followed by speciation in the symbiote? How often has the symbiosis arisen?

Conflicts of interest arise because each partner is a separate reproducing entity. The interests of one partner cannot be identical to the other partner because of this separation of genomes. If we consider that conflicts of interest have been observed between different genes within the genome of a single individual, it is not surprising that such conflicts would be found between two or more species.

Herre et al. (1999) list several factors that begin the formation of a framework for the alignment of mutualisms:

1. The vertical transmission of symbionts from parents to offspring
2. Genotypic uniformity of symbiotes within individual hosts
3. Spatial structure of populations that leads to repeated interactions between would-be mutualists
4. Restricted survival options outside of the relationship between the mutualists

Transfer of symbiotes between unrelated hosts by horizontal transfer, the presence of multiple genotypes of the symbiote within an individual host, and the ability to survive and reproduce outside the host (varied options) all work to prevent or undo mutualistic interactions.

An examination of each of the four factors listed earlier may provide insight into how they reduce conflicts of interest between partners in a symbiosis. In a successful mutualism, the reproductive success of the host ensures reproductive success of the symbiote. Reproductive success of the symbiote cannot be at odds temporally or spatially with the reproduction of the host. Any negative effect generated by the symbiote that affects the reproductive success of the host will directly affect the reproduction of the symbiote. Vertical transfer of the symbiote by successful reproduction of the host must, over evolutionary time, reduce the genetic diversity of the symbiote. Only genotypes that promote or are neutral toward host reproductive success will be favored. Over time, there may be an increased homogeneity in the symbiotes genetic background. If symbiote survival depends on successful vertical transfer by host reproduction, it is critical that there be increased interaction between the host and symbiont lineages. Increased interactions among the mutualists to the exclusion of life stages that are independent of one another would reduce the possibility and viability of nonsymbiotic alternatives.

This framework proposed by Herre et al. (1999) is logically appealing, but there are exceptions that can be observed from known mutualisms. For example, research on coral reef mutualisms has shown that there might be several genotypes and species of symbiotic algae associated with specific coral species. For many years, it was thought that a single species of dinoflagellate algae (*Symbiodinium*) was involved in the symbiosis. With the advent of modern molecular genetic techniques, it has been found that single species of coral have multiple species of symbiotic algae and that single colonies of coral might have multiple genetically different symbiotes. Coral symbionts are transferred horizontally although the free-living forms have not been clearly identified. Despite the absences of vertical transfer, it should be clear that this association is of great value to the coral reefs. Loss of coral reefs in part has resulted from the break up of these complex symbiotic relationships.

The discussion of conflicts of interest becomes more intense as we consider symbioses that have multiple species in a single host. Ecological theory predicts that these co-occupying species should not be closely related because of the increased levels of competition. However, molecular analyses have shown that these species are closely related. Why does competition not destabilize the mutualism and cause it to disintegrate? Herre et al. (1999) propose that the ecological flexibility that results from symbiote species and genetic diversity may counterbalance any negative effects of increased competition. An interesting question may be to determine how many and what kinds of mutualisms (ecto or endo) demonstrate patterns similar to the coral reef symbiosis (i.e., many species of symbionts).

Microbial symbiotic relationships can be found everywhere. We will consider only a few of these to give some appreciation for the diversity and extent of these relationships.

**Ectosymbioses**

Microbes are found in many ectosymbiotic relationships with various levels of interactions. These relationships can be found with protozoa, algae, fungi, plants, and animals in a variety of habitats and ecological conditions. Nardon and Charles (2003), in a paper entitled "Morphological Aspects of Symbiosis," give numerous examples of various symbiotic relationships involving microorganisms in association with many other types of organisms. We consider some of those examples.

Lichens have been discussed previously in some detail. Lichens are an example of increasing interactions among ectosymbionts. In the simplest cases, the fungus and algae have minimal interaction, whereas in the more highly evolved interactions, the fungus has developed various special devices that penetrate the gonidia. The fungus produces spores that find an algal cell and produce a new lichen. The symbiosis has produced a structure (i.e., thallus) that is stable between generations and whose development is controlled by the genetics of the fungus and the genetics of the algae (i.e., *epigenetics*).

Fungi can be associated with plant roots with the fungus mycelium coating the roots. If the fungus does not penetrate any root cells, this association is ectosymbiotic and is known as *ectomycorrhizae*. In this association, the fungal hyphae, which have a higher surface-to-volume ratio than the plant roots, increase the nutrient uptake capacity of the roots. Morphological changes are induced in some hosts, and the root cap and meristem is not infected by the fungus.

Fungi also form symbioses with animals. In Chapter 15, we discuss the ant-fungus interaction. Similar associations can be found with termites, wood-wasps, and some species of beetles. In the ant example, the fungus is grown by the ants, and no permanent contact is established between the invertebrate and the fungus. However, various devices have evolved in the hosts that allow the transfer or carrying of the fungus or its spores.

Certain ectosymbioses can be found involving protozoa or algae. For example, some benthic and planktonic diatoms are colonized by ciliates, dinoflagellates, cyanobacteria, and some other diatoms species. However, these relationships are not overly strong, and there is little fidelity between diatoms and the epibionts that are colonizing them, and any interactions are probably weak. In one curious association, an obligate endosymbiotic protozoan found in the guts of termites has an extensive ectosymbiotic assemblage attached to its surface. These ectosymbionts include small spirochetes, a few large spirochetes, and some short, rod-shaped bacteria attached to the host. These ectosymbiotic bacteria are attached in rows and have coordinated undulations that aid in the motility of the protozoan.

Another example of ectosymbiotic relationships is provided by the pogonophoran worms. Pogonophoran worms are found in the deep-sea hydrothermal vents. This worm, whose closest relative may be the annelids, does not have any mouth parts or digestive system, and it has no internal bacteria. On the external surface of the worm can be found four different types of bacteria belonging to the epsilon-Proteobacteria: rods attached to the cuticle by filaments; small, spiral-curved bacteria; some bacteria with appendages; and filamentous bacteria. The nature and strength of the interactions between these bacteria and the worm have not yet been worked out. However, it is presumed that the bacteria provide nutrition for the worm.

### Endosymbiosis

Endosymbionts are frequently found in the intestines of animals, but they are also found in specialized organs or in association with invertebrates. In endosymbiotic associations, the symbiote has penetrated into the organism but remains outside the cells. Nardon and Charles (2003) describe endosymbiotes of termites, mammals, Echinoderms, and sponges.

Within the mammals, there are two main groups that can be separated based on their gastric arrangement. *Monogastric* (i.e., single stomach) types include mice, humans, and organisms that are *nonruminant*. In contrast, *ruminants* are *polygastric* (i.e., many stomachs), and the anterior intestinal tract has been modified through evolution to facilitate mutualisms.

There are more than $10^{14}$ cells in a normal human (Savage, 1977; Nardon and Charles, 2003). Of this incredible number of cells, only about 10% ($10^{13}$) are animal cells, or what the average person would consider to be "human" cells. That is admittedly a large number of human cells, but it still means that 90% of all cells in a human are not human cells. From a community perspective, a normal human is a consortium of more than 300 different species, mostly bacteria that live in the gastrointestinal tract. Some of these species are found only in the guts of humans and nowhere else. Comparisons of the gastrointestinal tract between *axenic* (deprived of normal microbial associations) monogastric mammals and mammals with their natural microbial assemblages have demonstrated that the presence of the microorganisms results in morphological changes to the gastrointestinal tract. For example, comparisons made of axenic and normal rats shows that the lining of the intestine is much thinner and weaker in the axenic rats than in normal rats.

The main gastric compartment in ruminant mammals is the rumen. This specialized structure has been studied in much detail and found to be an incredibly complex, interactive organ. More than 200 species of endosymbiotes have been isolated from the rumen, of which 30% have been found only in the rumen. They include bacteria, protozoa (primarily ciliates), and some fungi. These organisms are obligate anaerobes that can survive in relatively warm temperatures (40°C) and circumneutral pH (6 to 7) of the rumen.

In contrast to mammals, termites have evolved numerous symbiotic relationships with several microbes, including bacteria, protozoa, and in some cases, yeasts. In the higher termites, only bacteria have been found as symbiotes. The hindgut of wood-eating termites is modified into a paunch that holds bacteria and protozoa. Studies have shown that the life span of termites fed antibiotics is reduced from 250 days to about 13 days. The termites are susceptible directly to the antibiotic, or the reduction in microbes has a major negative impact on termite survival. The hindgut of the alimentary tract has the highest densities and most diverse assemblages of symbionts. These diverse species are able to establish stable populations and obtain energy and nutrients, and in return, these microbes allow the host to consume a wider range of food resources. Many arthropods have firmly attached bacterial populations, but they also have free-living protozoa and spirochetes that persist without attachment. Both methanogens and nitrogen-fixing bacteria have been isolated from termite guts.

### Nitrogen Fixation and Symbiotic Gut Flora

The total number of species of arthropods that harbor nitrogen-fixing bacteria in their guts is not known (Nardi et al., 2002). Nitrogen fixation in the guts of these organisms has been demonstrated only in a few instances with rigorous tests but these studies suggest that symbiotic nitrogen-fixing bacteria may be important to arthropod nutrition. Nardi et al. (2002) estimated the possible global impact of nitrogen fixation in arthropods. The estimated annual global nitrogen fixation is approximately 8 kg/ha/yr for nonlegume crops, 10 kg/ha/yr for forests and woodlands, 15 kg/ha/yr for meadows and grasslands, and 2 kg/ha/yr for all other vegetated land. This gives a total estimate of 35 kg/ha/yr of fixed nitrogen. Using a conservative estimate of termite biomass of about $1 g/m^3$ in one Costa Rican forest, an estimate of 0.12 kg/ha/day of termite-associated fixed nitrogen was made. Over one annual cycle and at continuous rates, the results would be an estimated 43.8 kg/ha of $N_2$ fixation! It is highly unlikely that the rates of nitrogen fixation could be maintained at these rates, but in theory, the minimal estimate of termite biomass within a single forest can produce more fixed nitrogen than all vegetated environments in the world. Even if the rates were grossly overestimated, the total amount of nitrogen fixation may be immense and of critical importance in the cycling of carbon and other nutrients.

Endosymbionts have been found in subcuticular tissues of Echinoderms at very high numbers ($10^9$ cells/g of ash-free dry weight of the host tissue). Sponges are an ancient group of invertebrates that have been shown to have endosymbionts. It has been estimated that sponges are only 21% animal cells and 38% extracellular bacteria. The remaining "sponge" biomass is intercellular substances. Some studies have shown that there are genus-specific and, in some cases, species-specific associations of bacteria and sponges. Even though sponges have a long evolutionary history, there are no known morphological modifications or specialized organs to facilitate the symbiotic associations.

### Endocytobiosis

In the examples of symbioses previously discussed, the symbiote has been on the surface of or close to the host, or the symbiote has been inside the host but not inside the cells of the host. We now discuss associations in which one organism penetrates the cells of the host. Endocytobiosis is widespread. Plasmids and mitochondria have been shown to have prokaryotic DNA and to be most genetically similar to bacteria and represent one of the oldest symbiotic relationships. The capture of these prokaryotes and the transformation into organelles has been described in detail elsewhere, but it is mentioned here to highlight the ancient nature of symbioses that involve penetration of the cell.

Endocytobioses can emerge over ecological time frames. For example, the amoeba, *Amoeba proteus* was infected with a pathogenic gamma-Proteobacteria. Over a single year of observation, the association went from pathogenic to an obligate relationship between the amoeba and the bacteria (Jeon, 1987; Nardi et al., 2002). The degree of integration between the host and the symbiote is not necessarily an indicator of length of time the association has been developing.

Endocytobioses of protozoa involve bacteria in the cytoplasm of the host generally in specialized structures called *symbiosomes*. However, they are not restricted to the cytoplasm, and examples can be found of symbiotes in the nucleus. Not all protozoan endocytosymbiotes are bacteria. Some ciliates have algae in their nucleus, and there are other examples of protozoa being symbiotes of other protozoa. For example, a flagellate may infect the protozoan *Stentor*, causing hypertrophy of the nucleus. Morphological changes of the nucleus caused by endosymbiotes include changes in size, shape, and location. Hosts may be infected with symbiotes in their cytoplasm and in the nucleus.

## Bacterial Symbionts of Other Microbes

Most symbiotic associations (>96%) that involve prokaryotes are between eukaryotes and prokaryotes, but in a few cases, prokaryotes interact with other prokaryotes. Symbioses between prokaryotes and eukaryotes are no older than 250 million years. Considering that microbes have been interacting for 3.8 billion years, it is surprising that so few known symbioses exist between prokaryotes. Why? The answer may lie in our ability to observe or measure these interactions. Complex and highly evolved interactions may exist and may explain why so few microbes can be cultured.

The spatial distribution of microbes influences the magnitude of possible interactions among individuals or species. Based on an average size of microbes, a random distribution would mean that, on average, each cell would be approximately 1 to $112\,\mu m$ distant from the next closest cell. Overmann and Schubert (2002) report that for small soluble compounds (the compounds most readily taken up by microbes), the efficiency of transfer is inversely proportional to the distance between cells. For example, only 25% of the flux in metabolites between partners that are $1\,\mu m$ distant reach partners $2\,\mu m$ apart. There is a 75% reduction in available material and presumably a similar reduction in possible interactions. This reduction increases to 0.01% when the cells are $10\,\mu m$ apart. Selection favors situations in which interacting microbes are fairly close.

A careful examination of microbial distributions indicates that they are not randomly distributed. Microbes, especially bacteria, instead have clumped distributions. Often, these clumps show the existence of conspicuous morphologically distinct associations (Figure 14.10). Bacteria occur in micro-colonies, aggregated in association with various particles (inorganic or organic), or in biofilms. If the distributions of the

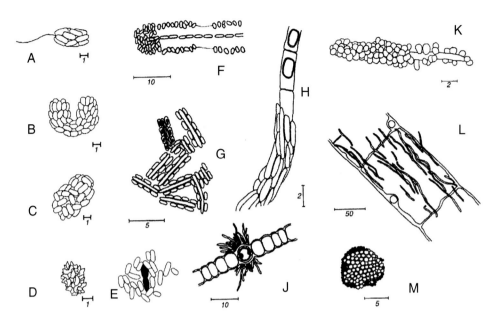

**Figure 14.10** **A:** Morphotype of *Chlorochromatium aggregatum* and *Pelochromatium roseum*.
**B:** *Chlorochromatium glebulum*. **C:** Morphology of *Chlorochromatium magnum* and *Pelochromatium roseo-viride*. **D:** Morphotype of *Chlorochromatium lunatum* and *Pelochromatium selenoides*.
**E:** *Chlorochromatium aggregatum* after disaggregation; the colorless central bacterium is visible.
**F:** *Cylindrogloea bacterifera*. Longitudinal transect on the right shows the central rod-shaped bacteria.
**G:** *Chloroplana vacuolata*. Gas vacuolation of colorless filaments and rod-shaped green sulfur bacteria is depicted for only a few cells. **H:** Consortium from the hindgut microbial community of the termite *Reticulitermes flavipes*. Upper portion shows a central chain of rod-shaped bacteria containing endospores. **J:** *Anabaena* sp. filament containing one heterocyst cell covered by chemotrophic bacteria.
**K:** Corn-cob formation from dental plaque. **I:** *Thioploca* sp. covered with filamentous, sulfate-reducing bacteria *Desulfonema* sp. **M:** Novel archaeal-bacterial consortia in which a central aggregate of archaeal cells (*open*) is surrounded by a few layers of sulfate-reducing bacteria (*solid*). Values on all bars are in micrometers (μm). (From Overmann J, Schubert K. Phototrophic consortia: model systems for symbiotic interrelations between prokaryotes. *Arch Microbiol* 177:201–208, 2002, with kind permission from Springer Science and Business Media.)

individual species found in these micro-colonies, aggregates, and biofilms are not haphazard, we might expect that species and individuals that are close enough have developed symbiotic interactions.

Overmann (2001) identified more than 15 different morphotypes of microbial consortia. These associations or consortia can be found in a wide number of habitats under very diverse environmental conditions. Overmann and Schubert (2002) list several examples, including phototrophic consortia that are found in freshwater and consortia consisting of Gram-negative rods or methanogenic archaea in association with filamentous, endospore-forming bacteria in termites occurring in the hindgut. Heterocysts of filamentous blue-green bacteria are colonized by chemotrophic bacteria, and plaque-forming streptococci form a "corn-cob" pattern when associated with a central *Bacterionema matruchotii* cell. Interesting consortia have been observed in marine upwelling sites, where filamentous sulfate-reducing bacteria are associated with sulfide-oxidizing *Thioploca*. A spherical consortium forms between sulfate-reducing bacteria and methanogenic archaea in certain marine sediments that are rich in methane hydrate.

Phototrophic consortia involve green sulfur bacteria surrounding a central core of chemotrophic bacteria (Overmann and Schubert, 2002). Some of these consortia are motile. Seven different morphotypes of motile phototrophic consortia have been identified that are distinguished based on the color and shape of the symbionts and the presence of intercellular gas vesicles. These consortia have been found and identified along the chemocline of stratified freshwater lakes worldwide, but until recently, all attempts to enrich, maintain, and isolate them have failed, even though the biomass of these consortia can exceed two-thirds of the total bacterial biomass in the chemocline.

Association is not definitive of symbiosis. In a symbiosis, cell-to-cell communication between the partners and products can be identified that benefit the partners because of the interaction. Do microbial associations meet this definition? Is the association specific—are certain species or strains always found in the association and no others? We will answer the second question first. Various phototrophic consortia were collected from European and North American lakes, and all microscopically identifiable morphotypes were isolated. Each isolated consortium was mechanically separated from accompanying bacteria by micromanipulation. DNA was extracted from the epibionts and analyzed using polymerase chain reaction (PCR) analysis of the 16S rRNA gene fragments of green sulfur bacteria. Amplified products were separated using denaturing gradient gel electrophoresis (DGGE) and sequenced. Amazingly, in each morphotype of consortia, only one, novel phylotype of green sulfur bacteria were identified. Unique epibionts were found in identical morphotypes that were sampled from different lakes. Such an observation suggests biogeographically isolated consortia. The phylogenetic variations observed in the epibionts were as diverse as that found in the free-living green sulfur bacteria, suggesting that the consortia developed after the diversification and radiation of the green sulfur bacteria.

Is there signaling or communication between the interacting partners in these phototrophic consortia? Overmann and Schubert (2002) present two lines of evidence in support of communication. First, the number of epibiont cells found in enrichment cultures and in consortia isolated from nature is consistent, and these cells are arranged in a "highly regular fashion." The central cell has been observed to be growing and multiplying while in association, suggesting that the cell division cycle of the partners is somewhat synchronized.

Second, when exposed to a continuous spectrum of light, the consortia accumulate at wavelengths between 730 and 742 nm, which is the action spectrum of the bacteriochlorophyll of the epibionts. Only the epibionts have the chlorophyll, and only the central cell is capable of motility. Somehow, a rapid signal transfer is made between the two different bacteria to facilitate movement to these adaptive wavelengths of light. Ecologically, the movement of these consortia toward sulfide (i.e., the electron donor) and the light response allows these partners to maintain a position in the chemocline that provides both needed light and resources. Free-living planktonic green sulfur bacteria rely on gas vesicles to maintain neutral buoyancy. Whereas the consortia can actively seek out and find suitable habitat.

The species interactions between nitrogen-fixing bacteria and various plant species (e.g., rhizobia and legumes) have been described in detail, but new aspects of the relationship continue to be discovered as researchers ask new questions. There is still much to be learned from this system in community, autecology, and population ecology, as well as the evolutionary relationships between and among interacting species. At its simplest level, this interaction involves the triggering of morphological changes in the roots of higher plants (legumes) that lead to the development of a nodule by certain soil bacteria. This complex system involves interesting and unique genes for signaling (chemical cues) between the participating partners. Many of the controlling genes have been identified. Several events lead to the formation of a nodule and the subsequent fixation of nitrogen gas into ammonia:

1. The higher plants release various compounds into the soil through their roots. These substances include amino acids, flavonoids, and other metabolic products. The release of plant compounds is not restricted to legumes. Many plants produce and release compounds into the soil. Some of these compounds are allelochemicals that inhibit the growth and development of other plant species. The release of substances that are needed for plant growth and reproduction is in itself quite interesting. Either the release is an uncontrollable phenomenon in which case the plants are losing some proportion of their production with no apparent benefit to the plants, or the plants have controlled release of the substances with an increased probability of higher fitness because of the release. Which is the case? For the legumes the evolutionary cost of losing metabolic products is paid for with interest by the readily available source of nitrogen available to these plants. Legumes release a variety of flavonoids that have an affect on specific bacteria by inducing symbiotic genes.

2. Soil bacteria such as *Rhizobium* organisms move toward these released compounds by *chemotaxis*. Before specific bacteria can move toward a substance, the bacteria must be present. There appears to be high levels of bacterial diversity in the rhizosphere of legumes. Chemotaxis is the first indicator of communication between the plant and the bacteria. The release of these compounds is not a species-specific invitation to the nitrogen-fixing bacteria. However, the ability of a bacterium to respond to a particular flavonoid is controlled by a transcriptional regulator that varies among strains and species of bacteria.

3. Nitrogen-fixing motile bacteria attach to the surface of roots. The attachment is not random, but rather selective. Only new root hairs are infected by the bacteria. Why are only young root hairs infected and not the other root tissue? Are there unique plant mechanisms that prevent infection of older tissue, or do the bacteria only infect the young tissue? In terms of breaching any plant defenses against infection, new root hairs offer the least defenses. The attachment of the bacteria is facilitated by lectins on the surface of the root hairs that bind to specific polysaccharides on the surface of the bacteria. Some of species of *Rhizobium* show species-specific infection patterns. Only certain species of legumes are infected by certain species of bacteria. However, there are other species of bacteria that can infect several species of plants. We can find generalists and specialists.

4. After they are attached to the root hairs, the bacteria secrete new compounds called *Nod factors* (nod refers to nodulation). Nod factors are lipochitooligosaccharides that cause deformation (i.e., curling) of the root hair (Figure 14.11). This communication then becomes the second direct interaction between the two partners. The diversity of Nod factors produced by the bacteria and the discrimination of these factors by the plants creates a second level of specificity between partners in

**Figure 14.11** The response of a growing root hair to a single spot application of 10⁻⁹ M purified Nod factor (NF). Fifteen minutes after NF application, reorientation of the root hair growth axis toward the site of application is already visible, and it becomes more pronounced at 30 minutes. Root hair growth is continuous during and after reorientation, and the root hair diameter does not change. Bar = 15 μm. (From Esseling JJ, Lhuissier GP, Emons AMC. Nod factor-induced root hair curling: continuous polar growth towards the point of nod factor application. *Plant Physiol* 132:1982–1988, 2003.)

the symbiosis. The Nod factors must be released by the bacteria, not by the plant. If the plant were releasing the factors, curling and deformation might occur before attachment of the bacteria. Deformation is critical.

5. The root tips curl under the influence of the Nod factors and entrap the attached bacteria within a pocket. Inside this pocket, a lesion of the root cell wall forms through hydrolysis.

6. The bacteria then enter into the roots by invagination of the plasma membrane of the root cells. Movement into the plant is done with help from the plant. This response of invagination may be a relict defense mechanism, in which infecting microbes were sequestered in invaginated vacuoles where degradative enzymes attacked the invaders. However, in this scenario, the captured cells cause the next phase of the interaction.

7. The bacterial cells promote a reaction in the plants that results in the formation of a tube made of cell wall material. This tube is called an *infection thread*. The infection thread is filled with actively reproducing and growing bacteria that are embedded in a matrix of mucopolysaccharide. At this point, the plant is providing a mechanism for movement of the bacteria to other locations and a suitable environment for growth to occur.

8. The bacteria are transported into the root cortex by the infection thread to the primordial root nodules. These primordial nodules function as plant meristem tissue, and their location depends on the species of plant. Once the infection thread reaches the primordial nodules, cell division is induced.

9. Cell division continues in the root nodules along with cell differentiation until a mature root nodule is formed that is composed of two tissue types: peripheral and central. Each nodule consists of four zones: meristem, invasion zone, infected zone, and degenerative zone. The invasion zone is the site of bacterial release. This occurs in special elongated cells that continue to elongate as the bacteria grow and reproduce. The host plant then provides energy to the bacteria in the forms of dicarboxylic acids and the plants maintain a low but steady oxygen flux through leghemoglobin (i.e., specialized molecule produced by the plant that behaves similar to hemoglobin and acts in the transport of oxygen).

We have discussed previously the formation of bacteroids from these invading bacteria and the evolutionary consequences of this formation. Bacteroids are incapable of further reproduction. However, the formation of bacteroids is consistent with the morphological changes induced by the bacteria on the plant. The plants are induced to form specialized structures that result in a supply of useable nitrogen by the bacteria housed in the structures. The morphological changes in the plant have a metabolic cost associated with them, but the return is sufficient to overcome these costs. The morphological changes to the bacteria result in non-reproducing bacteria whose distribution in the plants is completely controlled by the plants and a continuous release of plant compounds to the environment. The free-living bacteria in the soil continue to receive a "free lunch" from the plants. This maintains a source of future inoculating bacteria, even at the expense of feeding bacteria species that provide no positive benefit to the plants.

### Wolbachia

Not all cytosymbiotes have positive effects on their host. A chief example of a negative interaction is the *Wolbachia*-arthropod infection. *Wolbachia* (Rickettsiales) are obligate intracellular bacteria that infect many different species of arthropods, nematodes, isopods, and arachnids. After a host is infected, the bacterium is maternally transmitted through the cytoplasm of eggs. Survival and fitness of the bacteria depend on the number of female hosts.

*Wolbachia* species have evolved a number of strategies that increase the fecundity of infected daughters and increase the number and survival of daughters. Spread of the infection is a function of manipulating reproduction of the host. There are two general classes of manipulation by the bacteria. First, *sex-ratio distortion* is the process whereby the bacteria increase the proportion of daughters through the killing of males, causing the feminization of genetic males or by inducing parthenogenesis. The second class involves strains of bacteria that induce *cytoplasmic incompatibility*. Cytoplasmic incompatibility is the failure of a cross between hosts to produce any offspring because of cytoplasmic factors. Cytoplasmic incompatibility can be either unidirectional or bidirectional.

Unidirectional incompatibility exists when crosses between infected males and uninfected females results in a reduction of viability or complete failure to produce

**Table 14.1** Effects of Antibiotic Treatment of Either or Both Parents Infected with *Wolbachia* on the Hatch Rate of Butterflies

| | Hatch Rate (Total) | |
| --- | --- | --- |
| | Infected Males | Cured Males |
| **Infected Females** | 0.56 (n = 20) | 0.69 (n = 6) |
| **Cured Females** | 0.00 (n = 26) | 0.54 (n = 20) |

Data from Jiggins FM, Bentley JK, Majerus MEN, Hurst GDD. How many species are infected with *Wolbachia*? Cryptic sex ratio distorters revealed to be common by intensive sampling. *Proc R Soc Lond* 268:1123–1126, 2001.

offspring whereas the reciprocal cross between an infected female and an uninfected male results in normal numbers and viability of offspring. Rokas (2000) states that evolutionary theory predicts natural selection should favor increased fecundity of the infected females rather than increasing the incompatibility levels between infected and uninfected hosts, which should result in selection for strains of bacteria with lower incompatibility levels and with lower costs in the fecundity of the host. In contrast, bidirectional incompatibility results when two individuals infected with different strains of the bacteria fail to produce offspring. Cytoplasmic incompatibility can be determined when infected hosts are treated with antibiotic to kill the bacteria, and various results can be observed (Table 14.1). In one example (Jiggins et al., 2001), antibiotic treatment of females made them incompatible with unrelated males. No offspring were produced from this cross, whereas all other crosses resulted in viable offspring.

Based on numerous surveys within and among groups of arthropods, it is estimated that *Wolbachia* has infected between 17% and 22% of all insects, and that estimate is probably very conservative. However, based on that estimate, more than 5 million species of insects are thought to be infected. These estimates do not include the other arthropods or nematodes that are infected. Regardless of what the true level of infection really is, *Wolbachia* have been suggested to be a major force driving evolutionary change in infected host species. Strains that cause cytoplasmic incompatibility are usually found at higher frequencies than sex-ratio distorters.

Based on molecular phylogenetic analysis, it appears that the common ancestor of *Wolbachia* evolved between 80 and 100 million years ago (Heath et al., 1999), whereas the common ancestor of arthropods evolved at least 200 million years earlier. Would we expect *Wolbachia* to be important in the speciation of arthropods? Most documented transmission of *Wolbachia* appears to be vertical, but there is an almost complete lack of concordance between the phylogeny of *Wolbachia* organisms and their hosts. This observation suggests that there is significant horizontal transfer of strains among host species.

In general, the distribution of *Wolbachia* within and among hosts is a function of three processes outlined by Shoemaker et al. (2002): the rate of spread and the duration of maintenance of an infection within a host species; the rate of interspecific horizontal transfer of infections across species; and the rate of mutation of the bacteria genes. The interaction of the effects of these processes and the host population structure determine the rate of spread within a species. The evolutionary ecology of the

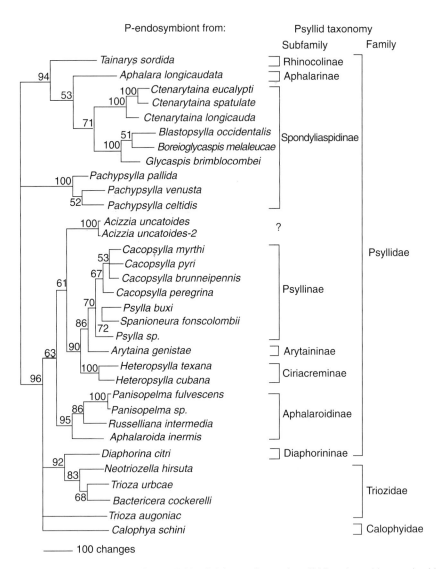

**Figure 14.12** Phylogenetic trees from neighbor-joining analyses of psyllid P-endosymbiont nucleotide sequences of combined 16S-23S rDNA and relationship of the results to the classification of psyllids based on morphology. Numbers at nodes are for bootstrap percentages from 1,000 replicates; only nodes supported by 50% or higher values are shown. Designations refer to psyllid hosts, *vertical lines* indicate assignments of psyllids to subfamilies and families, and the *question mark* indicates uncertain taxonomic affiliation. (From Thao ML, Moran NA, Abbot P, Brennan EB, Burkhardt DH, Baumann P. Cospeciation of psyllids and their primary prokaryotic endosymbionts. *Appl Environ Microbiol* 66:2898–2905, 2000.)

host species should have some effect on the spread of the infection. For example, unrelated species of hosts that live in ecologically close conditions or that have intimately connected life histories (e.g., parasitoids) may be expected to have the same or similar strains of *Wolbachia*. Highly inbred hosts should have lower spread and proportion of infections than other populations. At least for fig wasps, no evidence was found to support frequent or occasional co-speciation between wasps and the *Wolbachia* that infect them. The evidence suggested that the bacteria could cross species boundaries of hosts rapidly. Based on molecular analysis, Shoemaker et al. (2001) demonstrated that strains of *Wolbachia* from the New World and the Old World and from distant insect orders were fairly similar, and they suggested that this was evidence of horizontal transfer over large geographic distance and taxonomic difference. They further suggest that the high levels of infection are the result of historical processes involving horizontal transfer.

## Buchnera *and Aphids*

Aphids and other homopteran insects obtain their nutrition by tapping into and siphoning plant saps. Sap is nutritionally a very poor resource. How do these eukaryotes get the vitamins and limiting amino acids needed for growth and reproduction? The answer lies in a well-studied endosymbiosis between the insects and bacteria. This symbiosis is obligate for the bacteria and the insects. Insects that are treated with antibiotics or heat lose their bacterial endosymbionts, and they have little or stunted growth, decreased reproduction (often sterility), and premature death (Moran and Baumann, 1994). Through careful research, it has been determined that the endosymbiont bacteria have genes that provide needed nutrients to the aphids, and in return, they obtain safe passage within the insects and a constant supply of plant metabolites. There is evidence that infected aphids are resistant to parasitic wasps. In these studies, it was shown that infected and uninfected aphids have an equal probability of being attacked by ovipositing parasitoid wasps, but the infected aphids were significantly less likely to support parasitoid development (Oliver et al., 2003).

Several adaptations within the aphids facilitate carrying and transmitting these bacteria. Specialized cells are found in the insects that harbor the bacteria. The bacteria are carried into new generations through eggs or developing embryos by infection from the mother's symbionts. Based on phylogenetic analysis of the insects and the bacteria, it can be shown that the sequence-based phylogeny of *Buchnera* is completely concordant with the morphology-based aphid phylogeny (Figure 14.12). Such a result is consistent with a hypothesis of a single infection in a common ancestor of the aphids examined and subsequent co-speciation of the bacteria and the aphids. This distant ancestor of the aphids has been estimated to be about 200 to 250 million years old.

*Buchnera* are found only in association with aphids (i.e., there are no free-living forms). Which free-living bacteria are they most closely related to? Results of 16S rDNA sequence analyses indicate that *Buchnera* belong to a single clade that is most similar to *E. coli* and related bacteria. However, *Buchnera* sequences have a more than 11% divergence in 16S rDNA from other sequenced bacteria, suggesting ancient isolation of the bacteria through the development of the endosymbiosis.

# Additional Topics in Species Interactions

15

## Cheating and Cheaters

In previous chapters, we briefly considered the concept of cheating. For example, some bacteria may become embedded in the extracellular polysaccharide (ECP) matrix and take advantage of the other bacteria that produce the ECP and extracellular enzymes by not producing any of these substances. These organisms appear to have a major advantage over the other bacteria in that they do not have to use energy to make these specialized substances, although they receive all of the benefit. Not a bad strategy, right? Are there other situations in which cheating would appear to be an evolutionary advantage?

Before we answer that question, let us take some time to understand this strategy. Cheating in evolutionary terms is any strategy whereby one organism obtains or takes advantage of the behavior, metabolism, or physiology of another organism such that the first organism's fitness is increased without the associated costs of the behavior, metabolism, or physiology. Cheaters can be found among many different groups of organisms. Some species of birds are nest parasites. These birds lay their eggs in the nests of unrelated birds, and their offspring fledge with no parental investment after egg laying.

There are numerous other examples from higher organisms of cheating or free-loading, but our ability to determine evolutionary or ecological affects of such behaviors is limited by the longevity of the organisms. Microbial groups offer a means for testing the effects of cheating on the contributing and the noncontributing groups involved. Much of what follows is based on two papers (Dugatkin et al., 2003, Velicer, 2003) that have examined through modeling and observation the effects of cheating in various situations involving microbes.

We first consider the models of Dugatkin et al. (2003). These simple models are based on two genotypes. First, some microbes produce substances that always provide the producer with some benefit and may provide a benefit to other microbes that are nearby or part of their group. There is always a metabolic or energetic cost associated with producing a substance or in the maintenance of the cell machinery that produces the substance. The second genotype is that of nonproducers. These organisms take advantage of the substances produced by others without contributing anything. Nonproducers do not have the costs of producing the substance, but they can receive benefits from the substance under certain circumstances. Substances can be molecules and materials necessary for growth and reproduction or materials released into the environment such as ECP and exoenzymes. These substances can also be warning or signal molecules, such as those used in cell-to-cell communication or quorum sensing.

Dugatkin et al. (2003) describe a scenario in which the producer cells have a gene, carried on a plasmid, that encodes for an antibiotic-resistance mechanism that protects them in the presence of the antibiotic. Through leakage from the cell or through active secretion, some of this antibiotic-resistance substance is available to nonproducers, especially when they are associated with many producer cells. There are at least two immediate costs to the producer. First, the cells have to carry the plasmid and must provide the cellular resources for replication and maintenance. Second, the cells must produce the substance from cellular reserves or through uptake of needed materials and then the subsequent synthesis of the substance. We will consider only two of the models presented by Dugatkin et al. (2003)—Models I and II—and only the graphic representations, not the mathematics.

For each model, the frequency of producers is represented along the x-axis and is denoted as $p$, and because the total frequency of producers and nonproducers is equal to 1 (or 100%), $1 - p$ represents the frequency of the nonproducers. For the nonproducers to receive any benefit, there must be some minimal frequency of producers, which is designated $p_{min}$. There is a limit on the benefits that can be received from the production of the substance, and this maximum is designated as $b$. The actual cost to the producers is $c$. The benefits must exceed the costs ($b > c$), or the trait would never have been selected.

The models assume that in every generation the frequency of producers and nonproducers is at mutational levels. The model predictions are based on equilibrium frequencies of cheaters and producers. The fitness of the producers is the difference between the benefit and the cost ($b - c$). For these two examples, the frequency of the producers does not have an affect on the fitness of the producers. However, the fitness of the cheater is directly related to the frequency of the producers. In Model I (Figure 15.1), cheaters do not get any benefits from the producers until the frequency of the producer reaches $p_{min}$. Above this value, the cheater gets the same benefits ($b$) as the producers, but there are no associated costs. This type of relationship is known as a *step function*. In this step function, the fitness of the cheater is zero, whereas $p_{min}$ is greater than $p$, and the benefit is at the maximum ($b$) when $p_{min} \leq p$. In the antibiotic-resistance scenario, this model would apply if there was some minimum frequency of antibiotic-resistant producers that was necessary to produce enough substance such that the cheaters did not die. Below this frequency, the cheaters are not protected, but above the frequency, all cheaters are protected. Model I also predicts that the equilibrium value of producers ($p^*$) is equal to $p_{min}$. In other words, the equilibrium value is the minimum frequency needed to provide protection for cheaters and producers.

In model II (Figure 15.2), the cheaters do not receive any benefit when $p_{min} > p$, but once $p_{min} < p$, the cheaters begin to get some benefit, which is less than that obtained by the producers. This is not an all-or-none model, and in this simple case, the benefit response is linear with $p$ such that the fitness of the cheaters approaches $b$ when $p$ approaches 1. Using the antibiotic scenario, this response translates to a condition in

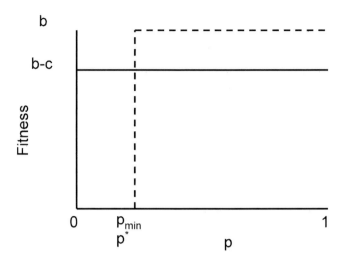

**Figure 15.1** Model I fitness function for producers and cheaters. (Adapted from Dugatkin LA, Perlin M, Atlas R. The evolution of group-beneficial traits in the absence of between-group selection. *J Theor Biol* 220:67–74, 2003, with permission from Elsevier.)

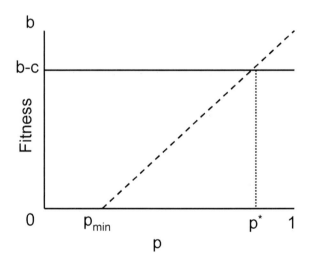

**Figure 15.2** Fitness function for producers (*bold solid line*) and cheaters (*bold dashed line*) using Model II of Dugatkin et al. (2003). (Adapted from Dugatkin LA, Perlin M, Atlas R. The evolution of group-beneficial traits in the absence of between-group selection. *J Theor Biol* 220:67–74, 2003, with permission from Elsevier.)

which cheaters die until a minimum frequency of producers is achieved, after which some protection is afforded the cheaters, although at a lower level than for the producers. As the frequency of the producers increases, so does the protection given the cheaters.

In this model, the equilibrium frequency of producers is much greater than $p_{min}$. If we set the fitness of the producers and cheaters equal,

$$b - c = \frac{\left(p^* - p_{min}\right)b}{1 - p_{min}}$$

and then solve for $p^*$,

$$p^* = \frac{(b-c)(1-p_{min})}{b} + p_{min}$$

We can see that by increasing $p_{min}$ we similarly increase $p^*$, or the equilibrium frequency of the producers. This equilibrium value of the producers can also be increased by increasing the benefit or decreasing the cost of production.

# Cooperation

Most of this discussion is based on the work of Velicer (2003). Several terms need defining before we proceed with this discussion (Box 15.1) (Velicer, 2003).

Mutualisms, symbioses, and cooperation are usually positive interactions between, within, and among species groups. There are different levels and types of cooperation. Velicer defines and gives examples of three distinct types of cooperation: minimal, density-enhanced, and group-limited. Organisms that engage in minimal cooperation are restrained from taking or using more of a group produced benefit when the amount or level of the benefit is not constrained by density-dependent factors. In general, a density-dependent factor is one that changes in intensity or effect with changes in the size of the population. Minimal cooperative traits do not show increased or decreased effects as a function of the population or group size. In other words, the effect is the same when an organism is isolated as when it is in a pure group of cooperators. Selection in minimal cooperative situations should favor organisms

---

**Box 15.1  Glossary**

**Altruism:** Any behavior that confers fitness benefits on other individuals while costing its performer a net reduction in evolutionary fitness.

**Cheating:** Obtaining benefits from a collectively produced public good that are disproportionately large relative to cheater's own contribution to that good.

**Cheating load:** The degree to which obligately defecting cheaters decreases the group-level benefits of cooperation in chimeric (mixed) social groups.

**Collective action:** The combined effect of individual behaviors within a group.

**Cooperation:** Proportional contribution by individuals to a collectively produced public good.

**Defection:** Disproportionately small contribution by individuals to a collectively produced public good. Biologically, defection does not necessarily entail cheating. Some mechanisms of defection may not enhance the relative ability of defectors to exploit a relevant public good.

**Public good:** Any fitness-enhancing resource that is accessible to multiple individuals within a local group. A preexisting public good originates independently of the group that benefits from it (e.g., rainwater). Alternatively, a collectively produced public good is generated by members of the group that use it.

that live alone or isolated from a group, because the fitness of cooperators that are close is not enhanced, and cheaters will decrease the fitness of localized cooperators.

In contrast to minimal cooperation traits, cooperation can have a synergistic effect when groups of individuals have a greater benefit together than they can obtain in isolation.

Such positive effects are density-dependent. Density-dependent traits provide a benefit for isolated individuals, but the same traits are greatly enhanced with increasing density of individuals within a group. In these situations, the benefit's effect is magnified as the group gets larger, but the trait has a minimal effect for an individual. Velicer provides the following example of a density-enhanced trait. Consider that individual isolated cells of *Myxococcus xanthus*, a predacious bacterium, can kill and obtain nutrition and energy from their prey. However, research has shown that the efficiency of predation by *M. xanthus* is greatly increased as the density of the bacteria increases. If this relationship is true, we would expect that the fitness of group associated individuals would be higher than isolated individuals because individuals found in the groups would be able to obtain more "food" and at faster rates than the isolated bacteria. Selection favors increased fitness and increased clustering of cells in cooperative groups.

Group-limited cooperation occurs when a trait increases fitness only in large and dense groups of cooperators but is a disadvantage when expressed at low density. This form of cooperation is the most complex. Natural selection is strongest at the individual level over selection of shared benefits, so we would expect higher stringency in the conditions that favor complex cooperative traits over simpler cooperation.

Examples of social interactions or cooperation among bacteria are primarily within populations. Cooperation within a population of the same species makes more sense than cooperation between unrelated species. However, cheating as a strategy can be beneficial to the cheater, whether in pure populations or in complex communities. The constraints on cheaters should be similar under either condition. If there are too many cheaters, there will be no resources available because there are no producers.

Recent evolutionary models consider when and under what conditions mutualisms are likely to persist. These models predict that if cheating is unconstrained, mutualisms should break down. Using the rhizobia and legume interaction as a model system, Simms et al. (2002, 2003) developed a theoretical framework for the maintenance of this ancient mutualism. How can plants constrain cheating bacteria? One model that may help to explain how plants can achieve this constraint is called *partner choice*. Partner choice occurs where there is a market of potential symbionts. Simms and Taylor list the following conditions that must apply to a partner-choice explanation:

1. A range of partners must be available. The genetic and species diversity of potential symbionts must be high to allow some "choice" to be made.
2. There must be a mechanism for effective choice among the potential partners.
3. The cost of evaluating partners is less than the benefit obtained from correctly identifying a good partner.

Through careful research of this system in agricultural and natural settings, it has been found that some legumes can control nitrogen fixation by restricting the phylogenetic groups of bacteria that colonize, limiting the number of infections, and controlling the numbers of nodules formed in response to environmental levels of nitrogen. Theory predicts that if the fitness status of a host and symbiont are tightly correlated from one generation to the next (i.e., reliance on vertical transmission of the symbiont), increased cooperation can be favored. Legumes and rhizobia have been shown to have high levels of horizontal transmission. Nevertheless, there appears to be high levels of cooperation in this system, and as such, the system is ripe for infection by cheaters.

In a partner-choice system, the individual partners enter what has been termed a *biological market of potential traders* and choose the best partners from this market. What is the basis of the choice? Individuals may choose a partner based on real signals (chemical or otherwise) that truly indicate the quality of the potential partner, or they may choose based on experience obtained through trial interactions (Simms and Taylor, 2002).

The fixation of dinitrogen gas is energetically expensive, and only a few organisms have the ability to perform this process. Free-living rhizobia bacteria rarely fix nitrogen outside of plants, indicating some positive benefit from the plant to the bacteria that allows them to fix nitrogen. From the plant's perspective, harboring and creating a suitable environment for the rhizobia to function is also energetically expensive. The plants must provide the energy needed to fix the nitrogen, maintain a complicated signal exchange system between the bacteria and the plant, produce the oxygen carrier molecule (i.e., leghemoglobin), and produce nodules, which are novel organs not found in uninfected plants. At first cut, it seems that the plants are bearing the major costs of the association. Can the plants regulate their costs? Research has shown that some legumes can reduce the formation of nodules when there is a supply of nitrogen available in the soil or when phosphorous is limiting.

Partner-choice models assume a biological market perspective for understanding the evolutionary ecology of the legume-rhizobia interaction. Biological market perspectives demonstrated in this interactions include the following:

1. An exchange of commodities to the mutual benefit of the partners. For example, plants produced energy molecules and bacteria produced ammonia.
2. Functional and phylogenetically diverse individuals within a trading class.
3. At least one of the trading partners (i.e., the plant) interacts simultaneously with diverse traders (i.e., the microbes).
4. The exchange value of the traded commodities is a source of conflict. At least one trader (i.e., the plant) has mechanisms that allow it to evaluate and choose the best symbiotic partners (Simms and Taylor, 2002).

In this interaction the plants, because of their large size relative to the bacteria, probably encounter more potential partners than do the bacteria. The number of plants is much less than the number of bacteria in a particular location. The plants are probably the members of the association that do the choosing. However, because the legumes and the bacteria disperse independent of each other, evolution will act on each member differently, resulting in incongruent trajectories (i.e., the fitness of one partner may be increased by selfish actions) (Simms and Taylor, 2002). From the plant's perspective, getting as much nitrogen at the lowest possible energetic cost may be accomplished through several mechanisms. The plants may be able to kill bacteroids or stop or reduce the formation of nodules. The plants may limit the rewards they pay the nodule bacteria. On the other hand, the bacteria may be selected to capture and store resources without concomitant nitrogen production, or they may break down structural carbohydrates found in the plant cell walls. Whether plants and rhizobia interact at these levels is unknown. Based on evolutionary theory and models, it seems likely that, given the time these organisms have been interacting, elegant mechanisms have evolved to detect and prevent cheating. Major research questions remain to be answered. For example, Simms and Taylor (2002) list five outstanding questions that they feel need to be researched:

1. Do legumes predictably encounter a bacterial market?
2. Can plants evaluate bacterial functions, and if so, how?
3. Are bacterial populations spatially structured in a way that facilitates accurate targeting of rewards and sanctions?
4. Do plant rewards and sanctions affect bacterial fitness?
5. What are the fitness costs to bacteria of being beneficial symbionts?

As with all aspects of science, the process of answering generates new questions. Limitations to exploration are restricted by the researcher, not the science.

# Evolutionary Arms Races

Antagonism between hosts and parasites is thought to result in between-population genetic differentiation of the parasites and the hosts. This differentiation is important in creating and maintaining genetic diversity or variation, which ultimately may result in speciation. Coupled with migration, differentiation may be the driver behind coevolution. These antagonistic interactions between the host and the parasite have been hypothesized to result in an evolutionary *arms race*. The designation of an arms race comes from the carefully choreographed dance between the world's superpowers during the height of the Cold War. The dance involved one country developing, producing, employing, and threatening another country with some new and improved weapon or counter-weapon system. As soon as the new system was detected, the other country sought to counter its perceived national security threat by devising, developing, producing, and employing a bigger and better system. Logic and prudence mandate limits on this sort of interaction. In the global-political system, the arms race was decided eventually by limitation of monetary and environmental resources.

Many biological systems appear to be antagonistic; one organism is seeking to exploit or kill another organism and thereby obtain resources to make more copies of itself and increase its fitness. Natural selection should favor any organism that can successfully counter the effects of an antagonist through phenotypic or genetic means. For example, green plants have a distinct disadvantage compared with animals in trying to escape their predators—they cannot move. Natural selection has in many plants selected modification of the architecture of the plant and produced protective spines, thorns, leathery leaves, hairs, and bark. Some plants produce secondary compounds that seem to have one function: to persuade predators not to feed on the plant. These compounds are distasteful, toxic, and in some cases, lethal to predators.

However, almost every plant has some organism that has successfully circumvented the plant's defense and exploits all or part of the plant. Some organisms are able to eat the toxic chemicals, store them, and use the chemicals in their own defense. In these examples, there are limits on the "arms" that can be selected in plants. Structures and chemicals that provide protection but do not increase photosynthetic capabilities are metabolically and evolutionarily costly to maintain. After a particular defense has been selected, the number of possible additional defenses becomes limited. For example, it is doubtful that plants with spines would also have thorns.

There are at least two conditions (Lenski and Levins, 1985) required for the perpetuation of an evolutionary arms race: exclusive antagonism and genetic complementarity. Exclusive antagonism in parasite-host systems depends on the mode of parasite transmission. Selection may favor interactions between hosts and parasites if any mutually beneficial characteristics increase the survival of both. The concept of genetic complementarity is based on a gene-for-gene relationship between the host and the parasite, such that for every genetic trait the host has to defend, the parasite has a gene to exploit.

Microbes with their limited DNA arsenal are nevertheless involved in evolutionary arms races. Bacteria are under continuous attack by bacteriophages or various microscopic and macroscopic predators. One difficulty in the study of coevolution is knowing the time scale of the coevolutionary change. How many generations are required before we can see or measure the effects of selection acting on a population? For higher organisms, generation times are much too long to allow observations in a single investigator's lifetime. Microbial populations on the other hand are ideal for studying and testing evolutionary hypotheses because of the high densities and very short generation times. However, careful observations of microbial interactions are best done under laboratory conditions, where environmental conditions can be maintained and controlled. Another plus for using microbes to test these hypotheses is that initial populations can be *isogenic*, or genetically identical. Any differences observed after treatments may be ascribed to mutations and not to initial genetic differences within or between communities.

Microbial hosts (i.e., bacteria) and microbial parasite (i.e., bacteriophages) interactions have been used in the study of antagonistic coevolution for several reasons. The virulent phage has to bind to the surface of the bacteria; the phage must inject its own genetic material into the bacteria; the viral genetic material must use the host cellular machinery to replicate; and the release of the newly synthesized phage occurs after lysis of the host cell. The interaction is entirely antagonistic (Buckling and Rainey, 2002).

To demonstrate the consequences of coevolution and the methods of observing the same, two studies are discussed. In the first study, cultures of *Pseudomonas fluorescens* SBW25 and a bacteriophage were found to persist for more than 300 bacterial generations, which suggested to the researchers that a possible evolutionary arms race was occurring (Buckling and Rainey, 2001). In follow-up laboratory experiments that eliminated or reduced environmental variability and by using isogenic cultures of the *P. fluorescens* to eliminate genetic variation, these scientists sought to observe between-population divergence and whether coevolutionary divergence resulted in local adaptations of the host or parasite. Did the parasites coevolve to be better at infecting hosts they had been grown with over hosts from different populations, and were the hosts better at resisting parasites they had been grown with than parasites that had evolved with other populations of hosts? They sought to determine which types of selection were predominantly driving any coevolution that took place.

Two different types of non–mutually exclusive selection may be acting on these populations. First, the selection may be directional, and the resulting hosts would be resistant to all parasite genotypes and parasites that are able to infect all host genotypes. Second, selection may be fluctuating, which results in different resistant ranges being alternately favored. According to Buckling and Rainey, directional selection appears to operate in laboratory studies, and fluctuating selection is assumed to operate in natural populations.

Reciprocal increases in phage infectivity and host resistance were observed over the time course of this experiment (Figure 15.3). There was a time lag in the antagonistic coevolution between host resistance and phage infectivity. At any point in time, phage collected two transfers in the future had higher levels of infectivity than did the phage collected at that particular time, and the contemporary bacteria (i.e., hosts) were more resistant to phage than were populations collected two transfers in the past.

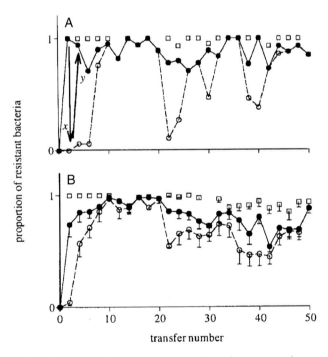

**Figure 15.3**  **A:** Proportion of bacteria resistant to ancestral phage (*open squares*), contemporary phage (*solid circles, solid lines*), and phage from two transfers in the future (*open circles, dashed lines*) in the replicate that underwent the greatest rate of coevolutionary change. Time-lagged evolution of phage infectivity to resistant bacteria is demonstrated by the difference between the resistance of contemporary bacteria to contemporary and future phage populations (*vertical arrow*, x). Subsequent increases in bacterial resistance in response to phage evolution is demonstrated by the difference between the resistance of past and contemporary bacteria, to contemporary phage (*diagonal arrow*, y). **B:** Mean (±SEM, *n* = 12) bacterial resistance of all populations. (From Figure 1 in Buckling A, Rainey PB. Antagonistic coevolution between a bacterium and a bacteriophage. *Proc R Soc Lond* 269:931–936, 2002.)

What about the mode of selection (directional or fluctuating)? Further analysis of these data demonstrated that bacteria at a particular time were resistant to phage collected at the same time and to all ancestral phage. However, phage, no matter when they were collected, were able to infect ancestral bacteria. The average bacterial resistance and the mean phage infectivity increased over time in all replicates. At the end of the study, there was considerable between-population divergence. Remember that each population was started with the same genetic background (i.e., isogenic). This divergence produced bacteria better able to resist the phage that had been grown in their microcosm than phage grown in other microcosms. However, the phages were less able to infect bacteria from their own microcosm than they were bacteria from other microcosms. The bacteria appear to be locally adapted or more resistant to their own phage.

Bacterial resistance to phage involves changes or loss of phage binding sites on the surface of the bacteria. Phage bind to specific sites on the cell surface, but these sites presumably have functions that are beneficial to the cell (e.g., metabolism). The binding sites have been exploited by the phage. Any change that alters the binding site configuration or results in the loss of the site would diminish the infectivity of the phage. However, if these sites are important, even marginally so, to the metabolism of the cell, change or loss of the sites should have a net cost associated with such change or loss, and bacteria with the altered sites should have lower growth rates than the unaltered bacteria. Competition between altered and unaltered cells should result in the unaltered succeeding. Phages that are able to infect bacteria with the altered sites may have reduced specificity and hence become more broad range. As the bacterial defense is radically modified, the range of phage genotypes that can infect would be reduced. Phages should evolve to be less specific for binding sites, which will result in a wider range of host genotypes they can infect. These mechanisms are suggested by Buckling and Rainey to "explain the long-term, predominantly directional, coevolution" they observed in this study.

Based on the extensive polymorphisms for infectivity and resistance observed in natural populations, it seems that fluctuating selection is driving coevolution. Why is there disparity between laboratory and field observations in the mode of selection? The study we have been discussing suggests that one reason may be that laboratory studies are not carried out over time frames needed to move selection from directional to fluctuating. Anecdotal evidence from the Buckling and Rainey study suggests that this may be possible. Increasing resistance and infectivity were associated with declining growth rates.

One strength of the previous example was that the initial starting populations were genetically identical or nearly so. This is also a weakness. Starting with pure cultures of genetically identical individuals allows certain hypotheses to be tested, but nature is usually not made up of pure genetically identical populations. Natural populations and communities are structured spatially and temporally into a number of geographically distinct subpopulations (Brockhurst et al., 2003). When two or more subpopulations mix, what are the effects on the coevolution between parasites and hosts? Mixing should increase the rate of parasite infections because of the increasing exposure of new susceptible hosts. Increased transmission of the parasite has a strong negative impact on the host fitness, which may result in stronger selection for increased defense against infection. Parasites within spatially mixed populations may be able to infect a wider range of hosts, and the hosts should be resistant to a wider range of parasite genotypes.

Brockhurst et al. (2003) sought to determine the effect of frequent population mixing had on the rate of coevolution between a bacterium and a parasitic phage. The bacterium and phage used in these experiments were identical to those used in the Buckling and Rainey experiments. This study was a follow-up study to that of Buckling and Rainey. For this study, initial populations consisted of the bacteria with phage added. One-half of the experimental units were in a static incubator (i.e., unmixed populations) at a constant temperature, and one-half of the experimental units were incubated at the same temperature but in orbital shakers (i.e., mixed populations) with alternating static and shaking cycles. Every 2 days, each culture was transferred to fresh medium for 16 transfers, which was approximately equivalent to 120 bacterial generations.

To estimate coevolution, the researchers measured the change in infectivity of phage to bacterial populations over time. At every second transfer, they determined the level of resistance of the bacteria to past, contemporary, and future phage populations. Past populations consisted of phage that had been saved from two transfers earlier, and the future populations were phage from two subsequent transfers that were then introduced into saved populations of bacteria. They predicted that if coevolution in this system was escalating over time, future phage should be better than contemporary phage, and contemporary phage should be better than past phage at infecting contemporary bacteria. A negative slope of the relationship between the proportions of resistant bacteria over time would be supportive of the predicted outcome.

The data from this experiment demonstrated oscillating cycles of bacterial resistance and phage infectivity (Figure 15.4) in the static and the mixed populations, but the rate of coevolution was nearly twice as great in the mixed populations as in the static populations. The bacteria from the mixed populations were more resistant to phage independent of where the phage had come from (mixed or unmixed populations) (Figure 15.5). Bacteria from the mixed populations were equally resistant to phage from either source, but the bacteria from the unmixed populations were more resistant to phage from the unmixed than to phage from mixed populations. The range of bacterial genotypes that phage could infect was increased by mixing; the number of phage genotypes that bacteria were resistant to increased from mixing. These data suggest that coevolution in this system (i.e., under the environmental conditions imposed by the researchers) escalates, with the product of selection being generalist phage and bacteria.

How long would we expect coevolution in this antagonistic system to continue? Are there limits imposed by ecology, or evolution that would keep this system or similar systems from escalating indefinitely? Selection favored generalist bacteria and phage, which ultimately means a reduction in genetic diversity. Reducing genetic diversity may reduce the ability of these organisms to respond favorably during periods of environmental change and accompanying change in selection pressures.

In nature, genetic diversity is ameliorated by migration of new individuals into these populations. Because the environment is not expected to be homogeneous, pockets of bacteria and phage may exist that are not coevolving at the rates as those observed. However, these data demonstrate that coevolution does occur and that varying

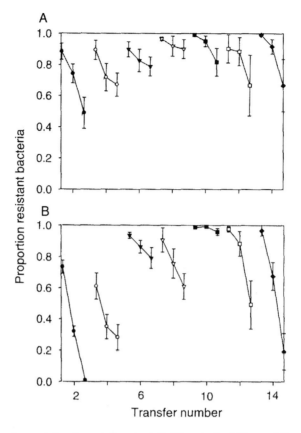

**Figure 15.4** Rates of coevolution through time, in static (**A**) and mixed (**B**) populations of bacteria and phage. Each set of lines shows (from left to right) the proportion of bacteria resistant to phage from two transfers in the past, contemporary phage, and phage from two transfers in the future. The slope of each line provides a measure of the rate of coevolution over each four-transfer period. (From Brockhurst MA, Morgan AD, Rainey PB, Buckling A. Population mixing accelerates coevolution. *Ecol Lett* 6:975–979, 2003, Blackwell Publishing.)

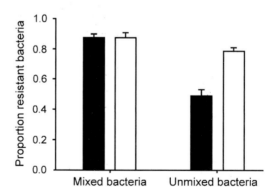

**Figure 15.5** Bacteria-phage interactions across all populations are shown as the mean ± SEM proportion of bacteria from mixed and unmixed populations resistant to phage from mixed (*black*) and unmixed (*blank*) populations. (From Brockhurst MA, Morgan AD, Rainey PB, Buckling A. Population mixing accelerates coevolution. *Ecol Lett* 6:975–979, 2003, Blackwell Publishing.)

environmental conditions (mixing or static) can produce unique responses from both interacting species.

To what extent do other microbial species coevolve? We have discussed antagonistic coevolution between two species, but what about other species interactions that are not negative or that involve more than two species? The various links in nutrient cycles probably evolved as independent processes. Is this really the case, or did species develop mutualistic relationships early in the formation of various metabolic pathways that benefited all the species more than if they had remained unassociated? Clearly, the evolution and ecology of microbes require understanding across a wide range of topics to begin to make generalizations that have any predictive power.

# Microbe Eukaryote Interactions

Microbial species interactions do not need to be restricted to be between two microbes. Species interactions transcend taxonomy. An example of a highly evolved mutualism is that of fungus-growing ants and their fungi. This is a classic example in which the ants care and tend the fungus, which is their primary food source. The ants maintain suitable growing conditions for the fungus, and it is presumed that the ants are able, through appropriate cultivation, to grow pathogen-free fungi. However, research has found that the ant fungal gardens are hosts to a virulent, highly specialized pathogen.

The presence of the fungal pathogen was predicted from theory before the discovery of the pathogen. Ants have been growing fungi for millions of years, and the fungal gardens should be a prime target for a parasite. Based on our previous discussion, the fungus (i.e., host) and the previously unidentified parasite have opposing interests: resistance and infectivity. However, no parasites were discovered or observed, even though thousands of research papers have been written on this system. To understand the importance of this lack of detection, it is necessary to describe in more detail the biology of the ant-fungus interaction.

This is a highly evolved form of mutualism. Only certain ant species have evolved the relationship, and it appears that these species have a single ancestral origin (i.e., monophyletic) from a presumably generalist foraging ant. These ants, all of which belong to the tribe Attini (subfamily Myrmicinae), are obligate feeders on their fungal gardens. The fungus provides all nutrition for the larvae and the queen. Workers obtain some nutrition from the fungus, but they often supplement the fungus with plant-derived products (e.g., sap).

The queen is responsible for transferring the fungus between generations by carrying a small ball of the fungus in a specialized cavity in her mouthparts, regurgitating the fungi, and tending to its cultivation through manuring with fecal fluids or suitable substrate until she reproduces the next generation of workers. These workers then perform numerous tasks for the colony, including tending the garden. Suitable substrate (e.g., leaves) for fungal growth is harvested by the ants, which promote decomposition by licking and masticating the leaf material to small pieces. Licking and chewing the leaf material are thought to reduce the resident microflora associated with the leaf surface and thereby reduce possible infections of the fungal garden or eliminate the appropriate fungus by competitors.

Some species of ants add a fecal droplet, which may add needed enzymes to help the fungus break down the pulp. The pulp is taken to the top of the garden and inoculated with a culture of fungus from an older part of the garden. All material completely decomposed by the fungus is removed to refuse chambers or completely removed from the nest to prevent hazardous infections of the ants and the fungus. The ants are also able to move proteolytic enzymes from areas of high concentration to areas that need the enzyme. These enzymes are thought to be produced by the fungus and are ingested by the ants and then defecated into appropriate locations where the enzymes are needed. This whole interaction is much more complicated and beautiful than I have described, but the previous discussion is sufficient to indicate the level of interaction.

Why are there not more infections from unwanted parasitic fungi or bacteria? The substrate used to grow the fungi is natural leaf material that must have fungi and bacteria from passive and active dispersal. After the ants make the pulp and start their fungal gardens, what keeps parasites from infecting the fungus? For years, researchers thought that the ants were able to physically cull unwanted microbes from the gardens, that the ants were able to secrete antibiotics that were selective for unwanted microbes, or that the fungus produced antibiotics that were similarly potent against other microbes. Although there is evidence in support of each of these methods of nuisance species protection, it appears that this mutualism is more complicated than just ants and fungi.

Several researchers had noticed that some of fungus-growing ants were covered completely or in certain locations with a white granular material (Currie, 2001), which was originally thought to be some sort of wax (Figure 15.6). This waxy material has since been identified as colonies of a filamentous bacterium, an actinomycete (Currie et al., 1999). The ant-actinomycete association was found in all 22 species of fungus-growing ants or with the colonies. The association, although crossing all species of ants, is unique for various groups in the location of the bacteria. In some species, it completely covers the surface of the ants, whereas in other species, it is restricted to specific locations on the surface of the ants. The bacteria are transferred exclusively from parent to offspring and are found on virgin queens but not on males that die before colony formation. In studies conducted by Currie et al. (2003), it has been established that the actinomycete produces a powerful antibiotic with specific activity against the predominant fungal garden parasite. The highly specific antibiotic suggests highly evolved mutualism between the ants, fungi, and the actinomycete (Figure 15.7). The actinomycete provides two other benefits. First, for some fungus-growing ants, there was a significant increase in growth rates of the fungus grown in broth containing a filtrate from the bacterium. Second, the bacteria may protect the ants from harmful pathogens.

For a relationship to be designated a mutualism, there must be some benefit that each species in the mutualism gets that it cannot get independently and that increases the fitness of each member. The ants are benefiting from the association. The ants obtain nutrition from the fungus, which is protected from infection by the actinomycete, and the ants may be protected from their own pathogens by the actinomycete. Although some of the fungal colony is consumed by the ants, the ants ensure that some of the fungus is maintained over long periods. The ants provide suitable conditions for fungal growth and bring suitable substrate to the fungus. What are the

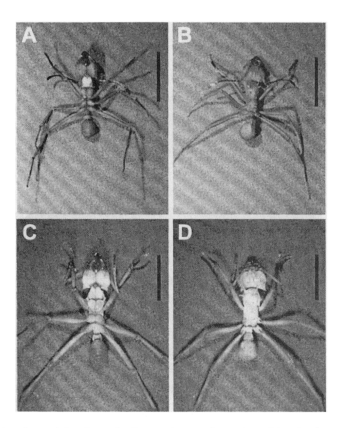

**Figure 15.6** Location and abundance of actinomycete on major workers of *A. octospinosus*. **A, B:** Ventral and dorsal views, respectively, of a worker representing individuals with none to a small amount of actinomycete present. **C, D:** Ventral and dorsal views, respectively, of a worker with thick coverage of actinomycete on the laterocervical plates and other locations on the body. Scale bar = 4 mm. (From Currie CR, Bot ANM, Boomsma JJ. Experimental evidence of a tripartite mutualism: bacteria protect ant fungus gardens from specialized parasites. *Oikos* 101:91–102, 2003, Blackwell Publishing.)

benefits to the actinomycete? Currie (2001) lists two possible benefits. First, the ants aid in the dispersal of the bacteria. Virgin queens carry the bacteria with them to new colonies. Second, the ants themselves provide a novel niche for the bacteria. Actinomycetes are not usually associated with insects. There is some evidence that the ants provide some level of nutrition to the bacteria, although this has not been definitively determined.

This example of a multiple species mutualism, which had been overlooked by numerous scientists, suggests that an extreme level of highly evolved relationships may exist among and between microbes and higher organisms. This section is on evolutionary arms races. In the previous example, the arms race was at first glance between a specialized parasite and the fungus. After careful examination, it was found that the arms race was between the parasite and an actinomycete. The arms race includes the evolution and ecology of ants, fungus, actinomycete, and the parasite. Any countermeasures evolved by the parasite must be met with countermeasures in the actinomycete that do not upset the mutualism among the ants, actinomycete, and the fungus.

There are probably innumerable other associations that have been overlooked by researchers and that are awaiting the careful examination by new students and

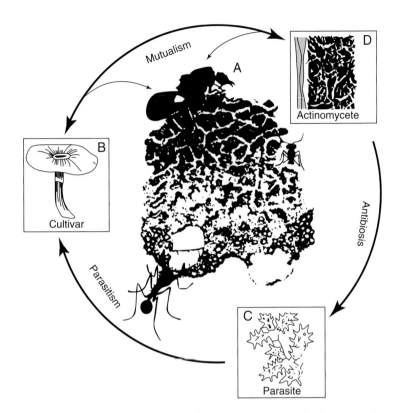

**Figure 15.7** Quadripartite symbiosis. **A:** The large-body queen represents the fungus-growing ants. **B:** The mushroom depicts the fungus that the ants cultivate and that has the appearance of free-living leucocoprineous fungi. **C:** The microfungus in the box represents the garden parasite *Escovopsis*. **D:** The actinomycete grows on the cuticle of fungus-growing ants and produces antibiotics that suppress the growth of *Escovopsis*. *Arrows* represent interacting components; *double-headed arrows* indicate mutual associations, and *single-headed arrows* indicate a negative interaction. (From Currie CR. A community of ants, fungi, and bacteria: a multilateral approach to studying symbiosis. *Annu Rev Microbiol* 55:357–380, 2001. Created by Cara Gibson. Reprinted with permission from the Annual Review of Microbiology, Volume 55, © 2001 by Annual Reviews.)

scientists and that will provide meaningful insights into the evolution and ecology of microbes and higher organisms.

# Biogeography

For organisms that can be found everywhere scientists have looked, it seems almost senseless to consider the biogeography of microorganisms. Microbes have been getting around longer than all other organisms combined. Their ability to disperse and traverse long distances and their ability of some types to survive extended periods of inhospitable conditions make them ideal for organisms that have worldwide distributions. However, perhaps everything is not everywhere. The ecology of specific microbes may be affected by the environmental conditions such that some microbes are able to survive and others to go locally extinct and not just remain in a quiescent or dormant state waiting for conditions to change so that they can grow. Many microbes are in a dormant state, and when given the appropriate conditions, they begin to grow and proliferate. For everything to be everywhere, all species of bacteria must be able to survive dispersal and be able to colonize new habitat. Is that a reasonable assumption? What, if any, are the limits on the geographic distributions of various microbial groups? It seems reasonable that bacteria that lack stages to effectively disperse and survive may have more limited ranges and distributions than microorganisms that do have such stages. How similar is the microbial fauna in one geographic location to the microbial fauna in a similar but distant geographic location? In this section, we consider the biogeography of microbes.

Organisms are distributed in space and time. The biosphere includes depth and height, as well as the latitude and longitude across the face of the globe. Organisms are distributed in time ecologically and historically. To understand the distribution of an organism, other factors must be considered, such as geology (i.e., movement of continents and rise and fall of mountains, basins, lakes, seas, oceans), phylogeny (i.e., how closely related are groups of organisms found in different regions or locations), climatology (i.e., glaciation, temperature, rainfall, and ultraviolet radiation), and ecology (i.e., competition, predation, production, availability of resources, dispersal, migration, and emigration). These factors are not exhaustive, however, they show that biogeography is influenced by biotic and abiotic processes.

The range of an organism expands and contracts based on the ability of the organism to establish into new areas or the ability to remain during times of biological or abiotic change. Often, there are ecological barriers such as scarcity of resources, physical conditions that prohibit survival, or the presence of other organisms that are better competitors or that are efficient predators on the species that prevent expansion. Sometimes, there are enormous distances between suitable habitats that limit the dispersal of an organism. For example, coral reefs around volcanic islands are haphazardly distributed along openings in the ocean floor and are often hundreds of miles from the next closest coral reef. All oceanic distance in between pairs of islands would be inhospitable to most of the organisms that make up the coral reef. All of these factors together or separately impact the distribution of organisms, especially higher organisms.

What about the distribution of microbes? Are there barriers to microbial dispersal? Are there disjunct distributions across space that require explanation? Are the distributions of microbes independent of large-scale geologic events such as continental drift and breakup? To what extent do local populations of microbes diverge genetically from the rest of the global gene pool?

We have discussed the dispersal ability of microbes, but our discussion was limited to microbes that could be wafted into the atmosphere through winds, waves, or sprays or that could be transported by attachment to other organisms. However, many microbes exist in habitats where such activities are not possible over ecological times and perhaps not possible over geologic time frames. For instance, consider bacteria that live in the deep subsurface environments. How long have the microbes been in the deep subsurface? Are they fairly recent colonizers, or do we have to go back to the time when these geologic layers were formed to understand observed patterns? If the microbes in the deep subsurface have been there for millions of years, the probability of differences between similar environments on different continents is great because of independent evolutionary events, different selection regimes, and the lack of lateral gene flow between the widely spaced communities. However, we do not have to go to extreme environments such as the deep subsurface to ask similar questions.

Consider obligate anaerobic bacteria living in the muck of some inland lake. Movement out of the anaerobic environment would result in death of the microbe, so how do obligate anaerobes get dispersed across the landscape? As with all things microbial, it is important to consider scale in our questions and possible explanations. It is possible and highly probable that higher animals wading through the muck could carry on their feet quantities of muck sufficient to maintain microzones of anaerobic conditions during transport between habitats. Alternatively, we may invoke the incredible amount of time bacteria have been on the planet. Anaerobes were presumably the first organisms to evolve, and their current distributions may be the remnants of ancient distributions. However, when new habitat is created, such as through the construction of a dam, vibrant anaerobic communities can be found in the sediments in short order. Were these anaerobes always present in the aerobic soils but in some dormant state? Were they transported in with the dammed river water and subsequently settled out over the increased wetted area of the new reservoir? Biogeography of microbes is experiencing a renewal of interest, and we may expect many different patterns to emerge with as many possible explanations. We will consider a few examples of spatial patterning in microbes.

Soil microbes may be distributed vertically within the soil profile and laterally among adjacent locations. Within this three-dimensional matrix, the microbes are generally not distributed homogenously because of physical and biological factors, such as availability of nutrients and food, soil moisture, plant root exudates, toxic compounds, allelochemicals, antimicrobials, predators, competitors, pH, $E_h$, and many others. Some studies have examined the distribution of bacteria within this heterogeneous milieu. In theory, we might expect communities or populations to be most similar to more closely adjacent sampling locations than from much more distant sampling location vertically or horizontally. For example, soil cores taken centimeters away from each other should have microbial communities more similar than cores taken meters apart and much more similar than samples taken kilometers apart. However, samples taken from locations with very similar ecological conditions that are distantly removed from each other would be expected to be more similar than samples taken from very different ecological conditions, even when such locations were close. For example, Franklin et al. (2000) found that microbial communities in the same shallow coastal plain aquifer differed depending on the groundwater chemistry at a specific location even when the locations were separated by short distance. There are several ways to determine similarity, including sequencing various genes and comparing sequence similarity. Another method is to determine the level of DNA:DNA hybridization among samples. A higher degree of reassociation suggests that the microbes are essentially from the same effective population.

Before we examine the biogeography of continents, let us begin with much smaller spatial scales. In the discussion on the size of bacteria and the inherent problems of sampling such a small organism, mention was made that scales as small as $1\,cm^2$ were equivalent to sampling almost $17\,km^2$ for a single mouse. Despite the coarseness of our sampling and observations, we should nevertheless expect that the closer the sampling stations are, the more similar the communities will be. Biogeographic analysis may focus on specific groups (e.g., species) of organisms, or it may seek to describe higher order taxonomic distributions (e.g., marsupials in Australia, placental mammals in North and South America). Species designations create some problems in microbial ecology (see Chapter 3) but some working definition of what is meant by a species or group must be made to compare taxonomic patterns across space or time.

Groups (species) of organisms that are found only at a single location are designated *endemic*, whereas widely dispersed organisms care called *cosmopolitan*. A higher organism is considered endemic if no others of that species can be found in any other location. The organism is endemic if it is localized, and the only way that that can be determined is through sampling all other similar locations. This is usually prohibitive. However, naturalists have been collecting and cataloging organisms for many years. Museums contain collections of species from many climes and habitats. Usually, if an organism is collected only from a small geographic area, it may be considered endemic if no record has ever been made of a capture outside that area. Does endemic have any meaning for microbes? If there are endemic bacteria, for instance, the evidence should be found in their traits. If all bacteria are everywhere, there should not be clustering of phylogenies with geography.

In a study of the purple nonsulfur bacterium *Rhodopseudomonas palustris*, the researchers sought to investigate the degree of relatedness among samples collected over increasing spatial scales from 1 cm up to 10 m apart (Oda et al., 2003). Samples were collected from five locations along a single transect. Thirty clones of phototrophic bacteria from each of the five sampling locations were isolated from phototrophically incubated agar plates. Genotypic comparisons were made on nearly complete 16S rRNA amplified genes. Operationally, the researchers defined groups based on a computer-assisted clustering analysis algorithm, and fingerprint patterns that had $r$ values of 0.8 were considered to be the same genotype. Based on this designation four distinct major genotypes were identified. Comparisons among the sites using the Morisita-Horns similarity coefficient showed that sites that were farther apart had decreasing similarity (Figure 15.8). This pattern is intriguing and suggests that multiple, small-scale, ecologically distinct habitats are found at this study site. If

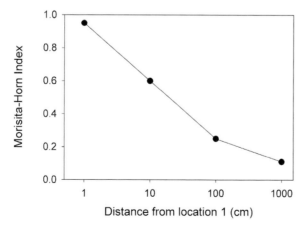

**Figure 15.8** Similarity between populations of *Rhodopseudomonas palustris* compared between sampling location 1 and four other sampling locations located 1, 10, 100, and 1,000 cm from the first location. (From Oda Y, Star B, Huisman LA, Gottschal JC, Forney LJ. Biogeography of the purple nonsulfur bacterium *Rhodopseudomonas palustris. Appl Environ Microbiol* 69:5186–5191, 2003.)

random neutral or nearly neutral mutations were occurring at each site, we would not expect to find any meaningful pattern between genotypes and distance. However, because the genotypes were more similar from geographically close samples, the data support the idea that selection regimes that are most similar produce the most similar genotypes.

Wise et al. (1995, 1996) investigated strain similarity among *Burkholderia cepacia* isolates collected spatially and temporally from a stream. The spatial study found results very similar to those described for the purple nonsulfur bacteria. Sites closest to each other had the most similar strains as determined by multilocus enzyme electrophoresis. Few studies have examined in the field how long specific strains of bacteria persist at a location. In the temporal study, Wise et al. (1996) collected samples from the same location in the stream over time. Because sampling was destructive, the same location could not be sampled, and this study therefore is also a spatial study showing the persistence of strains within a very limited sampling area. Strain similarity was fairly constant over time to about 16 to 32 days after initial sampling. Between day 16 and day 32, the strains changed significantly (Figure 15.9). This change could have been caused by some disturbance, such as a flood (a significant storm event occurred between days 16 and 32), or the pattern may result from sampling what was thought to be similar habitat (close to the initial samples) but actually was very ecologically different.

In another study, the distribution of nitrogen-fixing bacteria was determined along a 40-km stretch of one river that spanned a salinity gradient from freshwater to mesohaline (Affourtit et al., 2001). Very different nitrogen-fixing assemblages were found along the river. These differences in assemblage were related to the differential watershed hydrological inputs, sedimentation, and environmental selection pressures, such as salinity along the river course. These unique distributions once again confirm that

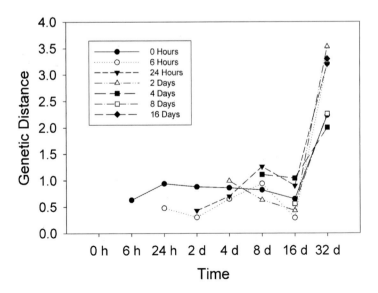

**Figure 15.9** Temporal stability of bacterial populations. (Adapted from Wise MG, McArthur JV, Wheat C, Shimkets LJ. Temporal variation in genetic diversity and structure of a lotic population of *Burkholderia* (*Pseudomonas*) *cepacia*. *Appl Environ Microbiol* 62:1558–1562, 1996.)

## Comparison of Bacterial Assemblages Using DNA/DNA Hybridizations

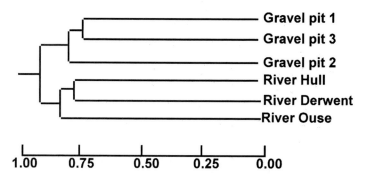

**Figure 15.10** Comparison of bacterial communities from riverine and adjacent gravel pits using DNA:DNA hybridization. (Adapted from Lambert DL, Taylor PN, Goulder R. Between-site comparison of freshwater bacterioplankton by DNA hybridization. *Microb Ecol* 26:189–200, 1993, with kind permission from Springer Science and Business Media.)

the environment selects for groups of organisms best able to survive under the prevailing conditions. Certain nitrogen-fixing genes were only found along certain reaches of the river, indicating that the genes and not necessarily the organism are being selected.

These studies were conducted over fairly short spatial scales within a habitat type. What happens when the spatial scales are increased to compare groups between different habitats that are separated by fairly short distances? Comparisons were made between bacteria collected from a flowing river and from gravel pits that were immediately adjacent to each of three rivers. In this study comparisons were made between the communities of bacteria using DNA:DNA hybridization (Figure 15.10). The bacteria were more similar within a habitat type than they were between habitat types. There were large differences within a habitat type, but these differences still were smaller than that found between the rivers and the gravel pits.

Yannarell and Triplett (2004) examined the composition of epilimnetic bacterial communities in several lakes in Minnesota that spanned differences in productivity and dissolved organic carbon, as well as different geographic locations. They found that lakes that were similar in physical characteristics had similar bacterial communities. The bacterial communities were characterized by ARISA. Samples collected at 10-m, 100-m, and between-lake scales differed by 13%, 17%, and 75%, respectively. Analytic profiles from northern lakes had fewer peaks and were less similar than southern lakes. These authors suggest that within-lake differences are much more ephemeral than between-lake differences.

What are the patterns observed between similar habitats that are extreme, and are even more isolated such as hyperthermal hot pools or between continents or poles? Hyperthermal pools occur throughout the Northern Hemisphere but are isolated from each other by geographic barriers. Stetter et al. (1993) compared hyperthermophilic archaea isolated from Alaskan oil reservoirs with cultures of known thermophilic archaea and found a high degree of DNA:DNA reassociation. In a similar study, Beeder et al. (1994) found 100% DNA:DNA reassociation between a strain of bacteria isolated from North Sea oil fields and an *Archaeoglobus fulgidus* isolated from an Italian hydrothermal environment. From these data, it appears that the adage may be right: Everything is everywhere!

*Sulfolobus*, a species of Archaea, inhabits hyperthermal pools and has an optimal growth temperature of 80°C and pH of 3. The cell cycle of these organisms is arrested abruptly with abrupt changes in temperature (Whitaker et al., 2003). *Sulfolobus* DNA begins to degrade after temperature change, and there are no known spore states for these organisms. However, *Sulfolobus* organisms are found in widely scattered geo-thermal pools. Whittaker et al. (2003) asked the obvious question: How can organisms with such specific growth requirements disperse across inhospitable distances and survive? To address this question, they sampled *Sulfolobus* from a nested hierarchy of geographic locations ranging from regions that were more than 250 km apart to areas within regions that were between 6 and 15 km apart with up to seven samples collected within an area. Comparisons were made based on the sequences of nine chromoso-mal loci using a maximum likelihood phylogenetic tree protocol that demonstrated that strains within a region share a common evolutionary history, which was distinct from the evolutionary history of the strains found in the other regions (Figure 15.11). These data suggest that there is little gene flow among the geographic regions. Either *Sulfolobus* do not disperse across these barriers, or dispersers are unable to establish in foreign hyperthermal habitats. Genetically isolated populations become adapted over time to the local conditions and increase differences among regions.

Fenchel (2003), commenting on the previous study, warned that although these data do show regional uniqueness, there were no differences in phenotypes among the strain so that the variation observed in the nine genes may be neutral or almost neutral mutations.

Among the most widely separated habitats on earth are the poles. Both poles have similar environmental conditions, and there are great expanses of very different habitat in between. Antarctica has been separated from all other land masses since the division of Gondwanaland and the formation of the Polar Front approximately 10 million years ago (Vincent, 2000). If there are examples of endemism in microor-ganisms, we might expect to find it in Antarctica because of this long separation. Iso-lated populations may, because of genetic drift and local mutations, be very genetically different from populations where gene flow occurs regularly.

Many Antarctic habitats have been "sealed" for millennia from the rest of the microbial world. Although gene flow into or out of Antarctica is restricted, there are several varied pathways for the transport of microbes, including large sea birds, fish, marine mammals, and in recent times, human vectors, including machinery, boats, and people themselves. Oceanic currents and atmospheric circulation can move microbes over extremely long distances. Vincent (2000) gives four lines of evidence that support the transfer of microbes between the rest of the world and Antarctica:

1. Beginning in the early 20th century, scientists monitored South American pollen in snow or in spore traps.
2. The Antarctic ice sheet contains species of microbes that are not present as living cells anywhere in Antarctica today and that were probably blown in (e.g., species of actinomycetes that have not been found elsewhere).

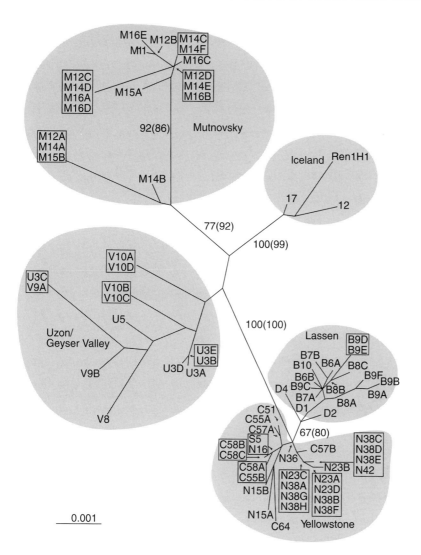

**Figure 15.11** Maximum likelihood tree for concatenated alignment of all loci for all strains. Numbers next to principal branches show bootstrap support of more than 70% using maximum likelihood and maximum parsimony (parenthetic information) algorithms. Individual strains are identified by site name. *Boxes* designate identical genotypes. *Shaded areas* highlight geographic regions. Scale bar = 1 substitution/1,000 sites. (From Whitaker RJ, Grogan DW, Taylor JW. Geographic barriers isolate endemic populations of hyperthermophilic archaea. *Science* 301:976–978, 2003.)

3. Geothermal habitats that are widely separated on the Antarctic continent and from the rest of the world have very similar assemblages that are similar to those found in the temperate regions. This observation needs to be examined using the same genetic analyses described for the hyperthermal pools *Sulfolobus*. Such an analysis would support or refute the conclusion drawn by Whittaker et al. (2003).

4. Many species of microbes, including algae, fungi, protozoa, and bacteria, in non-marine habitats of Antarctica are similar to those found throughout the world (i.e., are cosmopolitan species).

Genetic comparisons between temperate microbes and Antarctic microbes have been made. For example, Franzmann (1996) compared 10 Antarctic bacteria with their most closely related taxa from the temperate latitudes. He found phenotypic and genotypic (16S rDNA) differences. All of the Antarctic bacteria had lower optimal growth temperatures (7°C to 24°C) and median sequence dissimilarity relative to the temperate strains of 4.5%. Vincent (2000), using this dissimilarity, estimates that phylogenetic divergence from the temperate latitudes began more than 100 million years ago. This date is well before the division of Gondwanaland and the isolation and cooling of Antarctica. The Antarctic, at least according to the Franzmann study, contains a relatively unique subset of prokaryotes. However, care should be taken because of the extremely small database used to draw these conclusions. Many additional temperate and Antarctic samples need to be investigated to determine how robust the observation really is. The many novel species of bacteria, algae, fungi, and protozoa found in Antarctica and nowhere else (yet), including the Arctic, imply a fairly high level of endemism. The level of research being conducted in the Arctic is much reduced compared with the Antarctic. The following discussion compares bacteria from Antarctica to the same groups of bacteria isolated from the Arctic.

In a study, gas-vacuolate sea ice bacteria were compared between sea ice samples collected in Antarctica and the Arctic. Gas-vacuolate bacteria have been found only in or near sea ice. These organisms are true psychrophiles and cannot survive above 20°C and would therefore be sensitive to transport across the tropics. Strains of these bacteria were isolated from both poles using the same techniques and media (Staley and Gosink, 1999). Approximately 200 strains were compared using whole-cell fatty acid analysis, which resulted in three groups. One of these groups contained two strains of *Octadecabacter* spp., an alpha-Proteobacteria, and was used for further characterization using 16S rDNA, DNA:DNA hybridizations and various phenotypic comparisons, including vitamin requirements, pH ranges, and carbon use patterns.

The *Octadecabacter* from both poles were phenotypically similar, with some differences in vitamin requirements, pH ranges, and carbon use. Based on the 16S rDNA sequences, the strains differed only by 11 to 13 nucleotide positions, which suggests that they are the same species. However, the DNA:DNA hybridizations indicated that they were only 42% related to one another.

A second set of four strains collected in the Staley and Gosink study were identified as *Polarobacter*, with the two most closely related strains differing in size (ranges, 0.25 to 0.5 μm versus 0.8 to 1.6 μm). Few other phenotypic differences were observed, except for temperature ranges and some carbon use patterns. Sequences of 16S rDNA among the four strains differed by 18 to 50 nucleotide bases. Sequences that differ by between 40 and 50 nucleotide bases usually warrant designation as new species. However, the DNA:DNA hybridization studies of these four strains showed that the two most similar species were only 34% related, much below the relatedness needed to suggest that they are the same species, and none of these four strains has a bipolar distribution.

Microbial distributions can also be used to help understand the biogeography of larger organisms. Specifically we can use certain species of bacteria to understand the migrations of human populations. *Helicobacter pylori*, a chronic gastric pathogen of humans, can be classified into seven populations and subpopulations, each with a unique geographic distribution (Falush et al., 2003). By sequencing fragments of seven housekeeping genes, these scientists were able to reconstruct relationships among the different strains. These relationships mapped with some confidence onto the distributional data for historic human population migrations and helped to explain the spread of some of the ancestral genes across continents and oceans.

So what can be made of these different studies? Some show that species are widely distributed and others that there are clear differences between sites. Some species show differences over very small geographic distances and others over very large geographic distances. At this time in the development of microbial evolutionary ecology, we do not have enough data! Many more studies are needed. Many more comparisons need to be made and continuous development of techniques and methods is needed to aid in comparisons. As scientists address these questions a few guiding postulates may help in establishing whether the observed patterns are real or biased. These postulates are based on those presented in Staley and Gosink (1999):

1. At least four strains should be isolated from different samples from each ecosystem (or hosts) to be compared. It is essential that the strains be isolated from different samples taken from ecosystem using techniques that will result in the greatest possible diversity among the isolated strains. Collection of the strains from a single sample may not represent the true diversity found within the ecosystem. Diversity of habitats sampled in one ecosystem must be matched with similar effort in the other ecosystems that are geographically removed.
2. It is important to show that these strains can survive and grow (indigenous) under the conditions imposed by the environment from which they presumably were isolated. These conditions might not be the ones the strains were isolated from but would be found at other times during diel and annual cycles. However, it is extremely important to determine that they do grow under environmentally meaningful conditions imposed by the respective ecosystems.
3. Two or more groups of strains from each ecosystem being compared must be subjected to phylogenetic analyses by sequencing at least two genes.
4. Any clustering of groups by geographic regions would be presumptive evidence of endemism for these particular strains. No such clustering patterns would suggest cosmopolitan distributions.

Biogeographic studies can never prove that a particular strain or species is endemic to do so would require knowing that our sampling efforts have not excluded any individual. For example, how many robins' eggs would we have to sample to determine that robins produce only blue eggs? Answer: all of them! In other words, not finding an organism does not mean that it is not there. The smaller the sample size the higher the variance and the lower our ability to detect differences when they actually exist. Microbial ecologists must take sufficient samples (true samples not aliquots taken from a single sample) to be able to determine whether the community and biogeographic patterns observed are real or artifactual (i.e., sampling error).

# Bibliography

Affourtit, J., J. P. Zehr, and H. W. Paerl. 2001. Distribution of nitrogen-fixing microorganisms along the Neuse River estuary, North Carolina. Microb. Ecol. 41: 114–123.

Akkermans, A. D. L., J. D. Van Elsas, and F. J. De Bruijn (eds). 1995. Molecular microbial ecology manual. Kluwer Academic Publishers, Dordrecht.

Alcock, J. 1979. Animal behavior: an evolutionary approach, 2nd ed. Sinauer Associates, Sunderland, MA.

Alexander, M. 1971. Microbial ecology. John Wiley & Sons, New York.

Amann, R., W. Ludwig, and K. Schleifer. 1994. Identification of uncultured bacteria: a challenging task for molecular taxonomists. ASM News 60:358–369.

Amann, R. I., W. Ludwig, and K. H. Schleifer. 1995. Phylogenetic identification and in situ detection of individual microbial cells without cultivation. Microbiol. Rev. 59:143–169.

Anderson, B., and J. J. Midgley. 2002. It takes two to tango but three is a tangle: mutualists and cheaters on the carnivorous plant *Roridula*. Oecologia 132:369–373.

Andrewartha, H. G., and L. C. Birch. 1954. The distribution and abundance of animals. University of Chicago Press, Chicago.

Andrews, J. H. 1991. Comparative ecology of microorganisms and macroorganisms. Springer-Verlag, New York.

Andrews, J. H. 1995. What if bacteria are modular organisms? ASM News 61: 627–632.

Anesio, A. M., J. Theil-Nielsen, and W. Granéli. 2000. Bacterial growth on photochemically transformed leachates from aquatic and terrestrial primary producers. Microb. Ecol. 40:200–208.

Arsuffi, T. L., and K. Suberkropp. 1989. Selective feeding by shredders on leaf-colonizing stream fungi: comparison of macroinvertebrate taxa. Oecologia 79:30–37.

Atlas, R. M., A. Horowitz, M. Krichevsky, and A. K. Bej. 1991. Response of microbial populations to environmental disturbance. Microb. Ecol. 22:249–256.

Avilés, L. 2002. Solving the freeloaders paradox: genetic associations and frequency-dependent selection in the evolution of cooperation among nonrelatives. Proc. Natl. Acad. Sci. USA 99:14268–14273.

Avise, J. C., and K. Wollenberg. 1997. Phylogenetics and the origin of species. Proc. Natl. Acad. Sci. USA 94:7748–7755.

Bachofen, R., and A. Schenk. 1998. Quorum sensing autoinducers: do they play a role in natural microbial habitats? Microbiol. Res. 153:61–63.

Bagwell, C. E., and C. R. Lovell. 2000. Microdiversity of culturable diazotrophs from the rhizoplanes of the salt marsh grasses *Spartina alterniflora* and *Juncus roemerianus.* Microb. Ecol. 39:128–136.

Baldy, V., E. Chauvet, J.-Y. Charcosset, and M. O. Gessner. 2002. Microbial dynamics associated with leaves decomposing in the mainstream and floodplain pond of a large river. Aquat. Microb. Ecol. 28:25–36.

Barkay, T., S. M. Miller, and A. O. Summers. 2003. Bacterial mercury resistance from atoms to ecosystems. FEMS Microbiol. Revs. 27:355–384.

Barton, N. H., and B. Charlesworth. 1984. Genetic revolutions, founder effects, and speciation. Annu. Rev. Ecol. Syst. 15:133–164.

Bassler, B. L. 1999. How bacteria talk to each other: regulation of gene expression by quorum sensing. Curr. Opin. Mol. Biol. 2:582–587.

Bassler, B. L. 2002. Small talk: cell-to-cell communication in bacteria. Cell 109:421–424.

Battin, T. J., L. A. Kaplan, J. D. Newbold, X. Cheng, and C. Hansen. 2003. Effects of current velocity on the nascent architecture of stream microbial biofilms. Appl. Environ. Microbiol. 69:5443–5452.

Bauer, W. D., and J. B. Robinson. 2002. Disruption of bacterial quorum sensing by other organisms. Curr. Opin. Biotechnol. 13:234–237.

Bautsch W., D. Grothues, and B. Tummler. 1988. Genome fingerprinting of *Pseudomonas aeruginosa* by two-dimensional field inversion gel electrophoresis. FEMS Microbiol. Lett. 52:255–258.

Begon, M., and M. Mortimer. 1981. Population ecology: a unified study of animals and plants. Sinauer Associates, Sunderland, MA.

Begon, M., J. L. Harper, and C. R. Townsend. 1986. Ecology: individuals, populations, and communities. Blackwell Scientific Publications, Oxford, UK.

Belas, R. 1997. *Proteus mirabilis* and other swarming bacteria. In: J. Shapiro and M. Dworkin (eds). Bacteria as multicellular organisms, pp. 183–219. Oxford University Press, New York.

Bell, G. 1988. Recombination and the immortality of the germ line and the extinction of small populations. Am. Nat. 146:489–518.

Bennett, A. F., K. M. Dao, and R. E. Lenski. 1990. Rapid evolution in response to high-temperature selection. Nature 346:79–81.

Bennett, A. F., R. E. Lenski, and J. E. Mittler. 1992. Evolutionary adaptation to temperature. I. Fitness responses of *Escherichia coli* to changes in its thermal environment. Evolution 46:16–30.

Benson, D. R. 1985. Consumption of atmospheric nitrogen. In: E. R. Leadbetter and J. S. Poindexter (eds). Bacteria in nature, Vol. 1: Bacterial activities in perspective, pp. 155–198. Plenum Press, New York.

Benton, M. J., and F. J. Ayala. 2003. Dating the tree of life. Science 300:1698–1700.

Bever, J. D., and E. L. Simms. 2000. Evolution of nitrogen fixation in spatially structured populations of *Rhizobium.* Heredity 85:366–372.

Blahova J., M. Hupkova, and V. Krcmery. 1994. Phage F-116 transduction of antibiotic resistance from a clinical isolate of *Pseudomonas aeruginosa.* J. Chemother. 6:184–188.

Blahova, J., M. Hupkova, K. Kralikova, and V. Krcmery. 1994. Transduction of imipenem resistance by the phage F-166 from a nosocomial strain of *Pseuodomonas aeruginosa* isolated in Slovakia. Acta Virol. 38:247–250.

Bohannan, B. J. M., and R. E. Lenski. 1999. Effect of prey heterogeneity on the response of a model food chain to resource enrichment. Am. Nat. 153:73–82.

Bonner, J. T. 1965. Size and cycle. Princeton University Press, Princeton, NJ.

Boschker, H. T., and J. J. Middelburg. 2002. Stable isotopes and biomarkers in microbial ecology. FEMS Microbial. Ecol. 40:85–95.

Boulygina, E. S., B. B. Kuznetsov, A. I. Marusina, T. P. Tourova, I. K. Kravchenko, S. A. Bykova, T. V. Kolganova, and V. F. Galchenko. 2002. A study of nucleotide sequences of *nifH* genes of some methanotrophic bacteria. Microbiology (English translation of Mikrobiologiya) 71:500–508.

Bourtzis, K., and S. O'Neill. 1998. *Wolbachia* infections and arthropod reproduction. Bioscience 48:287–293.

Brisson-Noël, A., M. Arthur, and P. Courvalin. 1988. Evidence for natural gene transfer from Gram-positive cocci to *Escherichia coli*. J. Bacteriol. 170:1739–1745.

Brockhurst, M. A., A. D. Morgan, R. B. Rainey, and A. Buckling. 2003. Population mixing accelerates coevolution. Ecol. Lett. 6:975–979.

Brookfield, J. F. Y. 1998. Quorum sensing and group selection. Evolution 52:1263–1269.

Brown, S. P., and R. A. Johnstone. 2001. Cooperation in the dark: signaling and collective action in quorum-sensing bacteria. Proc. R. Soc. London B 268:961–965.

Brugger, A., B. Wett, I. Kolar, B. Reitner, and G. J. Herndl. 2001. Immobilization and bacterial utilization of dissolved organic carbon entering the riparian zone of the alpine Enns River, Austria. Aquat. Microb. Ecol. 24:129–142.

Buckley, D. H., and T. M. Schmidt. 2001. The structure of microbial communities in soil and the lasting impact of cultivation. Microb. Ecol. 42:11–21.

Buckling, A., and P. B. Rainey. 2002. Antagonistic coevolution between a bacterium and a bacteriophage. Proc. R. Soc. Lond. 269:931–936.

Buckling, A., M. A. Willis, and N. Colegrave. 2003. Adaptation limits diversification of experimental bacterial populations. Science 302:2107–2109.

Bull, J. J., and H. A. Wichman. 2001. Applied evolution. Annu. Rev. Ecol. Sys. 32:183–217.

Burkey, T. V. 1997. Metapopulation extinction in fragmented landscapes: using bacteria and protozoa communities as model ecosystems. Am. Nat. 150:568–591.

Burlage, R. S., R. Atlas, D. Stahl, G. Geesey, and G. Sayler (eds). 1998. Techniques in microbial ecology. Oxford University Press, New York.

Burns, T. P., B. C. Patten, and M. Higashi. 1991. Hierarchical evolution in ecological networks—environs and selection. In: M. Higashi and T. Burns (eds). Theoretical studies of ecosystems—the network prospective, pp. 211–239. Cambridge University Press, New York.

Buss, L. W. 1983. Evolution and development and the units of selection. Proc. Natl. Acad. Sci. USA 80:1387–1391.

Butcher, S. S., R. J. Charlson, G. H. Orians, and G. V. Wolfe (eds). 1992. Global biogeochemical cycles. Academic Press, New York.

Buu-Hoï, A., and T. Horodniceanu. 1980. Conjugative transfer of multiple antibiotic resistance markers in *Streptococcus pneumoniae*. J. Bacteriol. 143:313–320.

Cachran-Stafira, D. L., and E. N. von Ende. 1998. Integrating bacteria into food webs: studies with *Sarracenia purpurea* inquilines. Ecology 79:880–898.

Cairns, J., J. Overbaugh, and S. Miller. 1988. The origin of mutants. Nature 335: 142–145.

Cairns, J., Jr. 1993. Can microbial species with a cosmopolitan distribution become extinct? Specul. Sci. Technol. 16:69–73.

Caldwell, D. E., G. M. Wolfaardt, F. R. Korber, and J. R. Lawrence. 1997. Do bacterial communities transcend Darwinism? Adv. Microb. Ecol. 15:105–191.

Carpenter, S. R., and K. L. Cottingham. 1997. Resilience and restoration of lakes. Conserv. Ecol. 1:2. Available online at: http://www.ecologyandsociety.org/vol1/iss1/art2/.

Carson, H. L., and A. R. Templeton. 1984. Genetic revolutions in relation to speciation phenomena: the founding of new populations. Annu. Rev. Ecol. Syst. 15:97–131.

Casamatta, D. A., and C. E. Wickstrom. 2000. Sensitivity of two disjunct bacterioplankton communities to exudates from the cyanobacterium *Microcystis aeruginosa* Kützing. Microb. Ecol. 40:64–73.

Casida, L. E., Jr. 1989. Protozoan response to the addition of bacterial predators and other bacteria to soil. Appl. Environ. Microbiol. 55:1857–1859.

Casida, L. E., Jr. 1992. Competitive ability and survival in soil of *Pseudomonas* strain 679–2, a dominant, nonobligate bacterial predator of bacteria. Appl. Environ. Microbiol. 58:32–37.

Casti, J. L. 1989. Paradigms lost: images of man in the mirror of science. William Morrow and Company, New York.

Caugant, D. A., B. R. Levin, and R. K. Selander. 1981. Genetic diversity and temporal variation in the *E. coli* population of a human host. Genetics 98:467–490.

Characklis, W. G., and K. Marshall. 1990. Biofilms: a basis for an interdisciplinary approach. In: W. G. Characklis and K. Marshall K (eds). Biofilms, pp. 3–15. John Wiley & Sons, New York.

Chee, Y. E. 2004. An ecological perspective on the valuation of ecosystem services. Biological Conservation 120:549–565.

Chesson, P. 1994. Multispecies competition in variable environments. Theor. Pop. Biol. 45:227–273.

Chitty, D. 1960. Population processes in the vole and their relevance to general theory. Can. J. Zool. 38:99–113.

Cho, J., and J. M. Tiedje. 2000. Biogeography and degree of endemicity of fluorescent *Pseudomonas* strains in soil. Appl. Environ. Microbiol. 66:5448–5456.

Claridge, M. F., H. A. Dawah, and M. R. Wilson. 1997. Practical approaches to species concepts for living organisms. In: M. F. Claridge, H. A. Dawah, and M. R. Wilson (eds). The units of biodiversity, pp. 1–15. Chapman & Hall, London.

Cochran-Stafira, D. L., and C. N. von Ende. 1998. Integrating bacteria into food webs: studies with *Sarracenia purpurea* inquilines. Ecology 79:880–898.

Codeço, C. T., and J. P. Grover. 2001. Competition along a spatial gradient of resource supply: a microbial experimental model. Am. Nat. 157:300–315.

Cody, M. L., and J. M. Diamond. 1975. Ecology and evolution of communities. The Belknap Press of Harvard University Press, Cambridge, MA.

Cohan, F. M. 2001. Bacterial species and speciation. Syst. Biol. 50:513–524.

Cohan, F. M. 2002. What are bacterial species? Annu. Rev. Microbiol. 56:457–487.

Cole, G. A. 1975. Textbook of limnology. CV Mosby, St. Louis.

Cracraft, J. 1983. Species concepts and speciation analysis. Curr. Ornithol. 1:159–187.

Cracraft, J. 1987. Species concepts and the ontology of evolution. Biol. Philos. 2: 329–346.

Crespi, B. J. 2001. The evolution of social behavior in microorganisms. Trends Ecol. Evol. 16:178–183.

Crow, J. F., and M. Kimura. 1965. Evolution in sexual and asexual populations. Am. Nat. 99:439–450.

Currie, C. R. 2001. A community of ants, fungi, and bacteria: a multilateral approach to studying symbiosis. Annu. Rev. Microbiol. 55:357–380.

Currie, C. R., A. N. M. Bot, and J. J. Boomsma. 2003. Experimental evidence of a tripartite mutualism: bacteria protect ant fungus gardens from specialized parasites. Oikos 101:91–102.

Curtis, T. P., and W. T. Sloan. 2004. Prokaryotic diversity and its limits: microbial community structure in nature and implications for microbial ecology. Curr. Opin. Microbiol. 7:221–226.

Curtis, T. P., W. T. Sloan, and J. W. Scannell. 2002. Estimating prokaryotic diversity and its limits. Proc. Natl. Acad. Sci. USA 99:10494–10499.

Danovaro, R., M. Armeni, A. Dell'Anno, M. Fabiano, E. Manini, D. Marrale, A. Pusceddu, and S. Vanucci. 2001. Small-scale distribution of bacteria, enzymatic activities, and organic matter in coastal sediments. Microbiol. Ecol. 42:177–185.

Davey, M. E., and G. A. O'Toole. 2000. Microbial biofilms: from ecology to molecular genetics. Microbiol. Mol. Biol. Rev. 64:847–867.

Davey, R. B., and D. C. Reanney. 1980. Extrachromosomal genetic elements and the adaptive evolution of bacteria. Evol. Biol. 13:113–147.

Davison, J. 1999. Genetic exchange between bacteria in the environment. Plasmid 42:73–91.

Dawkins, R. 1989. The selfish gene. Oxford University Press, Oxford, UK.

Day, R. L., K. N. Laland, and J. Odling-Smee. 2003. Rethinking adaptation: the niche–construction perspective. Perspect. Biol. Med. 46:80–95.

de Rosa, R., and B. Labedan. 1998. The evolutionary relationships between the two bacteria *Escherichia coli* and *Haemophilus influenza* and their putative last common ancestor. Mol. Biol. Evol. 15:17–27.

Degens, B. P., L. A. Schipper, G. P. Sparling, and L. C. Duncan. 2001. Is the microbial community in a soil with reduced catabolic diversity less resistant to stress or disturbance? Soil Biochem. 33:1143–1153.

Denison, R. F., C. Bledsoe, M. Kahn, F. O'Gara, E. L. Simms, and L. S. Thomashow. 2003. Cooperation in the rhizosphere and the "free rider" problem. Ecology 84:838–845.

Denny, T. P. 1988. Phenotypic diversity in *Pseudomonas syringae* pv. *Tomato*. J. Gen. Microbiol. 134:1939–1948.

Denny, T. P., M. N. Gilmour, and R. K. Selander. 1988. Genetic diversity and relationships of two pathovars of *Pseudomonas syringae*. J. Microbiol. 134: 1949–1960.

Dial, K. P., and J. M. Marzluff. 1988. Are the smallest organisms the most diverse? Ecology 69:1620–1624.

Dini, F., and D. Nyberg. 1999. Growth rates of marine ciliates on diverse organisms reveal ecological specializations within morphospecies. Microb. Ecol. 37:13–22.

Dobzhansky, Th. 1970. Genetics of the evolutionary process. Columbia University Press, New York.

Dolfing, J. 2001. The microbial logic behind the prevalence of incomplete oxidation of organic compounds by acetogenic bacteria in methanogenic environments. Microb. Ecol. 41:83–89.

Doolittle, R. F., D. F. Feng, K. L. Anderson, and M. R. Alberro. 1990. A naturally occurring horizontal gene transfer from a eukaryote to a prokaryote. J. Mol. Evol. 31:383–388.

Drake, J. W. 1969. Comparative rates of spontaneous mutation. Nature 221:1132.

Drake, J. W. 1991. A constant rate of spontaneous mutation in DNA-based microbes. Proc. Natl. Acad. Sci. USA 88:7160–7164.

Dugatkin, L. A., M. Perlin, and R. Atlas. 2003. The evolution of group-beneficial traits in the absence of between-group selection: a model. J. Theor. Biol. 220: 67–74.

Duncan, K. E., C. A. Istock, J. B. Graham, and N. Ferguson. 1989. Genetic exchange between *Bacillus subtilis* and *Bacillus licheniformis*: variable hybrid stability and the nature of bacterial species. Evolution 43:1585–1609.

Dunn, A. K., and J. Handelsman. 2002. Toward an understanding of microbial communities through analysis of communication networks. Antonie van Leeuwenhoek 81:565–574.

Dykhuizen, D. E. 1990. Mountaineering with microbes. Nature 346:15–16.

Dykhuizen, D. E. 1990. Experimental studies of natural selection in bacteria. Annu. Rev. Ecol. Syst. 21:373–398.

Dykhuizen, D. E. 2000. *Yersinia pestis*: an instant species. Trends Microbiol. 8: 296–298.

Dykhuizen, D. E., and D. L. Hartl. 1983. Selection in chemostats. Microbiol. Rev. 47:150–168.

Dykhuizen, D. E., and L. Green. 1991. Recombination in *Escherichia coli* and the definition of biological species. J. Bacteriol. 173:7257–7268.

Eckschmitt, K., and B. S. Griffiths. 1998. Soil biodiversity and its implications for ecosystem functioning in a heterogeneous and variable environment. Appl. Soil Ecol. 10:201–215.

Edwards, R. T., J. L. Meyer, and S. E. G. Findlay. 1990. The relative contribution of benthic and suspended bacteria to system biomass, production, and metabolism in a low-gradient blackwater river. J. North Am. Benthol. Soc. 9:216–228.

Ehrlich, P. R., and P. H. Raven. 1969. Differentiation of populations. Science 165:1228–1232.

Elena, S. F., V. S. Cooper, and R. E. Lenski. 1996. Punctuated evolution caused by selection of rare beneficial mutations. Science 272:1802–1804.

Embley, T. M., and E. Stackebrandt. 1997. Species in practice: exploring uncultured prokaryote diversity in natural samples. In: M. F. Claridge, H. A. Dawah, and M. R. Wilson (eds). The units of biodiversity, pp. 61–81. Chapman & Hall, London.

Emlen, J. M. 1973. Ecology: an evolutionary approach. Addison-Wesley, Reading, MA.

Engelberg-Kulka, H., and R. Hazan. 2003. Cannibals defy starvation and avoid sporulation. Science 301:467–468.

Falush, D., T. Wirth, B. Linz, J. K. Pritchard, M. Stephens, M. Kidd, M. J. Blaser, D. Y. Graham, S. Vacher, G. I. Perez-Perez, Y. Yamaoka, F. Mégraud, K. Otto, U. Reichard, E. Katzowitsch, X. Wamg, M. Achtman, and S. Suerbaum. 2003. Traces of human migrations in *Heliobacter pylori* populations. Science 299: 1582–1585.

Fenchel, T. 2003. Biogeography for bacteria. Science 301:925–926.

Fiegna, F., and G. J. Velicer. 2003. Competitive fates of bacterial social parasites: persistence and self-induced extinction of *Myxococcus xanthus* cheaters. Proc. R. Soc. Lond. B 270:1527–1534.

Fitch, W. M., and E. Margoliash. 1970. The usefulness of amino acid and nucleotide sequences in evolutionary studies. Evol. Biol. 4:67–109.

Fitter, A. H., D. Atkinson, D. J. Read, and M. B. Usher. 1985. Ecological interactions in soil: plants, microbes and animals. Blackwell Scientific Publications, Oxford, UK.

Fogel, G. B., C. R. Collins, J. Li, and C. F. Brunk. 1999. Prokaryotic genome size and SSU rDNA copy number: estimation of microbial relative abundance from a mixed population. Microb. Ecol. 38:93–113.

Ford, T. E. (ed). 1993. Aquatic microbiology: an ecological approach. Blackwell Scientific Publications, Oxford, UK.

Fouace, J. 1981. Mixed cultures of *Staphylococcus aureus*: some observations concerning transfer of antibiotic resistance. Ann. Microbiol. (Paris) 132 B(3): 375–386.

Fox, J. L. 1994. Microbial diversity: low profile, immense breadth. ASM News 60:533–536.

Franklin, R. B., D. R. Taylor, and A. L. Mills. 2000. The distribution of microbial communities in anaerobic and aerobic zones of a shallow coastal plain aquifer. Microb. Ecol. 38:377–386.

Franzmann, P. D. 1996. Examination of Antarctic prokaryotic diversity through molecular comparisons. Biodivers. Conserv. 5:1295–1306.

Fromm, O. 2000. Ecological structure and functions of biodiversity as elements of its total economic value. Environmental and Resource Economics 16:303–328.

Fry, J. C., and M. J. Day (eds). 1990. Bacterial genetics in natural environments. Chapman & Hall, New York.

Futuyma, D. J. 1998. Evolutionary biology, 3rd ed. Sinauer Associates, Sunderland, MA.

Futuyma, D. J., and M. Slatkin (eds). 1983. Coevolution. Sinauer Associates, Sunderland, MA.

García-Moreno, J., M. D. Matocq, M. S. Roy, E. Geffen, and R. K. Wayne. 1996. Relationships and genetic purity of the endangered Mexican wolf based on analysis of microsatellite loci. Conserv. Biol. 10:376–389.

Garland, J. L., K. I. Cook, J. L. Adams, and L. Kerkhof. 2001. Culturability as an indicator of succession in microbial communities. Microb. Ecol. 42:150–158.

Gause, G. F. 1934. The struggle for existence. Hafner, New York (reprinted 1964).

Gavrilets, S., H. Li, and M. D. Vose. 1998. Rapid parapatric speciation on holey adaptive landscapes. Proc. R. Soc. Lond. B 265:1483–1489.

Ghiselin, M. 1987. Species concepts, individuality, and objectivity. Biol. Philos. 2: 127–143.

Gilpin, M., and I. Hanski (eds). 1991. Metapopulation dynamics: empirical and theoretical investigation. Academic Press, New York.

Gingerich, P. D. 1983. Rates of evolution: effects of time and temporal scaling. Science 222:159–161.

Giovannoni, S. J., T. B. Britschgi, C. L. Moyer, and K. G. Field. 1990. Genetic diversity in Sargasso Sea bacterioplankton. Nature 345:60–63.

Girvan, M. S., J. Bullimore, J. N. Pretty, A. M. Osborn, and A. S. Ball. 2003. Soil type is the primary determinant of the composition of the total and active bacterial communities in arable soils. Appl. Environ. Microbiol. 69:1800–1809.

Gliddon, C. J., and P.-H. Gouyon. 1989. The units of selection. Trends Ecol. Evol. 4:204–208.

Golovlev, E. L. 2001. Ecological strategy of bacteria: specific nature of the problem. Microbiology 70:379–383.

González-Pastor, J., E. C. Hobss, and R. Losick. 2003. Cannibalism by sporulating bacteria. Science 301:510–513.

Goodfellow, M., G. P. Manfio, and J. Chun. 1997. Towards a practical species concept for cultivable bacteria. In: M. F. Claridge, H. A. Dawah, and M. R. Wilson (eds). The units of biodiversity. Chapman & Hall, London.

Gosink, J. J., R. L. Irgens, and J. T. Staley. 1993. Vertical distribution of bacteria in arctic sea ice. FEMS Microbiol. Ecol. 102:85–90.

Gottschalk, G. 1979. Bacterial metabolism. Springer Verlag, New York.

Gray, M. W. 1994. One plus one equals one: the making of a *Cryptomonas* alga. ASM News 60:423–427.

Griffiths, A. J., and R. Lovitt. 1980. Use of numerical profiles for studying bacterial diversity. Microb. Ecol. 6:35–43.

Gunderson, L. H. 2000. Ecological resilience—in theory and application. Annu. Rev. Ecol. Syst. 31:425–439.

Hahn, M. W., and M. G. Höfle. 2001. Grazing of protozoa and its effect on populations of aquatic bacteria. FEMS Microbiol. Ecol. 35:113–121.

Hahn, M. W., E. R. B. Moore, and M. G. Höfle. 1999. Bacterial filament formation, a defense mechanism against flagellate grazing, is growth rate controlled in bacteria of different phyla. Appl. Environ. Microbiol. 65:25–35.

Hahn, M. W., E. R. B. Moore, and M. G. Höfle. 2000. Role of microcolony formation in the protistan grazing defense of the aquatic bacterium *Pseudomonas* sp. MWHI. Microb. Ecol. 39:175–185.

Hairston, N. G., Jr., S. Ellner, and C. M. Kearns. 1996. Overlapping generations: the storage effect and the maintenance of biotic diversity. In: O. E. Rhodes, R. K. Chesser, and M. H. Smith (eds). Population dynamics in ecological space and time, pp. 109–145. University of Chicago Press, Chicago.

Hall, B. G. 1982. Evolution on a Petri dish: the evolved β-galactosidase system as a model for studying acquisitive evolution in the laboratory. Evol. Biol. 15:85–150.

Hall, B. G. 1983. Evolution of new metabolic functions in laboratory organisms. In: M. Nei and R. K. Koehn (eds). Evolution of genes and proteins, pp. 234–257. Sinauer Associates, Sunderland, MA.

Hall, B. G. 1989. Selection, adaptation, and bacterial operons. Genome 31:265–271.

Hall, B. G., S. Yokoyama, and D. H. Calhoun. 1983. Role of cryptic genes in microbial evolution. Mol. Biol. Evol. 1:109–124.

Hamilton, W. D., and T. M. Lenton. 1998. Spora and Gaia: how microbes fly with their clouds. Ethol. Ecol. Evol. 10:1–16.

Harder, W., J. G. Kuenen, and A. Matin. 1977. A review: microbial selection in continuous culture. J. Appl. Bacteriol. 43:1–24.

Harper, J. L. 1985. Modules, branches, and the capture of resources. In: J. B. C. Jackson, L. W. Buss, and R. E. Cook (eds). Population biology and evolution of clonal organisms, pp. 1–33. Yale University Press, New Haven, CT.

Hartl, D. L., and D. E. Dykhuizen. 1984. The population genetics of *Escherichia coli*. Annu. Rev. Genet. 18:31–68.

Hartwig, A. 1998. Carcinogenicity of metal compounds: possible role of DNA repair inhibition. Toxicol. Lett. 102–103:235–239.

Hastings, J. W., and E. P. Greenberg. 1999. Quorum sensing: the explanation of a curious phenomenon reveals a common characteristic of bacteria. J. Bacteriol. 181:2667–2668.

Heath, B. D., R. D. J. Butcher, W. G. Whitfield, and S. F. Hubbard. 1999. Horizontal transmission of *Wolbachia* between phylogenetically distant insect species by a naturally occurring mechanism. Curr. Biol. 9:313–316.

Hegeman, G. 1985. The mineralization of organic materials under aerobic conditions. In: E. R. Leadbetter and J. S. Poindexter (eds). Bacteria in nature, Vol. 1: Bacterial activities in perspective. Plenum Press, New York.

Herdman, M. 1985. The evolution of bacterial genomes. In: T. Cavalier-Smith (ed) The evolution of genome size. John Wiley & Sons, London.

Hernandez, M. E., and D. K. Newman. 2001. Extracellular electron transfer: CMLS. Cell. Mol. Life Sci. 58:1562–1571.

Herre, E. A., N. Knowlton, U. G. Mueller, and S. A. Rehner. 1999. The evolution of mutualisms: exploring the paths between conflict and cooperation. Trends Ecol. Evol. 14:49–52.

Hirsch, A. M., H. I. McKhann, A. Reddy, J. Liao, Y. Fang, and C. R. Marshall. 1995. Assessing horizontal transfer of nifHDK genes in eubacteria: nucleotide sequence of nifK from Frankia strain HFPCcI3. Mol. Biol. Evol. 12:16–27.

Hoffman, A. A., and M. J. Hercus. 2000. Environmental stress as an evolutionary force. Bioscience 50:217–226.

Honda, D., A. Yokota, and J. Sugiyama. 1999. Detection of seven major evolutionary lineages in cyanobacteria based on the 16S rRNA gene sequence analysis with new sequences of five marine *Synechococcus* strains. J. Molec. Evol. 48:723–739.

Hsung, J., and A. Haug. 1975. Intracellular pH of *Thermoplasma acidophila*. Biochim. Biophys. Acta 389:477–482.

Hugenholtz, P., C. Pitulle, K. L. Hershberger, and N. R. Pace. 1998. Novel division level bacterial diversity in a Yellowstone hot spring. J. Bacteriol. 180:366–376.

Hughes, J. B., J. J. Hellmann, T. H. Ricketts, and B. J. M. Bohannan. 2001. Counting the uncountable: statistical approaches to estimating microbial diversity. Appl. Environ. Microbiol. 67:4399–4406.

Hull, D. L. 1992. Individual. In: E. F. Keller and E. A. Lloyd (eds). Keywords in evolutionary biology. Harvard University Press, Cambridge, MA.

Hull, D. L. 1997. The ideal species concept—and why we can't get it. In: M. F. Claridge, H. A. Dawah, and M. R. Wilson (eds). The units of biodiversity, pp. 355–379. Chapman & Hall, London.

Hulot, F. D., P. J. Morin, and M. Loreau. 2001. Interactions between algae and the microbial loop in experimental microcosms. Oikos 95:231–238.

Hunt, S. M., E. M. Werner, B. Huang, M. A. Hamilton, and P. S. Stewart. 2004. Hypothesis for the role of nutrient starvation in biofilm detachment. AEM 70:7418–7425.

Hurst, G. D. D., and M. Schilthuizen. 1998. Selfish genetic elements and speciation. Heredity 80:2–8.

Huston, M. A. 1994. Biological diversity: the coexistence of species on changing landscapes. Cambridge University Press, Cambridge, UK.

Hutchinson, G. E. 1959. Homage to Santa Rosalia or why are there so many kinds of animals? Am. Nat. 93:145–159.

Hutchinson, G. E. 1978. An introduction to population ecology. Yale University Press, New Haven, CT.

Hyder S. L., and M. M. Streitfeld. 1978. Transfer of erythromycin resistance from clinically isolated lysogenic strains of *Streptococcus pyogenes* via their endogenous phage. J. Infect. Dis. 138:281–286.

Ippen-Ihler, K. 1989. Bacterial conjugation. In: S. B. Levy and R. V. Miller (eds). Gene transfer in the environment. McGraw-Hill, New York.

Jassby, A. D. 1975. The ecological significance of sinking to planktonic bacteria. Can. J. Microbiol. 21:270–274.

Jeon, K. W. 1987. Change of cellular 'pathogens' into required cell components. Ann. N. Y. Acad. Sci. 503:359–371.

Jiggins, F. M., J. H. Von der Schulenburg, G. D. Hurst, and M. E. Majerus. 2001. Recombination confounds interpretations of *Wolbachia* evolution. Proc. R. Soc. Lond. B 268:1423–1427.

Jiggins, F. M., J. K. Bentley, M. E. N. Majerus, and G. D. D. Hurst. 2001. How many species are infected with *Wolbachia*? Cryptic sex ratio distorters revealed to be common by intensive sampling. Proc. R. Soc. Lond. 268:1123–1126.

Johannesson, K. 2001. Parallel speciation: a key to sympatric divergence. Trends in Ecology and Evolution 16:148–153.

Kamil, A. C., J. R. Krebs, and H. R. Pulliam (eds). 1987. Foraging behavior. Plenum Press, New York.

Kaufman, M. G., E. D. Walker, T. W. Smith, R. W. Merritt, and M. J. Klug. 1999. Effects of larval mosquitoes (*Aedes triseriatus*) and stemflow on microbial community dynamics in container habitats. Appl. Environ. Microbiol. 65:2661–2673.

Keeney, D. R., R. L. Chen, and D. A. Graetz. 1971. Importance of denitrification and nitrate reduction in sediments to the nitrogen budgets of lakes. Nature 233:66.

Keller, E. F., and E. A. Lloyd. 1992. Keywords in evolutionary biology. Harvard University Press, Cambridge, MA.

Keller, L., C. Liautard, M. Reuter, W. D. Brown, L. Sundström, and M. Chapuisat. 2001. Sex ratio and *Wolbachia* infection in the ant *Formica exsecta*. Heredity 87:227–233.

Kellogg, C. A., J. B. Rose, S. C. Jiang, J. M. Thurmond, and J. H. Paul. 1995. Genetic diversity of related vibriophages isolated from marine environments around Florida and Hawaii, USA. Mar. Ecol. Prog. Ser. 120:89–98.

Kirk, J. L., L. A. Beaudette, M. Hart, P. Moutoglis, J. N. Klironomos, H. Lee, and J. T. Trevors. 2004. Methods of studying soil microbial diversity. J. Microbiol. Methods 58:169–188.

Kjelleberg, S., B. A. Humphrey, and K. C. Marshall. 1982. Effect of interfaces on small, starved marine bacteria. Appl. Environ. Microbiol. 43:1166–1172.

Knowlton, N., and F. Rohwer. 2003. Multispecies microbial mutualisms on coral reefs: the host as a habitat. Am. Nat. 162 (Suppl.):S51–S62.

Kolenbrander, P. E. 2000. Oral microbial communities: biofilms, interactions, and genetic systems. Annu. Rev. Microbiol. 54:413–437.

Konopka, A., T. Bercot, and C. Nakatsu. 1999. Bacterioplankton community diversity in a series of thermally stratified lakes. Microb. Ecol. 38:126–135.

Krebs, C. J. 1978. Ecology: the experimental analysis of distribution and abundance, 2nd ed. Harper & Row, New York.

Krebs, J. R., and N. B. Davies (eds). 1978. Behavioural ecology: an evolutionary approach. Blackwell Scientific Publications, Oxford, UK.

Kushner, D. J. 1993. Growth and nutrition of halophilic bacteria. In: R. H. Vreeland and L. I. Hochstein (eds). The biology of halophilic bacteria, pp. 87–103. CRC Press, Boca Raton, FL.

Laland, K. N., F. J. Odling-Smee, and M. W. Feldman. 1996. The evolutionary consequences of niche construction: a theoretical investigation using two-locus theory. J. Evol. Biol. 9:293–316.

Laland, K. N., F. J. Odling-Smee, and M. W. Feldman. 1999. Evolutionary consequences of niche construction and their implications for ecology. Proc. Natl. Acad. Sci. USA 96:10242–10247.

Laland, K. N., F. J. Odling-Smee, and M. W. Feldman. 2000. Niche construction, biological evolution, and cultural change. Behav. Brain Sci. 23:131–175.

Lawrence, J. R., B. Scharf, G. Packroff, and T. R. Neu. 2002. Microscale evaluation of the effects of grazing by invertebrates with contrasting feeding modes on river biofilm architecture and composition. Microb. Ecol. 44:199–207.

Lazazzera, B. A. 2000. Quorum sensing and starvation: signals for entry into stationary phase. Curr. Opin. Microbiol. 3:177–182.

Leadbetter, E. R., and J. S. Poindexter (eds). 1986. Bacteria in nature, Vol. 1: Bacterial activities in perspective. Plenum Press, New York.

LeCleir, G. R., A. Buchan, and J. T. Hollibaugh. 2004. Chitinase gene sequences retrieved from diverse aquatic habitats reveal environment-specific distributions. AEM 70:6977–6983.

Leff, L. G., J. V. McArthur, and L. Shimkets. 1992. Information spiraling: movement of bacteria and their genes in streams. Microb. Ecol. 24:11–24.

Leff, L. G., J. R. Dana, J. V. McArthur, and L. J. Shimkets. 1993. Detection of Tn5-like sequences in kanamycin-resistant stream bacteria and environmental DNA. Appl. Environ. Microbiol. 59:417–421.

Leff, L. G., J. V. McArthur, J. L. Meyer, and L. J. Shimkets. 1994. Effect of macroinvertebrates on detachment of bacteria from biofilms in stream microcosms. J. North Am. Benthol. Soc. 13:74–79.

Leff, L. G., A. A. Leff, and M. J. Lemke. 1998. Seasonal changes in the planktonic bacteria assemblages of two Ohio streams. Freshwater Biol. 39:129–134.

Legan, J. D., J. D. Owens, and G. A. Chilvers. 1987. Competition between specialist and generalist methylotrophic bacteria for intermittent supply of methylamine. J. Gen. Microbiol. 133:1061–1073.

Lemke, M. J., and L. G. Leff. 1999. Bacterial populations in an anthropogenically disturbed stream: comparison of different seasons. Microb. Ecol. 38:234–243.

Lenski, R. E. 1984. Coevolution of bacteria and phage: are there endless cycles of bacterial defenses and phage counterdefenses? J. Theor. Biol. 108:319–325.

Lenski, R. E. and B. R. Levin. 1985. Constraints on the coevolution of bacteria and virulent phage: a model, some experiments, and predictions for natural communities. Am. Nat. 125:585–602.

Lenski, R. E. 1988. Experimental studies of pleiotropy and epistasis in *Escherichia coli*. I. Variation in competitive fitness among mutants resistant to virus T4. Evolution 42:425–440.

Levin, B. R. 1981. Periodic selection, infectious gene exchange and the genetic structure of *E. coli* populations. Genetics 99:1–23.

Levins, R. 1968. Evolution in changing environments: some theoretical explorations. Princeton University Press, Princeton, NJ.

Levy, S. B., and R. V. Miller (eds). 1989. Gene transfer in the environment. McGraw-Hill, New York.

Liebert, C. A., M. Hall, and A. O. Summers. 1999. Transposon Tn21: flagship of the floating genome. Microbiol. Mol. Biol. Rev. 63:507–522.

Lighthart, B., K. Prier, G. M. Loper, and J. Bromenshenk. 2000. Bees scavenge airborne bacteria. Microb. Ecol. 39:314–321.

Linström, E. S. 2000. Bacterioplankton community composition in five lakes differing in trophic status and humic content. Microb. Ecol. 40:104–113.

Lock, M. A. 1981. River epilithon: a light and organic energy transducer. In: M. A. Lock and D. D. Williams (eds). Perspectives in running water ecology. Plenum Press, New York.

Lock, M. A. 1993. Attached microbial communities in rivers. In: T. E. Ford (ed). Aquatic microbiology, pp. 113–138. Blackwell Scientific Publications, Cambridge, MA.

Lock, M. A., and T. E. Ford. 1985. Microcalorimetric approach to determine relationships between energy supply and metabolism in river epilithon. Appl. Environ. Microbiol. 49:408–412.

Lock, M. A., R. R. Wallace, J. W. Costerton, R. M. Ventullo, and S. E. Charlton. 1984. River epilithon: toward a structural functional model. Oikos 42:10–22.

Loomis, W. F. 1988. Four billion years: an essay on the evolution of genes and organisms. Sinauer Associates, Sunderland, MA.

López-Archilla, A. I., I. Marin, and R. Amils. 2001. Microbial community composition and ecology of an acidic aquatic environment: the Tinto River, Spain. Microb. Ecol. 41:20–35.

Loreau, M. 2001. Microbial diversity, producer–decomposer interactions and ecosystem processes: a theoretical model. Proc. R. Soc. Lond. B 268:303–309.

Loreau, M., S. Naeem, P. Inchausti, J. Bengtsson, J. P. Grime, A. Hector, D. U. Hooper, M. A. Huston, D. Raffaelli, B. Schmid, D. Tilman, and D. A. Wardle. 2001. Biodiversity and ecosystem functioning: current knowledge and future challenges. Science 294:804–808.

Lorenz, M. G., and W. Wackernagel. 1994. Bacterial gene transfer by natural genetic transformation in the environment. Microbiol. Rev. 58:563–602.

Lynch, M., and J. S. Conery. 2003. The origins of genome complexity. Science 302:1401–1404.

MacLeod, R. A. 1968. On the role of inorganic ions in the physiology of marine bacteria. In: M. R. Droop and E. J. F. Wood (eds). Advances in microbiology of the sea I, pp. 95–126. Academic Press, London.

Manefield, M., A. S. Whiteley, R. I. Griffiths, and M. J. Bailey. 2002. RNA stable isotope probing, a novel means of linking microbial community function to phylogeny. Appl. Environ. Microbiol. 68:5367–5373.

Manz, W., K. Wendt-Potthoff, T. R. Neu, U. Szewzyk, and J. R. Lawrence. 1999. Phylogenetic composition, spatial structure, and dynamics of lotic bacterial biofilms investigated by fluorescent in situ hybridization and confocal laser scanning microscopy. Microb. Ecol. 37:225–237.

Margulis, L., and D. Sagan. 1986. Origins of sex: three billion years of genetic recombination. Yale University Press, New Haven, CT.

Marshall, K. C. 1992. Biofilms: an overview of bacterial adhesion, activity, and control at surfaces. ASM News 58:202–207.

Martin, M. O. 2002. Predatory prokaryotes: an emerging research opportunity. J. Mol. Microbiol. Biotechnol. 4:467–477.

Matsui, K., N. Ishii, and Z. Kawabata. 2003. Release of extracellular transformable plasmid DNA from *Escherichia coli* cocultivated with algae. Appl. Environ. Microbiol. 69:2399–2404.

Mayden, R. L. 1997. A hierarchy of species concepts: the denouement in the saga of the species problem. In: M. F. Claridge, H. A. Dawah, and M. R. Wilson (eds). The units of biodiversity. Chapman & Hall, London.

Mayden, R. L., and R. M. Wood. 1995. Systematics, species concepts, and the evolutionarily significant unit in biodiversity and conservation biology. Am. Fish. Soc. Symp. 17:58–113.

Maynard Smith, J. 1976. Group selection. Q. Rev. Biol. 51:277–283.

Maynard Smith, J. 1978. The evolution of sex. Cambridge University Press, Cambridge, UK.

Maynard Smith, J., C. G. Dowson, and B. G. Spratt. 1991. Localized sex in bacteria. Nature 349:29–31.

Maynard Smith, J., N. H. Smith, M. O'Rourke, and B. G. Spratt. 1993. How clonal are bacteria? Proc. Natl. Acad. Sci. USA 90:4384–4388.

Mayr, E. 1966. Animal species and evolution. Belknap Press of Harvard University Press, Cambridge, MA.

McArthur, J. V., and R. C. Tuckfield. 1997. Information length: spatial and temporal parameters among stream bacterial assemblages. J. North Am. Benthol. Soc. 16:347–357.

McArthur, J. V., J. R. Barnes, B. J. Hansen, and L. G. Leff. 1988. Seasonal dynamics of leaf litter breakdown in a Utah alpine stream. J. North Am. Benthol. Soc. 7:44–50.

McArthur, J. V., D. A. Kovacic, and M. H. Smith. 1988. Genetic diversity in natural populations of a soil bacterium across a landscape gradient. Proc. Natl. Acad. Sci. USA 85:9621–9624.

McEwen, H. A., and L. G. Leff. 2001. Colonization of stream macroinvertebrates by bacteria. Arch. Hydrobiol. 151:51–65.

McGenity, T. J., R. T. Gemmell, W. D. Grant, and H. Stan-Lotter. 2000. Origins of halophilic microorganisms in ancient salt deposits. Environ. Microbiol. 2:243–250.

Meeûs, Y., Y. Michalakis, and F. Renaud. 1998. Santa Rosalia revisited: or why are there so many kinds of parasites in 'The garden of earthly delights'? Parasitol. Today 14:10–13.

Merkley, M., R. B. Rader, and J. V. McArthur. 2004. Bacteria as bioindicators in wetlands: bioassessment in the Bonneville Basin of Utah, USA. Wetlands 24:600–607.

Merrell, D. J. 1981. Ecological genetics. University of Minnesota Press, Minneapolis.

Miao, E. A., and S. I. Miller. 1999. Bacteriophages in the evolution of pathogen–host interactions. Proc. Natl. Acad. Sci. USA 96:9452–9454.

Milkman, R. 1973. Electrophoretic variation in *Escherichia coli* from natural sources. Science 182:1024–1026.

Miller, M. B., and B. L. Bassler. 2001. Quorum sensing in bacteria. Annu. Rev. Microbiol. 55:165–199.

Mills, G. L., J. V. McArthur, and C. P. Wolfe. 2003. Lipid composition of suspended particulate matter (SPM) in a southeastern blackwater stream. Water Res. 37: 1783–1793.

Minshall G. W., K. W. Cummins, R. C. Peterson, C. E. Cushing, D. A. Bruns, J. R. Sedell, and R. L. Vannote. 1985. Developments in stream ecosystem theory. Can. J. Fish. Aquat. Sci. 42:1045–1055.

Mira, A., and N. A. Moran. 2002. Estimating population size and transmission bottlenecks in maternally transmitted endosymbiotic bacteria. Microb. Ecol. 44: 137–143.

Molin, S., and M. Givskov. 1999. Application of molecular tools for in situ monitoring of bacterial growth activity. Environ. Microbiol. 1:383–391.

Monogold, J. A. 1993. DNA repair and the evolution of sex in bacteria. ASM News 59:397–400.

Moran, N. A. 2002. The ubiquitous and varied role of infection in the lives of animals and plants. Am. Nat. 160 (Suppl.):S1–S8.

Moran, N. A., and J. J. Wernegreen. 2000. Life style evolution in symbiotic bacteria: insights from genomics. Trends Ecol. Evol. 15:321–326.

Moran, N., and P. Baumann. 1994. Phylogenetics of cytoplasmically inherited microorganisms of arthropods. Trends Ecol. Evol. 9:15–20.

Morita, R. Y. 1982. Starvation-induced survival of heterotrophs in the marine environment. Adv. Microbiol. Ecol. 6:171–198.

Morita, R. Y. 1997. Bacteria in oligotrophic environments: starvation–survival lifestyle. Chapman & Hall, New York.

Muela, A., J. M. Garcia-Bringas, I. Arana, and I. Barcina. 2000. The effect of simulated solar radiation on *Escherichia coli*: the relative roles of UV-B, UV-A, and photosynthetically active radiation. Microb. Ecol. 39:65–71.

Muyzer, G. 1999. DGGE/TGGE a method for identifying genes from natural ecosystems. Curr. Opin. Microbiol. 2:317–322.

Nardi, J. B., R. I. Mackie, and J. O. Dawson. 2002. Could microbial symbionts of arthropod guts contribute significantly to nitrogen fixation in terrestrial ecosystems? J. Insect Physiol. 48:751–763.

Naylor, B. G., and P. Handford. 1985. In defense of Darwin's theory. Bioscience 35:478–484.

Newbold, J. D., J. W. Elwood, R. V. O'Neill, and W. Van Winkle. 1981. Measuring nutrient spiraling in streams. Can. J. Fish. Aquat. Sci. 38:860–863.

Nielsen, K. M., M. D. van Weerelt, T. N. Berg, A. M. Bones, A. N. Hagler, and J. D. Van Elsas. 1997. Natural transformation and availability of transforming DNA to *Acinetobacter calcoaceticus* in soil microcosms. Appl. Environ. Microbiol. 63:1945–1952.

Nieminen, J. K., and H. Setälä. 2001. Bacteria and microbial-feeders modify the performance of a decomposer fungus. Soil Biol. Biochem. 33:1703–1712.

Nuttall, D. 1982. The populations, characterization, and activity of suspended bacteria in the Welsh River Dee. J. Appl. Bacteriol. 53:49–59.

Ochman, H., and A. C. Wilson. 1987. Evolution in bacteria: evidence for a universal substitution rate in cellular genomes. J. Mol. Evol. 26:74–86.

Oda, Y., B. Star, L. A. Huisman, J. C. Gottschal, and L. J. Forney. 2003. Biogeography of the purple nonsulfur bacterium *Rhodopseudomonas palustris*. Appl. Environ. Microbiol. 69:5186–5191.

Odling-Smee, F. J., K. N. Laland, and M. W. Feldman. 1996. Niche construction. Am. Nat. 147:641–648.

Ohta, T. 1992. The meaning of natural selection revisited at the molecular level. Trends Ecol. Evol. 7:311–312.

Oliver, K. M., J. A. Russell, N. A. Moran, and M. S. Hunter. 2003. Facultative bacterial symbionts in aphids confer resistance to parasitic wasps. PNAS 100:1803–1807.

Olivieri, I., and S. A. Frank. 1994. The evolution of nodulation in Rhizobium: altruism in the rhizosphere. J. Hered. 85:46–47.

O'Toole, G., H. B. Kaplan, and R. Kolter. 2000. Biofilm formation as microbial development. Annu. Rev. Microbiol. 54:49–79.

Otte, D., and J. A. Endler. 1989. Speciation and its consequences. Sinauer Associates, Sunderland, MA.

Overmann, J. 2001. Phototrophic consortia: a tight cooperation between non-related eubacteria. In: J. Seckbach (ed). Symbiosis: mechanisms and model systems. Kluwer Academic Publishers, Dordrecht.

Overmann, J., and K. Schubert. 2002. Phototrophic consortia: model systems for symbiotic interrelations between prokaryotes. Arch. Microbiol. 177:201–208.

Pace, M. L., and J. J. Cole. 1996. Regulation of bacteria by resources and predation tested in whole-lake experiments. Limnol. Oceanogr. 41:1448–1460.

Pace, N. R. 1996. New perspective on the natural microbial world: molecular microbial ecology. ASM News 62:463–470.

Paerl, H. W. 1993. Emerging role of atmospheric nitrogen deposition in coastal eutrophication—biochemical and rophic perspectives. Can. J. Fish. Aquat. Sci. 50:2254–2269.

Palleroni, N. 1994. Some reflections on bacterial diversity. ASM News 60:537–540.

Panikov, N. S. 1999. Understanding and prediction of soil microbial community dynamics under global change. Appl. Soil Ecol. 11:161–176.

Parker, M. A., and J. M. Spoerke. 1998. Geographic structure of lineage associations in plant-bacterial mutualism. J. Evol. Biol. 11:549–562.

Pedrós-Alió, C. 1993. Diversity of bacterioplankton. Trends Ecol. Evol. 8:85–90.

Perdue, E. M., and E. T. Gjessing (eds). 1990. Organic acids in aquatic ecosystems. John Wiley & Sons, New York.

Pernthaler, J., T. Posch, K. Simek, J. Vrba, R. Amann, and R. Psenner. 1997. Contrasting bacterial strategies to coexist with a flagellate predator in an experimental microbial assemblage. Appl. Environ. Microbiol. 63:596–601.

Peters, R. H. 1983. The effect of body size on animal abundance. Oecologia 30:89–96.

Pettibone, G. W., J. P. Mear, and B. M. Sampsell. 1996. Incidence of antibiotic and metal resistance and plasmid carriage in *Aeromonas* isolated from brown bullhead (*Ictalurus nebulosus*). Lett. Appl. Microbiol. 23:234–240.

Pianka, E. R. 1970. On r- and K-selection. Am. Nat. 104:592–597.

Pierson, B. K., V. M. Sands, and J. L. Frederick. 1990. Spectral irradiance and distribution of pigments in a highly layered marine microbial mat. Appl. Environ. Microbial. 56:2327–2340.

Poindexter, J. S., and E. R. Leadbetter. 1986. Bacteria in nature, Vol. 2: Methods and special applications in bacterial ecology. Plenum Press, New York.

Poindexter, J. S., and E. R. Leadbetter. 1986. Bacteria in nature, Vol. 3: Structure, physiology, and genetic adaptability. Plenum Press, New York.

Pomeroy, L. R. 2001. Caught in the food web: complexity made simple? Scientia Marina 65 (Suppl. 2):31–40.

Posch, T., J. Jezbera, J. Vrba, K. Šimek, J. Pernthaler, S. Andreatta, and B. Sonntag. 2001. Size selective feeding in *Cyclidium glaucoma* (Ciliophora, Scuticociliatida) and its effects on bacterial community structure: a study from a continuous cultivation system. Microb. Ecol. 42:217–227.

Postgate, J. R. 1976. Death in microbes and microbes. In: T. R. G. Gray and J. R. Postgate (eds). The survival of vegetative microbes, pp. 1–19. Cambridge University Press, Cambridge, UK.

Poulsen, M., A. N. M. Bot, C. R. Currie, M. G. Nielsen, and J. J. Boomsma. 2003. Within-colony transmission and the cost of mutualistic bacterium in the leaf-cutting ant *Acromyrmex octospinosus*. Funct. Ecol. 17:260–269.

Provorov, N. A. 1998. Coevolution of Rhizobia with legumes: facts and hypotheses. Symbiosis 24:337–368.

Pulliam, H. R. 1988. Sources, sinks, and population regulation. Am. Nat. 132: 652–661.

Pulliam, H. R. 1996. Sources and sinks: empirical evidence and population consequences. In: O. E. Rhodes, Jr., R. K. Chesser, and M. H. Smith (eds). Population dynamics in ecological space and time, pp. 45–70. University of Chicago Press, Chicago.

Rader, R. B., D. P. Batzer, and S. A. Wissinger (eds). 2001. Bioassessment and management of North American freshwater wetlands. John Wiley & Sons, New York.

Rashidan, K. K., and D. F. Bird. 2001. Role of predatory bacteria in the termination of a cyanobacterial bloom. Microb. Ecol. 41:97–105.

Rayssiguier, C., D. S. Thaler, and M. Radman. 1989. The barrier to recombination between *Escherichia coli* and *Salmonella typhimurium* is disrupted in mismatch-repair mutants. Nature 342:396–401.

Reanney, D. C. 1976. Extrachromosomal elements as possible agents of adaptation and development. Bacteriol. Rev. 40:552–590.

Reanney, D. C. 1978. Coupled evolution: adaptive interactions among the genomes of plasmids, viruses, and cells. Int. Rev. Cytol. (Suppl.) 8:1–68.

Redfield, R. J. 2002. Is quorum sensing a side effect of diffusion sensing. Trends Microbiol. 10:365–370.

Reice, S. R. 1994. Nonequilibrium determinants of biological community structure. Am. Scientist 82:424–435.

Rheinheimer, G. 1992. Aquatic microbiology, 4th ed. John Wiley & Sons, West Sussex, UK.

Rhodes, O. E., Jr., R. K. Chesser, and M. H. Smith (eds). 1996. Population dynamics in ecological space and time. University of Chicago Press, Chicago.

Rice, L. B. 1998. Tn916 family conjugative transposons and dissemination of antimicrobial resistance determinants. Antimicrob. Agents and Chemother. 42: 1871–1877.

Rickard, A. H., A. J. McBain, A. T. Stead, and P. Gilbert. 2004. Shear rate moderates community diversity in freshwater biofilms. AEM 70:7426–7435.

Rodríquez-Valera, F. 2002. Approaches to prokaryote biodiversity: a population genetics perspective. Environ. Microbiol. 4:628–633.

Rokas, A. 2000. Wolbachia as a speciation agent. Trends Ecol. Evol. 15:44–45.

Rosas, I., E. Salinas, A. Yela, E. Calva, C. Eslava, and A. Craviota. 1997. Escherichia coli in settled-dust and air samples collected in residential environments in Mexico City. Appl. Environ. Microbiol. 63:4093–4095.

Rose, M. R., and W. F. Doolittle. 1983. Molecular biological mechanisms of speciation. Science 220:157–162.

Rosenberg, S. M., and R. J. Hastings. 2003. Modulating mutation rates in the wild. Science 300:1382–1383.

Rosenzweig, M. L. 1978. Competitive speciation. Biol. J. Linnaean Soc. 10:275–289.

Rounick, J. S., and M. J. Winterbourn. 1983. Leaf processing in two contrasting beech forest streams: effects of physical and biotic factors on litter breakdown. Arch. Hydrobiol. 96:448–474.

Salyers, A. A., A. Reeves, and J. D'Elia. 1996. Solving the problem of how to eat something as big as yourself: diverse bacterial strategies for degrading polysaccharides. J. Ind. Microbiol. 17:470–476.

Sarathchandra, S. U., G. Burch, and N. R. Cox. 1997. Growth patterns of bacterial communities in the rhizoplane and rhizosphere of white clover (Trifolium repens L.) and perennial ryegrass (Lolium perenne L.) in long-term pasture. Appl. Soil Ecol. 6:293–299.

Satfish, N., T. Krugman, O. N. Vinogradova, E. Nevo, and Y. Kashi. 2001. Genome evolution of the cyanobacterium Nostoc linckia under sharp microclimatic divergence at "Evolution Canyon" Israel. Microb. Ecol. 42:306–316.

Savage, D. C. 1977. Microbial ecology of the gastrointestinal tract. Annu. Rev. Microbiol. 31:107–133.

Schauder, S., and B. L. Bassler. 2001. The languages of bacteria. Genes Dev. 15:1468–1480.

Scheffer, M., and S. R. Carpenter. 2003. Catastrophic regime shifts in ecosystems: linking theory to observation. Trends Ecol. Evol. 18:648–656.

Scheffer, M., S. Carpenter, J. A. Foley, C. Folke, and B. Walker. 2001. Catastrophic shifts in ecosystems. Nature 413:591–596.

Schidlowski, M. 2001. Carbon isotopes as biogeochemical recorders of life over 3.8 Ga of earth history: evolution of a concept. Precambr. Res. 106:117–134.

Schupp, T., R. Hutter, and D. A. Hopwood. 1975. Genetic recombination in *Norcardia mediterranei*. J. Bacteriol. 121:128–136.

Sela, S., D. Yogev, S. Razin, and H. Bercovier. 1989. Duplication of the *tuf* gene: a new insight into phylogeny of eubacterial. J. Bacteriol. 171:581–584.

Semenov, A. M. 1991. Physiological bases of oligotrophy of microorganisms and the concept of microbial community. Microb. Ecol. 22:239–247.

Service, R. F. 1997. Microbiologists explore life's rich, hidden kingdoms. Science 275:1740–1742.

Shaffer, R. T., F. Widmer, L. A. Porteous, and R. J. Seidler. 2000. Temporal and spatial distribution of the *nifH* gene of $N_2$ fixing bacteria in forests and clearcuts in western Oregon. Microb. Ecol. 39:12–21.

Shapiro, J. A. 1991. Multicellular behavior of bacteria. ASM News 57:247–253.

Shapiro, J. A., and M. Dworkin (eds). 1997. Bacteria as multicellular organisms. Oxford University Press, New York.

Shibata, M., H. Tanaka, and T. Nakashizuka. 1998. Causes and consequences of mast seed production of four co-occurring *Carpinus* species in Japan. Ecology 79:54–64.

Shlegel, H. G. 1988. General microbiology, 6th ed. Cambridge University Press, Cambridge, UK.

Shoemaker, D. D., C. A. Machado, D. Molbo, J. H. Werren, D. M. Windsor, and E. A. Herre. 2001. The distribution of *Wolbachia* in fig wasps: correlations with host phylogeny, ecology, and population structure. Proc. R. Soc. Lond. 269: 2257–2267.

Siefert, J. L., K. A. Martin, F. Abdi, W. R. Widger, and G. E. Fox. 1997. Conserved gene clusters in bacterial genomes provide further support for the primacy of RNA. J. Mol. Evol. 45:467–472.

Sigler, W. V., S. Crivil, and J. Zeyer. 2002. Bacterial succession in glacial forefield soils characterized by community structure, activity and opportunistic growth dynamics. Microb. Ecol. 44:306–316.

Sillman, C. E., and L. E. Casida, Jr. 1986. Isolation of nonobligate bacterial predators of bacteria from soil. Can. J. Microbiol. 32:760–762.

Simek, K., P. Hartman, J. Nedoma, J. Pernthaler, D. Springmann, J. Vrba, and R. Psenner. 1997. Community structure, picoplankton grazing and zooplankton control of heterotrophic nanoflagellates in a eutrophic reservoir during the summer phytoplankton maximum. Aquat. Microb. Ecol. 12:49–63.

Sime-Ngando, T., S. Demers, and S. K. Juniper. 1999. Protozoan bacterivory in the ice and water column of a cold temperate lagoon. Microb. Ecol. 37:107–115.

Simms, E., and D. L. Taylor. 2002. Partner choice in nitrogen-fixation mutualisms of legumes and rhizobia. Integr. Comp. Biol. 42:369–380.

Simonsen, L. 1991. The existence condition for bacterial plasmids: theory and reality. Microb. Ecol. 22:187–205.

Simpson, G. G. 1961. Principles of animal taxonomy. Columbia University Press, New York.

Sinsabaugh, R. L., D. Repert, T. Weiland, S. W. Golladay, and A. E. Linkins. 1991. Exoenzyme accumulation in epilithic biofilms. Hydrobiologia 222:29–37.

Sinsabaugh, R. L., S. W. Golladay, and A. E. Linkins. 1991. Comparison of epilithic and epixylic biofilm development in a boreal river. Freshwater Biol. 25:179–187.

Smith C. L., and C. R. Cantor. 1987. Purification, specific fragmentation, and separation of large DNA molecules. In: R. Wu (ed). Methods in enzymology, Vol. 155, pp. 449–467. Academic Press, San Diego.

Smith, M. E., and A. T. Bull. 1976. Studies of utilization of coconut water for production of food yeast *Saccharomyces fragilis*. J. Appl. Bacteriol. 41:81–95.

Smith, R. L. 1980. Ecology and field biology, 3rd ed. Harper & Row, New York.

Sneath, P., and R. R. Sokal. 1973. Numerical taxonomy. W. H. Freeman, San Francisco.

Sniegowski, P. 1998. Mismatch repair: origin of species? Curr. Biol. 8:R59–R61.

Solbrig, O. T. 1970. Principles and methods of plant biosystematics. Macmillan, Toronto.

Son, R., G. Rusul, A. M. Sahilah, A. Zainuri, A. R. Raha, and I. Salmah. 1997. Antibiotic resistance and plasmid profile of *Aeromonas hydrophila* isolates from cultured fish, *Telapia* (*Telapia mossambica*). Lett. Appl. Microbiol. 24:479–482.

Sørheim, R., V. L. Torsvik, and J. Goksøyr. 1989. Phenotypical divergences between populations of soil bacteria isolated on different media. Microb. Ecol. 17: 181–192.

Southwood, T. R. E. 1977. Habitat, the templet for ecological strategies. J. Anim. Ecol. 46:337–365.

Southwood, T. R. E. 1988. Tactics, strategies and templets. Oikos 52:3–18.

Sparrow, A. H., and A. F. Nauman. 1976. Evolution of genome size by DNA doublings. Science 192:524–529.

Stackebrandt, E., and B. M. Goebel. 1994. Taxonomic note: a place for DNA:DNA reassociation and 16S rRNA sequence analysis in the present species definition in bacteriology. Int. J. Syst. Bacteriol. 44:846–849.

Stackebrandt, E. F., A. Rainey, and N. Ward-Rainey. 1996. Anoxygenic phototrophy across the phylogenetic spectrum: current understanding and future perspectives. Arch. Microbiol. 166:211–223.

Stahl, D. A., R. Key, B. Flesher, and J. Smit. 1992. The phylogeny of marine and freshwater Caulobacters reflects their habitat. J. Bacteriol. 174:2193–2198.

Staley, H. T., and J. J. Gosink. 1999. Poles apart: biodiversity and biogeography of sea ice bacteria. Annu. Rev. Microbiol. 53:189–215.

Staley, J. T., and A. Reysenbach (eds). 2002. Biodiversity of microbial life: foundation of Earth's biosphere. John Wiley & Sons, New York.

Stearns, S. C. 1989. Trade-offs in life-history evolution. Functional Ecology 3:259–268.

Stephens, D. W., and J. R. Krebs. 1986. Foraging theory. Princeton University Press, Princeton, NJ.

Stetter, K. O. 1996. Hyperthermophiles in the history of life. Evolution of hydrothermal ecosystems on Earth (and Mars?), pp. 1–18. Ciba Foundation Symposium 202. Wiley, Chichester, UK.

Stetter, K. O., A. Hoffmann, and R. Huber. 1993. Microorganisms adapted to high-temperature environments. In: R. Guerrero and C. Petrós-Alió C (eds). Trends in microbial ecology. Spanish Society for Microbiology, Barcelona, pp. 25–28.

Stevenson, B. S., and T. M. Schmidt. 2004. Life history implications of rRNA gene copy number in *Escherichia coli*. AEM 70:6670–6677.

Stewart, G. J. 1989. The mechanism of natural transformation. In: S. B. Levy and R. V. Miller (eds). Gene transfer in the environment. McGraw-Hill, New York.

Stewart, G. J., and C. A. Carlson. 1986. The biology of natural transformation. Annu. Rev. Microbiol. 40:211–235.

Stewart, G. J., and C. D. Sinigalliano. 1991. Exchange of chromosomal markers by natural transformation between the soil isolate, *Pseudomonas stutzeri* JM300 and the marine isolate *Pseudomonas stutzeri* strain ZoBell. Antonie Van Leeuwenhoek 59:19–25.

Stock, M. S., and A. K. Ward. 1989. Establishment of a bedrock epilithic community in a small stream: microbial (algal and bacterial) metabolism and physical structure. Can. J. Fish. Aquat. Sci. 46:1874–1883.

Stoodley, P., K. Sauer, D. G. Davies, and J. W. Costerton. 2002. Biofilms as complex differentiated communities. Annu. Rev. Microbiol. 56:187–209.

Stotzky, G., L. R. Zeph, and M. A. Devanas. 1991. Factors affecting the transfer of genetic information among microorganisms in soil. In: L. R. Ginzburg (ed). Assessing ecological risks of biotechnology, pp. 95–122. Butterworth-Heinemann, Boston.

Strobeck, C. 1975. Selection in a fine-grained environment. Am. Nat. 109:419–425.

Suzuki, M. T., and E. F. DeLong. 2002. Marine prokaryote diversity. In: J. T. Staley and A.-L. Reysenbach (eds). Biodiversity of microbial life, pp. 209–234. Wiley-Liss, New York.

Swift, S., J. P. Throup, P. Williams, G. P. C. Salmond, and G. S. A. B. Stewart. 1996. Quorum sensing: a population-density component in the determination of bacterial phenotype. Trends Biochem. Sci. 21:214–219.

Tate, R. L. III (ed). 1986. Microbial autecology: a method for environment studies. John Wiley & Sons, New York.

Templeton, A. R. 1981. Mechanisms of speciation—a population genetic approach. Annu. Rev. Ecol. Syst. 12:23–48.

Templeton, A. R., and E. D. Rothman. 1981. Evolution in fine grained environments. II. Habitat selection as a homoeostatic mechanism. Theor. Popul. Biol. 19: 326–340.

Teske, A., and D. A. Stahl. 2002. Microbial mats and biofilms: evolution, structure, and function of fixed microbial communities. In: J. T. Staley and A.-L. Reysenbach (eds). Biodiversity of microbial life, pp. 49–100. Wiley-Liss, New York.

Thimm, T., A. Hoffmann, I. Fritz, and C. C. Tebbe. 2001. Contribution of the earthworm *Lumbricus rubellus* (Annelida, Oligochaeta) to the establishment of plasmids in soil bacterial communities. Microb. Ecol. 41:341–351.

Thingstad, T. F., A. Hagstrom, and F. Rassoulzadegan. 1997. Accumulation of degradable DOC in surface waters: is it caused by a malfunctioning microbial loop? Limnol. Oceanogr. 42:398–404.

Thompson, J. N. 1999. The raw material for coevolution. Oikos 84:5–16.

Thompson, W., and A. D. M. Rayner. 1982. Structure and development of mycelial cord systems of *Phanerochaete laevis* in soil. Trans. Br. Mycol. Soc. 78:193–200.

Tilman, D. 1997. Biodiversity and ecosystem functioning. In: G. C. Daily (ed). Nature's services: societal dependence on natural ecosystems, pp. 93–112. Island Press, Washington, DC.

Torsvik, V., and L. Øvreås. 2002. Microbial diversity and function in soil: from genes to ecosystems. Curr. Opin. Microbiol. 5:240–245.

Tuomi, J., and T. Vuorisalo. 1989. Hierarchical selection in modular organisms. Trends Ecol. Evol. 4:209–213.

Vannote, R. L., G. W. Minshall, K. W. Cummins, J. R. Sedell, and C. E. Cushing. 1980. The river continuum concept. Can. J. Fish. Aquat. Sci. 37:130.137.

Velicer, G. J. 2003. Social strife in the microbial world. Trends Microbiol. 11:330–337.

Velicer, G. J., and R. E. Lenski. 1999. Evolutionary trade-offs under condition of resource abundance and scarcity: experiments with bacteria. Ecology 80: 1168–1179.

Vellai, T., A. L. Kovács, G. Kovács, C. Ortutay, and G. Vida. 1999. Genome econo-mization and a new approach to the species concept in bacteria. Proc. R. Soc. Lond. B 266:1953–1958.

Vincent, W. F. 2000. Evolutionary origins of Antarctic microbiota: invasion, selection and endemism. Antarctic Sci. 12:374–385.

Vulic, M., R. E. Lenski, and M. Radman. 1999. Mutation, recombination, and incip-ient speciation of bacteria in the laboratory. Proc. Natl. Acad. Sci. USA 96: 7348–7351.

Wade, M. J., and S. Kalisz. 1990. The causes of natural selection. Evolution 44: 1947–1955.

Wallace, D. C., and H. J. Morowitz. 1973. Genome size and evolution. Chromosoma 40:121–126.

Walvoord, M. A., F. M. Phillips, D. A. Stonestrom, R. D. Evans, P. C. Hartsough, B. D. Newman, and R. G. Striegl. 2003. A reservoir of nitrate beneath desert soils. Science 302:1021–1024.

Ward, B. B. 2002. How many species of prokaryotes are there? Proc. Natl. Acad. Sci. USA 99:10234–10236.

Ward, B. B., M. A. Voytek, and K.-P. Witzel. 1997. Phylogenetic diversity of natural populations of ammonia oxidizers investigated by specific PCR amplification. Microb. Ecol. 33:87–96.

Webb, C. O., D. D. Ackerly, M. A. McPeek, and M. J. Donoghue. 2002. Phylogenies and community ecology. Annu. Rev. Ecol. Systematics 33:475–505.

Werren, J. H., and J. D. Bartos. 2001. Recombination in *Wolbachia*. Curr. Biol. 11:431–435.

Wetzel, R. G. 1983. Limnology, 2nd ed. Saunders College Publishing, New York.

Wheeler, Q. D., and R. Meier. 2000. Species concepts and phylogenetic theory: a debate. Columbia University Press, New York.

Whitaker, R. J., D. W. Grogan, and J. W. Taylor. 2003. Geographic barriers isolate endemic populations of hyperthermophilic Archaea. Science 301:976–978.

Whitehead, N. A., A. M. L. Barnard, H. Slater, N. J. L. Simpson, and G. P. C. Salmond. 2001. Quorum-sensing in Gram-negative bacteria. FEMS Microbiol. Revs. 25:365–404.

Whittaker, R. H. 1973. Ordination and classification of communities. Part V. Hand-book of vegetation science. W. Junk Publishers, The Hague.

Wiley, E. O. 1978. The evolutionary species concept reconsidered. Syst. Zool. 27:17–26.

Wiley, E. O. 1981. Phylogenetics: the theory and practice of phylogenetic systematics. John Wiley & Sons, New York.

Wiley, E. O., and R. L. Mayden. 1997. The evolutionary species concept. In: Q. D. Wheeler and R. Meier (eds). Species concepts and phylogenetic theory: a debate. Columbia University Press, New York.

Wilkinson, D. M. 1999. Bacterial ecology, antibiotics, and selection for virulence. Ecol. Lett. 2:207–209.

Williams, D. M., and T. M. Embley. 1996. Microbial diversity: domains and kingdoms. Annu. Rev. Ecol. Syst. 27:569–595.

Williams, G. C. 1974. Adaptation and natural selection. Princeton University Press, Princeton, NJ.

Williams, H. G., M. J. Day, J. C. Fry, and G. J. Stewart. 1996. Natural transformation in river epilithon. Appl. Environ. Microbiol. 62:2994–2998.

Williams, P., M. Camara, A. Hardman, S. Swift, D. Milton, V. J. Hope, K. Winzer, B. Middleton, D. I. Pritchard, and B. W. Bycroft. 2000. Quorum sensing and the population-dependent control of virulence. Phil. Trans. R. Soc. Lond. 355:667–680.

Wilson, D. S. 1997. Altruism and organism: disentangling the themes of multilevel selection theory. Am. Nat. 150 (Suppl.):S121–S134.

Wilson, E. O., and W. II. Bossert. 1971. A primer of population biology. Sinauer Associates, Sunderland, MA.

Wise, M., L. J. Shimkets, and J. V. McArthur. 1995. Genetic structure of a lotic population of *Burkholderia* (*Pseudomonas*) *cepacia*. Appl. Environ. Microbiol. 61:1791–1798.

Wise, M. G., J. V. McArthur, and L. J. Shimkets. 2001. *Methylosarcina fibrata* gen. Nov., sp. Nov. and *Methylosarcina quisquiliarum* sp. Nov., novel type I methanotrophs. Int. J. Syst Evol. Microbiol. 51:611–621.

Wise, M. G., L. G. Shimkets, C. Wheat, and J. V. McArthur. 1996. Temporal variation in genetic diversity and structure of a lotic population of *Burkholderia* (*Pseudomonas*) *cepacia*. Appl. Environ. Microbiol. 62:1558–1562.

Woese, C. R., O. Kandler, and M. L. Wheelis. 1990. Towards a natural system of organisms: proposal for the domains Archaea, Bacteria, and Eucarya. Proc. Natl. Acad. Sci. USA 87:4576–4579.

Wohl, D. L., and J. V. McArthur. 2001. Aquatic actinomycete–fungal interactions and their effects on organic matter decomposition: a microcosm study. Microb. Ecol. 42:446–457.

Wolf, A., J. Wise, G. Jost, and K. Witzel. 2003. Wide geographic distribution of bacteriophages that lyse the same indigenous freshwater isolate (*Sphingomonas* sp. Strain B18). Appl. Environ. Microbiol. 69:2395–2398.

Wommack, K. E., and R. R. Colwell. 2000. Virioplankton: viruses in aquatic ecosystems. Microbiol. Mol. Biol. Rev. 64:69–114.

Wu, Q. L., J. Boenigk, and M. W. Hahn. 2004. Successful predation of filamentous bacteria by a nanoflagellate challenges current models of flagellate bacteriovory. Appl. Environ. Microbiol. 70:332–339.

Yair, S., D. Yaacov, K. Susan, and E. Jurkevitch. 2003. Small eats big: ecology and diversity of *Bdellovibrio* and like organism, and their dynamics in predator–prey interactions. Agronomie 23:433–439.

Yannarell, A. C., and E. W. Triplett. 2004. Within- and between-lake variability in the composition of bacterioplankton communities: investigations using multiple spatial scales. Appl. Environ. Microbiol. 70:P214–P223.

Young, J. P. W. 1989. The population genetics of bacteria. In: D. A. Hopwood and K. F. Chater (eds). Genetics of bacterial diversity, pp. 417–438. Academic, London.

Young, J. P. W., and K. E. Haukka. 1996. Diversity and phylogeny of rhizobia. New Phytol. 133:87–94.

Zavilgelsky, G. B., and I. V. Manukhov. 2001. Quorum sensing or how bacteria "talk" to each other. Mol. Biol. 35:224–232.

Zehr, J. P., M. T. Mellon, and W. D. Hiorns. 1997. Phylogeny of cyanobacterial *nifH* genes: evolutionary implications and potential applications to natural assemblages. Microbiology 143:1443–1450.

Zeph, L. R., and L. E. Casida, Jr. 1986. Gram-negative versus Gram-positive (Actinomycete) nonobligate bacterial predators of bacteria in soil. Appl. Environ. Microbiol. 52:819–823.

Zubkov, M. V., and M. A. Sleigh. 1999. Growth of amoebae and flagellates on bacteria deposited on filters. Microb. Ecol. 37:107–115.

Zund, P., and G. Lebek. 1980. Generation time-prolonging R plasmids: correlation between increases in the generation time of *Escherichia coli* caused by R plasmids and their molecular size. Plasmid 3:65–69.

# Glossary

**Acidophiles:** Microorganisms that love acidic conditions. These organisms can tolerate and even propagate at pH < 2.0.

**Adhesion:** The attraction or holding together of molecules of different substances. For example, the attraction of water molecules to surfaces.

**Alkalophiles:** Microorganisms that love basic or alkaline conditions. These organisms can tolerate and propagate at pH > 12.

**Allochthonous:** Material that is produced or arises from outside a system. For example, leaves from deciduous trees that fall into streams and rivers are allochthonous organic matter, and bacteria that wash into rivers during storms are allochthonous bacteria.

**Allopatric speciation:** Speciation that occurs because of geographic isolation or separation of a population of organisms.

**Ammonification:** Generation of ammonia as a primary end product of decomposition of organic matter by heterotrophic bacteria either directly from proteins or from other nitrogen-rich organic compounds.

**Anastomoses:** The union or fusion of two or more structures to form a branching network, such as the mycelia of fungi.

**Apoptosis:** Programmed cell death.

**Assortative mating:** Choice of mates based on phenotype; nonrandom mating.

**Autecology:** The study of individual organisms or species within an ecosystem.

**Autochthonous:** Material that is produced within a system. Organic matter produced by organisms within an ecosystem. For example, algae or aquatic macrophytes result from autochthonous production in a river ecosystem.

**Autotrophy:** The synthesis of all needed organic molecules from inorganic substances with the use of some energy source (e.g., sunlight).

**Axenic:** Culturing of organisms in the absence of other organisms.

**Azote:** Dinitrogen gas ($N_2$); literally, it means *no life*.

**Bacterioplankton:** Bacteria that live in the open-water column of lentic and lotic aquatic habitats.

**Bacteroids:** Modified rhizobia cells that infect plant tissue and are incapable of further reproduction.

**Barophilic:** Pressure-loving organisms; microbes that inhabit and reproduce only at extreme barometric pressures, such as those found in the deep sea.

**Barophobic:** Organisms that cannot reproduce or survive at extreme barometric pressures.

**Barotolerant:** Organisms that can survive but not necessarily reproduce at high pressure.

**Beltian bodies:** Specialized protein-rich structures found at the tips of leaflets of some types of acacia trees that provide nutrition for symbiotic ants.

**Binomial nomenclature:** Classification system in which all living organisms are categorized based on genus and species names.

**Bioelements:** Elements essential for life.

**Biofilms:** Growth of microorganisms on surfaces, usually with the production of extracellular polysaccharides.

**Biogeography:** Distributions of organisms across the landscape over short or long spatial scales.

**Biological species concept:** The classification of organisms into distinct species based on reproductive isolation.

**Bioreactors:** Engineered environments in which the growth of specific microbes is promoted to perform a specific function or to produce a specific cellular product.

**Brownian motion:** Random movement of inanimate particles in solutions due to thermal motion of the molecules of the fluid.

**Carrying capacity:** The theoretical maximum number of organisms that can be supported by a specific habitat.

**Catabolite repression:** The repression of metabolic pathways through the accumulation of breakdown products of the same or different metabolic pathways.

**Catastrophic:** Sudden, severe event that usually covers a large geographic area.

**Chemotaxis:** Directed movement by bacteria toward (i.e., positive) or away from (i.e., negative) some stimulus.

**Chemotrophy:** The use of energy released from specific inorganic reactions to synthesize organic molecules and to perform life processes.

**Cline:** A series of graded changes within a species that is correlated with some gradual change in an environmental variable such as climate or geography.

**Coacervates:** Small droplets that form when certain oily liquids are mixed with water. When certain enzymes are present along with sugars, the droplets sequester the sugar in the form of starch, and they grow and divide. The localized systems were thought by Aleksandr Oparin to be precursors of cells.

**Codon:** Three adjacent nucleotides on the same DNA or mRNA molecule that code for a specific amino acid or polypeptide chain termination.

**Cohesion:** The attraction or holding together of molecules of the same substance, such as water molecules forming drops.

**Cometabolism:** The fortuitous breakdown of some substance while another substance is being degraded. Often, the microbes performing the breakdown obtain little benefit in terms of carbon or energy.

**Competent:** The ability to take up naked DNA from the environment.

**Competitive exclusion principle:** The ecological principle that states that two organisms that have the same basic requirements cannot stably coexist in the same location. The species that is more efficient in using available resources will out-compete or exclude the other species from the location.

**Conjugation:** Novel evolutionary mechanism found in bacteria; one-way sharing of genetic material between two bacteria by direct contact through pili. In conjugation, one bacterium is the donor, and the other is the recipient. Movement of genetic material is always unidirectional.

**Conservative:** Not subject to change. In genetics, the term is used to describe regions of the chromosome (DNA or RNA) that has not been changed through random mutations over long periods. Conservative sequences are usually regions that code for essential cellular functions.

**Consolidation:** Solidification into a firm, dense mass; movement of bacterial cells back into a single, large colony.

**Conspecifics:** Members of the same species group.

**Continuous environment:** An environment in which there are no large changes in physical conditions or barriers.

**Controlled experiments:** Experiments in which various aspects (i.e., variables) of the environment are maintained (i.e., control) while other aspects are altered (i.e., treatment) to elucidate cause and effect.

**Cosmopolitan:** An organism that has a very broad distribution; found in many locations.

**Covalent bond:** Chemical bond formed as the result of the sharing of one or more pairs of electrons.

**Cryosphere:** Part of the environment in which cold temperatures are the norm.

**Cryptic growth:** Growth of bacteria when no nutrients are added; thought to be at the expense of conspecifics that die and release their cellular contents.

**Cytoplasmic incompatibility:** The failure of a cross between hosts to produce any offspring because of cytoplasmic factors.

**Deaminated:** Removal of an amine group through enzymatic activity.

**Deleterious mutation:** Most mutations in which a substitution, deletion, or rearrangement of the bases in a DNA molecule results in a significant change in the gene product, which is nonfunctional or lethal.

**Density dependence:** Change in the density of a population that is the result of the initial density, such as a disease that spreads more rapidly when the population is dense.

**Density independence:** Change in the density of a population that is caused by factors that are independent of the initial density, such as sudden climatic changes that kill individuals independent of whether there are hundreds or thousands.

**Density:** The number of individuals of a species that is found in some unit area. It usually is measured and reported as numbers per unit volume or numbers per unit area.

**Diazotroph:** Nitrogen-fixing bacteria.

**Dienes phenomenon:** Thin line of demarcation between different strains of *Proteus mirabilis* when inoculated onto a solid medium.

**Differentiation:** The process of *Proteus mirabilis* cells changing from swimming to swarming cell forms.

**Directional selection:** Net change in the mean level of a phenotypic characteristic; an increase or decrease in the mean due to selection.

**Disruptive selection:** Form of natural selection in which the two extreme types in a population are increased at the expense of intermediate forms. This type selection results in two phenotypically unique populations.

**Dissimilatory food web:** Microbial food web based on the breakdown of some complex molecule. In each step in the process, new microbes may be involved in the subsequent degradation or mineralization.

**Dissociation:** Breaking up of compounds into simpler forms through a reversible reaction.

**Diurnal:** Periodic biological event that repeats on a daily basis; activity of organisms during daylight hours as opposed to nighttime activity.

**Ecological succession:** Change in community structure over time due to the modification of the environment by the current community that promotes or allows other species to establish.

**Ectomycorrhizae:** Fungi that have a symbiotic association with plant roots. The fungi surround but do not penetrate the root cells.

**Ectosymbiosis:** Symbiosis in which one member of the association lives on the other member.

**Electrolyte:** A substance whose aqueous solutions conduct electricity.

**Endemic:** An organism that is found only in a single location.

**Endocytobiosis:** Symbiotic relationship in which one organism lives within the cells of another organism.

**Endospores:** A specialized, resistant, resting bacterial cell that develops from a vegetative cell through a series of biochemical reactions called *sporulation*.

**Endosymbiosis:** Symbiotic relationship in which one organism lives within the other organism.

**Energy efficiency:** The efficiency of energy transfer within communities of organisms. Efficiency may be measured within a trophic level or between trophic levels.

**Environmental grain:** The way an organism perceives its environment. The term relates the size of a patch to the size of the individual and the space in which the individual is active.

**Ephemeral:** Short-lived event or organism; fleeting event in time and space.

**Epidemic population structure:** Microbial population structure characterized by frequent recombination in which a particular successful clone may increase, predominate for a time, and then disappear as a result of recombination.

**Epigenetics:** Developmental effects and interactions above the level of the gene.

**Evolutionary arms race:** Coevolution between two or more species in which selection for increased resistance or avoidance by one species is countered with an increased virulence of ability to pursue and capture by the other.

**Exogenous:** Developed or originating outside the organism.

**Extrinsic trait:** External to or not a basic part of, such as an extrinsic isolating mechanism.

**Facultative:** A microbial trait or process that is turned on in response to environmental conditions. For example, facultative anaerobes are normally aerobic and use oxygen, but under anaerobic conditions, these organisms continue to reproduce in the absence of oxygen by switching to other metabolic pathways.

**Favorableness:** Environmental conditions that promote growth and reproduction.

**Fecundity:** The potential to produce eggs or offspring, which is usually measured as the total number of eggs produced per female.

**Fitness:** The genetic legacy of an organism in succeeding generations relative to other individuals in the population.

**Food chain:** Ecological sequence of organisms that are related to one another as prey or predator.

**Food web:** The set of interactions among organisms in a specific location that includes producers, decomposers, herbivores, predators through energy moves and matter is cycled within a community or ecosystem.

**Founder effect:** The reduction in genetic diversity that results from a small number of individuals colonizing a new habitat.

**Gene duplication:** Condition in which entire sequences of DNA are duplicated and maintained in an organism. Gene duplication may be one way that organisms increase metabolic diversity.

**Generalist:** An organism that is capable of survival under a variety of conditions and able to feed on a wide variety of materials.

**Genet:** A genetic individual resulting from the growth of a zygote; usually applied to unitary organisms, such as most mobile organisms.

**Genetic conservation:** Maintenance of genetic integrity over time. The same or nearly the same sequences of DNA are found between and within species over time.

**Genetic drift:** Evolution or change in gene frequencies due to random or chance events and not resulting from selection.

**Genomic species:** Strains that show DNA:DNA relatedness values greater than some specified value and thermal denaturation values less than some specific rating.

**Geographic barrier:** Any physical barrier that prevents gene flow from occurring between populations on either side of a barrier, such as rivers, oceans, or mountains. At the microbial scale, geographic barriers may include chemical or thermal barriers.

**Geographic isolate:** A population of organisms separated from other populations by a geographic barrier.

**Geographic speciation:** Speciation that occurs between populations due to isolation by some geographic barrier that prevents gene flow between the populations. Because environmental conditions are most likely different in the isolated locations, selection will probably favor different combinations of genes, resulting in the inability of the populations to breed if and when the barrier is breached.

**Geostatistics:** Branch of statistics that examines the distributions of objects over spatial gradients. It can be used to determine the distribution of organisms, genes, and nutrients and other materials.

**Germination:** Resumption of growth or development from a seed or spore.

**Glycoproteins:** Proteins with a carbohydrate attached.

**Guerrilla growth form:** Growth form characterized by long internodes, infrequent branching, spaced modules, and minimum overlap of resource depletion zones.

**Halophilic:** Organisms that require high-saline conditions to grow; literally, the term means *salt loving*. Organisms are found in brackish, marine, and other high-saline environments.

**Halophobic:** Organisms that cannot grow under saline conditions.

**Heterocyst:** Specialized cell of many cyanobacteria in which nitrogen fixation takes place. Fixed nitrogen is supplied to neighboring vegetative cells in return for the products of photosynthesis.

**Heteroduplex:** Double-stranded DNA in which the two DNA strands do not show perfect base complementarity.

**Heterotrophy:** Any organism that must feed on organic matter that has been produced by other organisms to obtain energy and carbon necessary for growth and reproduction.

**Homeostasis:** Ability of an organism to maintain a stable internal physiologic environment or internal equilibrium.

**Homoplasy:** Similarity in phenotypic characteristics among different species or populations that does not accurately represent patterns of common descent. It is produced by parallel evolution or convergence.

**Hydrogen bond:** A weak molecular or chemical bond that links a hydrogen atom that is covalently bonded to a larger atom, such as oxygen or nitrogen.

**Hydrolysis:** Splitting of one molecule into two molecules by addition of the $H^+$ and $OH^-$ ions of water.

**Hyphal growth unit:** The ratio of the total length of the mycelium to the number of branches.

**Hyporheic:** Aquatic habitat that is found below the surface sediments of rivers and streams.

**Infection thread:** A tube made of plant cell wall material that is promoted by bacteria. The infection thread is filled with actively reproducing and growing bacteria that are embedded in a matrix of mucopolysaccharide.

**Information length:** The theoretical maximum distance a bacterium or its genes are distributed along a stream course.

**Information spiraling:** The movement of bacteria or their genes along streams or rivers.

**Integron:** A mobile DNA element that can capture and carry genes, particularly those responsible for antibiotic resistance. Integrons do this by site-specific recombination.

**Interspecific interaction:** Biological interaction, including competition, predation, or mutualism, between two individuals that are not of the same species.

**Intraspecific interaction:** Biological interaction, including interaction between two or more individuals of the same species.

**Intrinsic trait:** Internal to, a basic part of some biological system.

**Isogenic:** Individuals that are genetically identical (except for sex); daughter cells arising from fission or coming from the same individual or from the same inbred strain.

**Isolated environment:** An environment that prevents immigration or emigration because of distance or some barrier.

**Isolating mechanisms:** Any biological or physical condition that prevents two organisms from mating.

**$K$:** The carrying capacity of an environment; the theoretical maximum number of individuals the environment can support.

**Leachate:** Soluble organic and inorganic materials that leak from plant tissues into aqueous substances after submersion.

**Lentic:** Aquatic habitats that do not have measurable flows, such as lakes or ponds; standing-water habitats.

**Limiting factor:** A necessary factor whose availability or concentration in the environment limits the amount of reproduction and growth of a population.

**Linkage disequilibrium:** When the observed frequencies of haplotypes in a population does not agree with haplotype frequencies predicted by multiplying the frequencies of individual genetic markers in each haplotype.

**Lithosphere:** Inorganic environment in which organisms can be found; literally, the term means *rock habitat*.

**Lithotrophy:** Use of inorganic substances as sources of energy.

**Logistic equation:** Mathematical equation that describes the ideal sigmoid growth of biological populations over time.

**Longevity:** Life span.

**Lotic:** Aquatic habitat with flowing water, such a streams and rivers.

**Masting:** Production of large numbers of seeds at infrequent intervals to satiate and otherwise overwhelm seed predators to ensure seedling success.

**Maternal effects:** Contribution of mothers to their offspring that ensures or promotes their survival. Examples include extra yolk in eggs and unequal cell division during binary fission, which would give a putative advantage to the daughter cell with the extra cytoplasmic material.

**Meiofauna:** Small invertebrate fauna that inhabit the interstices between sand particles in marine and freshwater environments.

**Melting:** Unwinding of a nucleic acid into two separate strands through the addition of heat or chemicals.

**Mesocosm:** An artificial experimental system that is larger than a microcosm and that more closely mimics aspects of a natural environment. Examples include cattle tanks, ponds, exclosures, or enclosures. Because they are larger than microcosms, more or larger species can be included, and more of the complexity of the natural system can be established.

**Metabiosis:** Reliance by an organism on another to produce a favorable environment.

**Metapopulation:** A population of populations. In ecology, it is the distribution of distinct populations on the landscape.

**Microbial loop:** A food web found in most pelagic systems with strong linkages between bacteria and algae.

**Microcosm:** An artificial experimental apparatus that attempts to mimic components of a natural system. Microcosms are small. Examples include beakers, flasks, and aquaria. Select species and environmental conditions are established by the researcher.

**Microsatellite:** Short sequences of dinucleotide or trinucleotide repeats of various lengths distributed widely throughout the genome.

**Miniaturization:** Cell division by bacteria without concomitant cell growth such that each division results in smaller cells, each with a full complement of genetic material but less and less cytoplasmic material.

**Monogastric:** An organism with only one stomach.

**Monophyletic:** Condition in which a taxon contains the most recent common ancestor of the group and all of its descendants.

**Mortality:** Death of an organism.

**Mutualism:** Biological interaction in which two or more species derive benefits from the association and in which the association is necessary for the species involved.

**Natality:** Birth of organisms.

**Natural history:** Observational study of organisms in their natural environments.

**Natural selection:** An evolutionary process that occurs if there is variation among individuals of a population in some trait, nonrandom mating based on the variation, and a relationship between the parents and offspring in trait. Natural selection is often confused with evolution. Natural selection does not always cause change in gene or trait frequencies because populations may be at equilibrium.

**Nectaries:** Specialized structures on plants in which nectar is produced.

**Net energy:** Amount of energy delivered minus the energy needed for maintenance, growth, and reproduction.

**Nitrification:** The biological conversion of organic and inorganic nitrogen-containing compounds from a reduced state to a more oxidized state.

**Nod factors:** Signal molecules produced by bacteria during the initiation of nodules on the roots of legumes. The compounds are lipochitooligosaccharides, which cause deformation (i.e., curling) of the root hairs of legumes.

**Numerical taxonomy:** Classification of organisms into taxonomic groups based on the measurement of several phenotypic characteristics. These measurements are combined into a single metric that is presumably diagnostic for specific taxons.

**Obligate:** Microbes that require certain conditions for growth and reproduction. For example, some microbes require anaerobic conditions to survive, and the addition of oxygen is lethal. These organisms are obligate anaerobes.

**Oligonucleotide:** A small DNA molecule composed of a few nucleotide bases.

**Oligotrophic:** An environment that has low levels of nutrients and organic carbon.

**Open system:** An ecological system that has inputs and exports of energy and matter. Examples include most ecosystems that receive thermal energy from the sun.

**Operational taxonomic unit:** One of the organisms being compared in a phylogenetic analysis.

**Operon:** A group of genes, including an operator, a common promoter, and one or more structural genes that are controlled as a unit to produce messenger RNA (mRNA).

**Optimal solution:** The process of minimizing costs and maximizing benefits or of obtaining the best possible compromise between the two.

**Organotroph:** An organism that gets its energy from organic compounds, usually by consuming or breaking down other organisms. Most organotrophs are also heterotrophs, meaning that they also get their carbon from that source.

**Ovipositing:** Laying of eggs in some medium such as water, soil, fruit, or other organisms.

**Oxidative phosphorylation:** The movement of electrons down the electron transport chain in the final stages of cellular respiration that provides the energy needed to phosphorylate (i.e., add a phosphate group) adenosine diphosphate (ADP), producing adenosine triphosphate (ATP).

**Oxidative reduction:** The balanced gain of oxygen, loss of hydrogen, or loss of an electron by an ion, atom, or molecule. Oxidation and reduction reactions occur simultaneously.

**Paralous:** Copies of a gene that issued from duplication of an ancestral gene, with each copy having diverged before any speciation event.

**Partner choice:** Differential response to partners (i.e., benefits are returned by specifically chosen partners).

**Patchy environment:** An environment in which there are abrupt changes in environmental conditions such that there are patches of unfavorable habitat for certain organisms.

**Paternal effects:** Advantages given by fathers to offspring that increase their probability of survival and reproduction.

**Peritrichous flagella:** Flagella occur all around the bacterial cell.

**Phalanx growth form:** Growth form that maximizes the consolidation of resources within a habitat, characterized by closed architecture and shorter spacers.

**Phenetic:** Use of a criterion of overall similarity to classify organisms into taxonomic units.

**Pheromone:** Chemical substance produced and released by one organism into the environment that causes a behavioral or physiological response in another organism.

**Photoperiodism:** The ability to measure and respond to changes in the length of periods of light. Examples include nocturnal and diurnal behaviors.

**Phototrophy:** The ability to use carbon dioxide ($CO_2$) in the presence of light as a source of metabolic energy.

**Phyllosphere:** Abundant population of microbes living on the surface of leaves. They are influenced by climate and the type of plant used as a habitat.

**Phytoplankton:** Photosynthetic organisms that are suspended within the water column of aquatic marine and freshwater systems.

**Polar molecule:** Covalently bonded molecule that has a slight charge due to the difference in the sizes of the bonded molecules. Water molecules are polar because the oxygen molecules is so much larger than the hydrogen that the shared electrons are more frequently associated with the oxygen molecule, giving that side of the water molecule a slight negative charge.

**Polygastric:** An organism with more than one stomach.

**Polyphyletic:** Condition of a taxon that does not contain the most recent common ancestor of the group and all of its descendants. This condition implies that there are multiple evolutionary origins of the group and that the group is not valid as a taxonomic group.

**Preadaptive:** Condition of an organism that has a mutation before being introduced into a new habitat where the mutation provides an immediate advantage. The condition preceded the selection, and the organism is called *predapted*.

**Predictable:** Environmental condition for which the magnitude and duration of changes in physical or biological circumstances are expected.

**Promiscuous:** Indiscriminate sharing of genes.

**Proteinoids:** Protein-like molecules formed inorganically by polymerization of amino acids that are capable of some nonspecific enzymatic activity.

**Protonmotive force:** Sum of chemical gradient of protons and membrane potential that provides the energy for active transport across cell membranes.

**Quorum sensing:** Presumptive communication among bacteria using chemical cues (i.e., hormones) that elicit various behaviors or physiologic responses as a group or quorum of bacteria.

**_r_:** The intrinsic rate of increase in a population.

**Ramet:** A colony or clone.

**Range of tolerance:** The range of some environmental factor over which an organism can grow and reproduce. Environmental conditions below or above this range result in the death or incapacitation of the organism.

**Recalcitrant:** Difficult to break down. In microbial ecology, the term primarily refers to organic substrates that are not easily broken down by microorganisms.

**Redox potential:** A measure in volts of the affinity of a substance for electrons; also called its electronegativity. It is measured relative to hydrogen, which has a redox potential set at 0.

**Refractory:** Substance that is resistant to microbial degradation; synonym for recalcitrant.

**Regrowth:** Growth of microbes, especially in water delivery systems after treatment to eliminate organisms.

**Regulatory gene:** A gene that controls or regulates some process, such as growth hormones and insulin.

**Reproductive isolation:** Any biological or physical mechanism that prevents two organisms from mating.

**Resource depletion zone:** The area around an organism that can be depleted of available resources through continued growth of the organism.

**Resources:** Inorganic or organic energy sources needed for the growth and reproduction of an organism.

**Respiratory chain phosphorylation:** The oxidation of metabolic intermediates using molecular oxygen that occurs through an ordered series of substances that act as hydrogen and electron carriers.

**Rhizopines:** Compounds found in some rhizobia-legume symbioses that are synthesized by bacteroids and used by free-living cells of the producing strain. Rhizopines may act as specific growth substrates and enhance the competitive ability of the producing strain in interactions with the diverse microbial community found within the rhizosphere.

**Rhizosphere:** Habitat immediately around the roots of plants.

**Riparian zone:** The area immediately adjacent to a stream or river.

**Ruminant:** Mammal with more than one stomach that relies on the symbiosis of microorganisms to break down the complex plant molecules consumed.

**Seasonal:** Events that recur in response to changes in climatic conditions brought on by seasons.

**Selection regime:** The total number of factors that are acting as selection agents on an organism.

**Senesce:** To grow old; to reach later stages of maturity.

**Serial replacement:** The periodic change in strains of bacteria grown in culture due to selection caused by metabolically induced physical and chemical differences in the culture media.

**Sink:** A suboptimal habitat where an organism may reside but where the organism is unable to reproduce.

**Source:** A optimal habitat where organisms can survive and reproduce. These locations are the source of organisms that move into suboptimal or sink habitats.

**Specialist:** An organism that has a restrictive diet or conditions that promote growth.

**Speciation:** The biological process of forming species from other species through natural selection.

**Species richness:** The total number of species found within a specific habitat.

**Specific heat:** The amount of heat required to raise the temperature of 1 gram of a substance 1°C.

**Sporogenesis:** The formation of spores.

**Stabilizing selection:** Selection that eliminates extreme individuals from both ends of a phenotypic distribution, resulting in most of the population having the mean phenotype.

**Stable polymorphism:** A polymorphism (i.e., one of several possible alleles of a gene) in a population that remains constant over time.

**Starvation state:** The presumed condition of many environmental bacteria because of the inherent oligotrophic state of many environments.

**Stomata:** Specialized structures on leaves that allow the exchange of gases.

**Structural gene:** A gene that produces a protein.

**Substrate:** Organic material that can be used for energy or carbon by microbes.

**Substrate-level phosphorylation:** The production of ATP from ADP by direct transfer of a high-energy phosphate group from a phosphorylated intermediate metabolic compound in an exergonic catabolic pathway.

**Surface tension:** The tautness of the surface of a liquid because of cohesion between the molecules of the liquid.

**Swarmer cell:** Morphologically different form of *Proteus mirabilis* when grown associated with a surface that allows the bacteria to move across solid surfaces.

**Symbiosis:** A long-term intimate association between two or more organisms of different species. The association may be neutral; mutualistic, in which all species benefit; or parasitic, in which only one species benefits at the expense of the others.

**Symbiosome:** Specialized structure found in endosymbiotic interactions; site where nitrogen fixation occurs in legumes.

**Symbiotes:** Organisms that live inside other organisms.

**Sympatric speciation:** Speciation that takes place without geographic isolation of the population.

**Synecology:** Study of groups of organisms that are associated together as a unit.

**Syngamy:** The union of two gametes to form a zygote.

**Synonomy:** Existence of multiple systematic names to label the same organism.

**Taxonomic resolution:** The taxonomic level an organism can be classified based on the criteria being used to classify.

**Taxospecies:** Naming of a microbial entity based on numerical taxonomy of various measurable characters.

**Tetraploid:** Having four copies of an organism's genome within a single cell.

**Trade-off:** Evolutionarily conflicting demands on an organism. Trade-offs may include organisms making decisions about the best sites for reproduction that are also the worst sites in terms of predation.

**Transconjugants:** Bacteria that have received novel DNA through conjugation.

**Transcription:** The process by which the information for the synthesis of a protein is transferred from the DNA strand on which it is carried to the messenger RNA strand involved in the actual synthesis.

**Transduction:** Evolutionary mechanism in bacteria by which novel DNA is moved into or out of bacterial cells through infection and lysis of cells by a bacteriophage.

**Transformation:** Evolutionary mechanism in bacteria by which foreign or novel naked DNA is taken up by bacteria directly from the environment. It usually involves chromosomal DNA.

**Transmissibility:** The ability of extrachromosomal elements (e.g., plasmids, transposons) to disseminate genes in an infectious manner and greatly accelerate an adaptive response.

**Transposition:** The movement of certain pieces of DNA from one location to another location on the chromosome or to some other mobile element.

**Transposon:** Mobile segments of DNA that cannot exist independent of a replicon (i.e., region of DNA that replicates as an individual unit) or chromosome.

**Uniparental:** Organisms that arise from a single parent. Most microbes and many parthenogenic organisms are uniparental.

**Valence:** The number of electrons needed to fill the outermost shell of an atom; the number of electrons with which a given atom generally bonds or the number of bonds an atom forms.

**Variogram:** Two-dimensional plot resulting from geostatistical analysis. The plot shows increasing and decreasing levels of spatial correlation between sites in terms of the variable of interest.

**Versatility:** The ability of an organism to do many things or the ability of an organism to master many different situations quickly and easily.

**Xenobiotic:** A chemical or mix of chemicals that is not a normal component of the organism exposed to it.

**Zooplankton:** Small, suspended invertebrates that live in the water column of lakes, ponds, and oceans.

# Index

# Index